rn
vai
fte
h

r conditions of l

Integrated circuits

Integrated circuits

Technology and applications

F. F. MAZDA

Manager, Component Engineering,
Rank Xerox Ltd

CAMBRIDGE UNIVERSITY PRESS

Cambridge
London · New York · Melbourne

Published by the Syndics of the Cambridge University Press
The Pitt Building, Trumpington Street, Cambridge CB2 1RP
Bentley House, 200 Euston Road, London NW1 2DB
32 East 57th Street, New York, NY 10022, USA
296 Beaconsfield Parade, Middle Park, Melbourne 3206, Australia

First published 1978

Phototypeset by Western Printing Services Ltd, Bristol
and printed in Great Britain by The Pitman Press, Bath

Library of Congress Cataloguing in Publication Data

Mazda, F. F.

Integrated circuits

Bibliography: p. 201

Includes index

1. Integrated circuits. I. Title
TK7874.M35 621.381'73 77-71418

ISBN 0 521 21658 3

Contents

Preface

Perhaps no other field of engineering has progressed so far and so fast as that of integrated circuits. Undoubtedly these developments have been of considerable benefit to mankind. However this rapid progress poses a problem to the would-be author of a book on integrated circuits. He must aim to cover as much of the field as possible in a text which is small enough to be saleable. At the same time he must ensure that further developments have not outdated much of his work before the book goes to print.

The present book aims to overcome both these difficulties by concentrating primarily on the newer technology areas. Well-known and established techniques are only briefly mentioned and the student is directed to the bibliography for further material. Comparatively new devices, however, such as memories and microprocessors are described in greater detail. Not only are these emerging technologies, whose use is likely to increase during the life of this book, but there are few books which cover these topics.

The emphasis throughout the book is on the applications of integrated circuits. However it is important that the user has some understanding of the processes which are involved in fabricating the devices, in order that he may select and utilize them correctly. Therefore the first chapter of the book provides an introduction to the manufacturing techniques used in integrated circuits, and describes the differences between the various types of commercial processes. Packaging technologies are also described since it is an area of vital interest to the user, as often the packaged device is all that he sees.

The next five chapters of the book describe digital applications of integrated circuits. Chapter 2 covers the different circuit techniques which are used to produce families of devices, each family having its own distinct characteristics. Developments in this area have meant that many of the once well-known families are now becoming obsolete and are available commercially only as maintainence types for existing designs. These families are briefly described in the chapter, the emphasis being on new devices.

Chapter 3 introduces mathematical techniques which are used in the design of digital circuits. The binary and octal number systems are described first, followed by sections on Boolean algebra and function minimization techniques. Also described are analytical methods for sequential circuits and a brief introduction to threshold logic circuits. The material covered in chapter 3 is used in subsequent chapters to explain the operation of various integrated circuits.

Integrated circuits are now largely bought as off-the-shelf, 'black-box', devices which have been designed to produce certain output signals for sets of input signals. These devices are available to perform a variety of functions and a selection of these are described in chapter 4. They can all be stated to have low to medium complexity, in terms of number of circuit elements on a die, and are therefore known as small and medium scale integrated components. Complex or large scale integrated devices are described in chapters 5 and 6. Chapter 5 covers the different types of semiconductor memories which are commercially available. The construction of a single cell is first described and this is then followed by a discussion of the different organizations which may be used for an array of static and dynamic memory cells. Applications covered include read only memories, random access memories, and content addressable memories.

Chapter 6 describes devices which may be

considered to be universal logic elements. These are primarily large scale integrated components which can be adapted, with very few modifications, to meet a variety of different applications. Such devices cover uncommitted logic arrays, programmable logic arrays and microprocessors. The chapter describes the construction and applications of all these devices, with the emphasis being placed on microprocessors. Since software forms an important part of a microprocessor system the operation of the device is explained with reference to a list of software commands.

Linear integrated circuits are described in chapter 7. The number of different component types available in this area is very large and ranges from the well-known operational amplifier, to the less well-known charge coupled device. Once again the chapter introduces all the devices, so as to give the reader a feel for the whole field, but the emphasis is placed on new technologies.

This book is primarily intended for practicing engineers and technicians in electronics and other allied disciplines, who need an introduction to modern integrated circuit usage. It also covers the needs of some introductory post-graduate and specialist undergraduate courses.

Many people contributed to the contents of this book, either directly by giving advice, or indirectly through their own publications. I am grateful to them all, and to the countless other engineers and technicians who have been too busy pushing forward the frontiers of our knowledge to have time to write. My special thanks also go to Dr H. Ahmed of the University of Cambridge for reading through the entire draft of this book and for making invaluable suggestions for changes and corrections.

F.F.M.

1978

1. The technology of integrated circuits

1.1. Introduction

Integrated circuits can be made with one of two technologies, *monolithic** or *hybrid*. In the monolithic circuit all the component parts are formed into a single silicon *die*. The hybrid circuit resembles a miniature printed circuit board. Conductor tracks are built onto a small glass or alumina *substrate* and miniature resistor, capacitor, or semiconductor *chips* can be added. Generally the resistors are not included as discrete devices but are formed as tracks on the substrate surface, the resistance value being determined by the dimensions of the track and the resistivity of the material.

In addition to their production technology, integrated circuits can also be classified according to their application. These are broadly divided into linear and digital circuits. In this chapter the production technology of integrated circuits is briefly introduced. Subsequent chapters develop its applications.

1.2 Monolithic integrated circuits

Monolithic integrated circuits can be *bipolar* or *unipolar*. The difference between these two is best explained by comparing the action of their transistors, as shown in Fig. 1.1. With the *bias* arrangement shown, the holes in the base region are attracted into the emitter region and electrons flow from the emitter to the base. Provided the base region is thin, most of the electrons cross over into the collector and reach the collector terminal. However, some of the electrons combine with holes in the base region, and this is compensated for by a flow of holes from the battery terminal.

* The first occurrence of words that are included in the glossary of terms (p. 203) are given in italic.

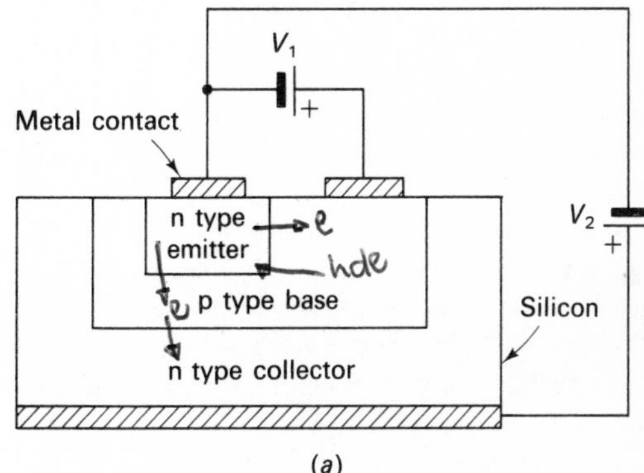

(a)

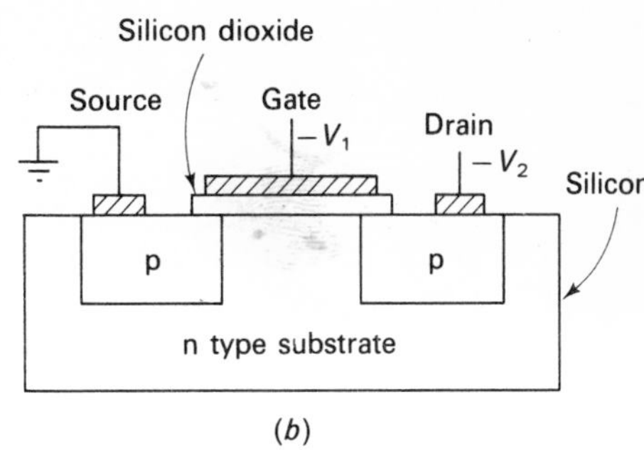

(b)

Fig. 1.1. Transistor action; (*a*) bipolar, (*b*) unipolar.

The bipolar transistor action involves both holes and electrons. By contrast the unipolar transistor operates due to a flow of holes or electrons, but not both in the same device. Fig. 1.1(*b*) shows a p-channel unipolar transistor. With no gate bias, current cannot flow between source and drain. If the gate is taken to a high enough negative voltage it will attract holes to the surface between the source and drain so that a conducting channel exists between the two. Similarly if the substrate is of p type material and the source and drain are of n type then a positive gate bias attracts electrons and forms the conducting channel.

Several *process* steps are used in the production of monolithic integrated circuits. The *wafer* is first prepared by pulling a seed crystal from the silicon melt in a Czochralski puller. p or n type impurities may be added to the melt. The ingot is then cut into slices with a diamond impregnated saw and polished with diamond powder to give a strain free, highly flat surface.

During the processing of an integrated circuit a layer of silicon oxide is often required to be grown over the silicon surface. This can be grown by heating the silicon wafer in a quartz tube and passing wet or dry oxygen, or steam, over it. Selective areas of the silicon dioxide can be removed using photolithography. In this process the whole surface is covered with a photoresist and exposed to ultraviolet light through a mask. If the material used was negative photoresist then the exposed areas harden and the non-exposed areas can be removed by a developer. The surface is now etched with hydrofluoric acid to remove the exposed silicon dioxide layer. *Impurities* can be introduced into the exposed regions by *diffusion*. The wafer is heated and nitrogen gas, which has previously been bubbled through a container of the impurity, is passed over it. At the high temperatures involved, in the region of 1200°C, the silicon atoms are very mobile so that the impurities from the gas rapidly move through the silicon crystal structure. No diffusion can occur in regions protected by a silicon dioxide layer since these absorb the impurities.

An alternative technique for introducing impurities into selected regions of the silicon surface is known as *ion implantation*. In this the ions are accelerated until they attain a high energy and they are then bombarded onto the desired silicon region. The *dopants* force their way into the silicon structure and their concentration and penetration depth can be closely controlled.

The interconnection pattern on the silicon surface is obtained by depositing a layer of aluminium over the surface and then etching it. Vacuum deposition techniques are commonly used in which the silicon *slices* are placed face down on the top of an evacuated bell jar and the metal source is evaporated onto it. The slices are heated so that the metal forms a strong bond with the silicon surface.

The silicon substrate, prepared by means of pulling, usually has a relatively wide tolerance in its impurity doping. It is usual to form a layer, called an *epitaxy*, over the substrate. This is a single crystal silicon structure and is a molecular extension of the original silicon. The epitaxy is built up by heating the silicon slices in an atmosphere of hydrogen carrying silicon tetrachloride. When the vapour reaches the hot silicon surface it dissociates and silicon atoms are deposited on the slice, where they establish themselves as part of the original silicon crystal structure, Dopants can be introduced into the vapour stream and the concentration in the epitaxy layer can be closely controlled.

The masks used in the preparation of integrated circuits can be made by several techniques. The layout drawing for each mask can be cut into a stable Mylar film which is then photographically reduced down to $\times 10$ size and finally to the required die size in a step-and-repeat camera, which produces many identical patterns of the die, covering the entire surface area of the silicon slice. Alternatively the mask can be prepared by writing directly onto the film by a beam of light or by building up the pattern by altering the position and dimensions of a slit and then repeatedly exposing the film through this. For very fine line circuits an electron beam is used to write onto the film since it has a lower wavelength than conventional light.

Fig. 1.2 shows the structure of a typical integrated circuit. An n^+ diffusion, called the buried diffusion, is first formed into the substrate, before the epitaxy is grown over it. This is used to provide a low impedance path for the collector current. p diffusions are made through the epitaxy to link up with the original substrate such that the individual components are formed in the n type epitaxy and are completely surrounded by p type material. If the substrate is now taken to the most negative voltage in the system then the components are all electrically separated from each other by

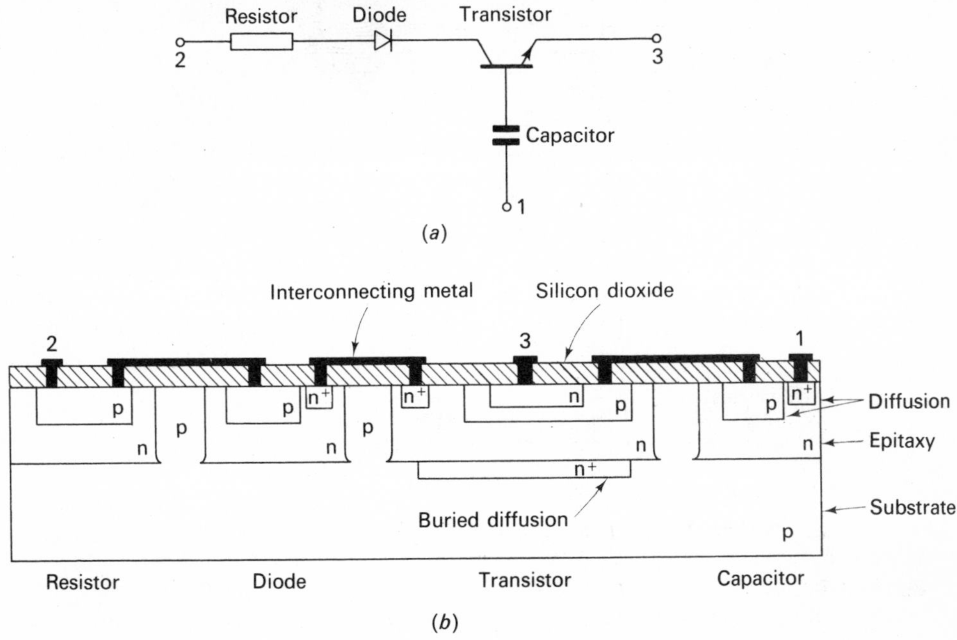

Fig. 1.2. An integrated circuit with diode isolation; (a) electronic circuit, (b) integrated circuit layout.

reverse biased pn junction. This is known as diode *isolation*.

The diodes used in integrated circuits can be formed using the base–emitter or collector–base junction of the transistors. The former is fast but has low breakdown voltage, whereas the latter has a higher voltage rating but is slower. Schottky diodes can also be used in which a metal conductor forms the p electrode and it is placed in contact with a lightly doped n type silicon material. Schottky diodes are very fast. In Fig. 1.2 a collector–base diode is shown and high impurity n^+ diffusions are used under all metal terminals to prevent unwanted Schottky action occurring. The value of the resistors is determined by their dimensions and the resistivity of the diffusion in which they are formed. Integrated circuit resistors are difficult to make to close absolute tolerances although between adjacent resistors on the same die the matching is good. Capacitors can only be built to low values. Fig. 1.2 shows a diffusion capacitor which uses a reverse biased pn junction. Better quality capacitors, having lower leakage, higher Q and a lower voltage coefficient, can be obtained with metal–oxide-semiconductor (MOS) capacitors. In these a thin layer of silicon dioxide acts as the dielectric and is sandwiched between a top metal conductor and a conducting bottom diffusion into the silicon, as illustrated in Fig. 1.4.

Fig. 1.3 illustrates several other techniques which have been used to isolate devices on a silicon chip. Process III is a very fast technology. It uses a very thin epitaxy and shallow diffusions to give narrow transistor bases and low *parasitic capacitance*. The CDI process uses a p dope epitaxy and a non-masked p diffusion over the whole surface. The collector diffusion is taken down to the buried layer and surrounds the whole element, so giving the isolation. The advantage of the system is that no separate isolation diffusion is required so that a larger number of devices can be accommodated on a die.

Silicon dioxide can be used to provide the isolation on the die. The silicon surface is etched to the required depth and oxide is then grown over it. The Isoplanar process gives a high die density since devices can be placed

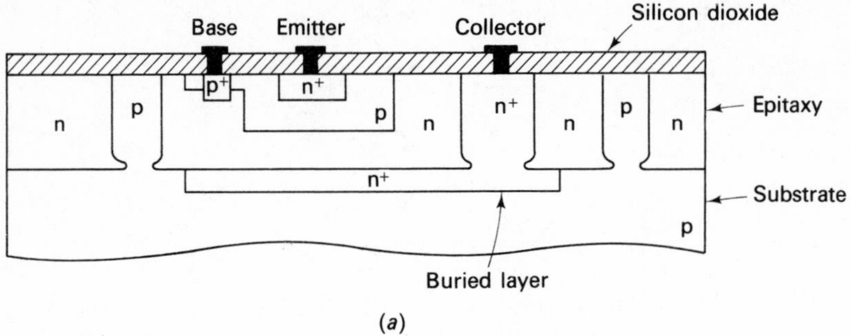

(*a*)

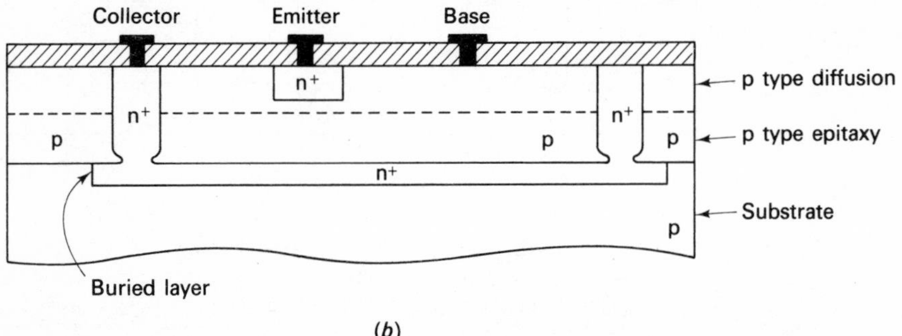

(*b*)

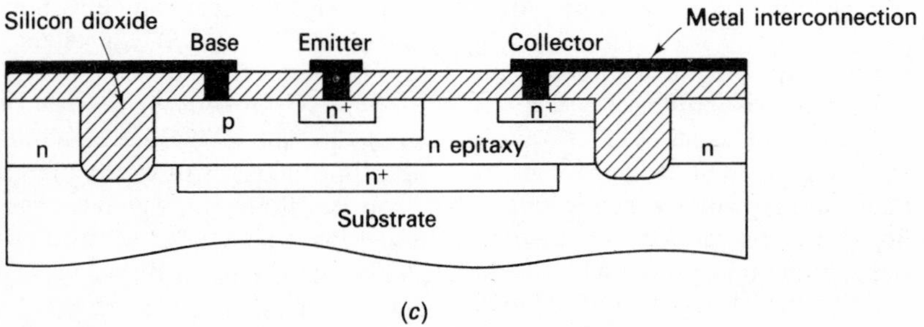

(*c*)

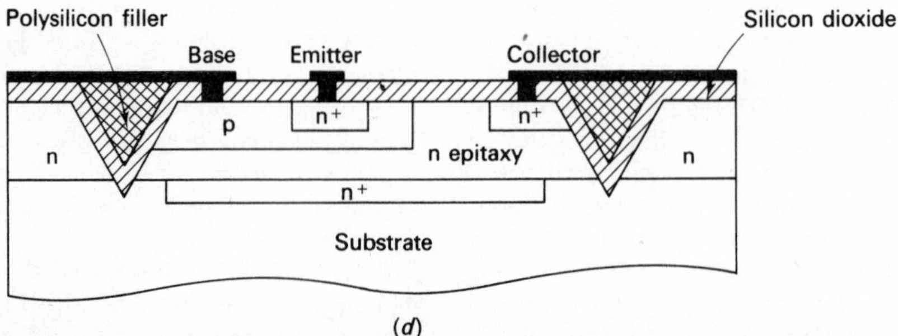

(*d*)

Fig. 1.3. Bipolar isolation techniques; (*a*) process III, (*b*) collector diffusion isolation (CDI), (*c*) isoplanar, (d) VIP and Polyplanar.

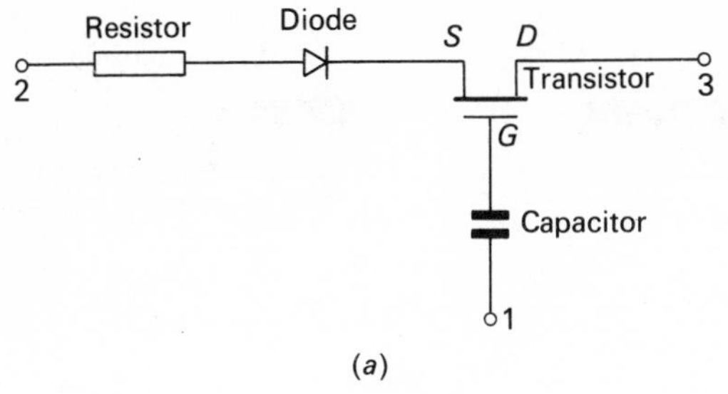

(a)

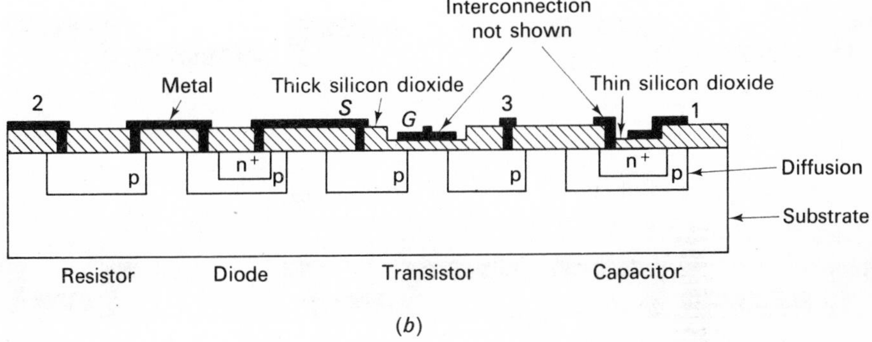

(b)

Fig. 1.4. Unipolar integrated circuit; (a) functional diagram, (b) chip construction.

close together. Speeds are also higher since parasitic capacitances, associated with isolation diffusions, are absent. Alternative oxide isolation techniques use V grooves between circuit elements which are then oxidized. Unfortunately the metallization which runs over these grooves is likely to crack due to the sharp bends involved. The V isolation with polysilicon backfill (VIP) process and the Polyplanar process overcome this disadvantage by filling the V groove with polycrystalline silicon before the metallization stage in order to avoid sharp bends in the metal.

Fig. 1.4 shows the structure of an integrated circuit constructed in unipolar technology. The p wells are self-isolating provided the n substrate is taken to the most positive voltage in the system, therefore unipolar circuits are constructionally simpler than bipolar and are capable of greater die densities.

The disadvantages of the unipolar circuit shown in Fig. 1.4 are that it requires a relatively large voltage to turn on the transistors and it is fairly slow due to parasitic capacitances primarily caused by the overlap of the gate electrode over the source and drain diffusions. Fig. 1.5 shows several modifications which have been made to overcome these disadvantages. The metal–nitride–oxide-semiconductor structure (MNOS) uses a nitride layer under the gate, which gives lower threshold voltages. Similarly the threshold can be reduced by using a polycrystalline silicon gate instead of a metal gate. A further advantage of this process is that since the silicon gate can withstand the high temperatures used during the diffusion stage it can be used as a mask during the source and drain diffusion so that the overlap between the gate and these regions is reduced, giving higher operating speeds. The metal gate can be used as a mask during ion implantation, which is a cold process. This again results in a structure which has very little gate overlap and is therefore fast.

Isoplanar techniques can be used in unipolar technology as was done for bipolar. As before, the advantage is reduced stray capacitance and higher speeds. The n channel unipolar device (NMOS) is inherently faster than a p channel

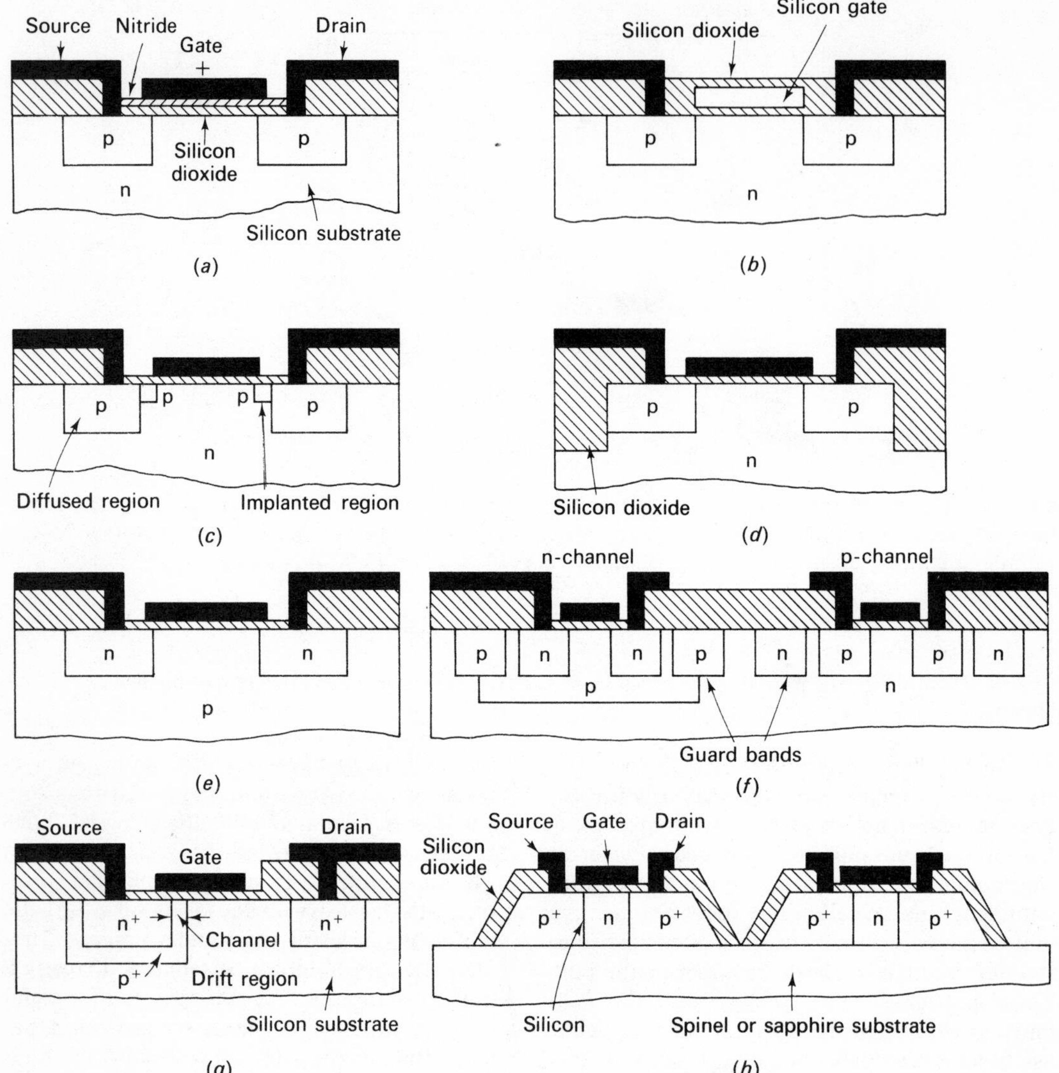

Fig. 1.5. Unipolar processes; (*a*) MNOS, (*b*) silicon gate, (*c*) ion implant, (*d*) isoplanar, (*e*) NMOS, (*f*) CMOS, (*g*) DMOS, (*h*) SOI.

device (PMOS) since the electrons carry the charge in NMOS and these have higher mobility than the holes which operate in PMOS transistors. Both p- and n-channel transistors can be combined to form complementary MOS devices or CMOS. These occupy a large area on the die, specially if guard bands are used to surround groups of like devices in order to prevent leakage between them. The advantages of CMOS are a low threshold voltage and low power consumption.

For very high speeds double diffused MOS or DMOS can be used. The gate channel is formed as the difference between the p and n diffusions and it can be made very narrow, which gives this device its high speed. Silicon on insulator (SOI) is another device which has been developed for high speed applications. The

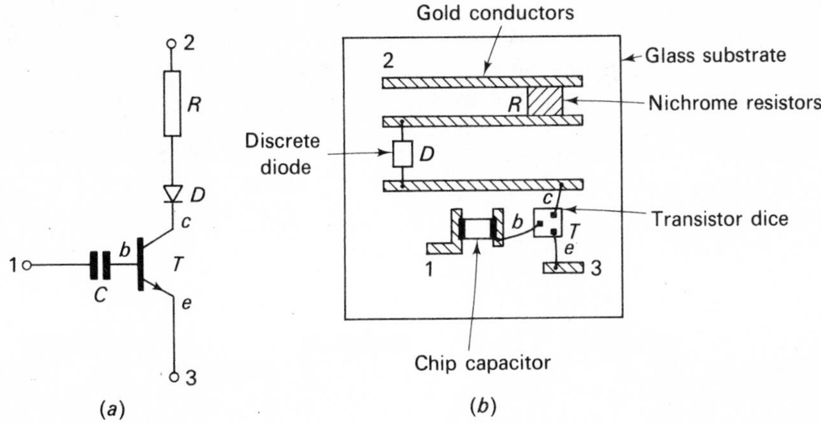

Fig. 1.6. Hybrid thin film circuit; (a) circuit schematic, (b) thin film circuit.

silicon is grown on the insulator, which must have a crystal structure similar to silicon. Sapphire and spinel are suitable materials for this. The silicon is then etched into islands and the unipolar devices are formed into these. The stray capacitance between devices is very low in this technique so that it is capable of high speeds.

1.3 Hybrid integrated circuits

Hybrid integrated circuits can be one of two types, thin film and thick film. Although these circuits look very similar they differ in their production techniques which give the thin film better performance characteristics but makes it more expensive than thick film. Fig. 1.6 shows the construction of a typical thin film circuit. The diode has been shown added in discrete form although unpackaged devices may be used as is done for the transistor. The capacitor is in the form of a miniature block and the conductors and resistor are thin film tracks on the glass substrate. A thick film circuit would look very similar to Fig. 1.6 but would use different substrate, conductor and resistor materials.

Glass is the most frequently used substrate material for thin film hybrid circuits. It must be flat and free from deformation, in order to prevent cracks in the top film, and it must also be chemically stable so as not to effect the film characteristics. Gold is the commonest conductor material since it has a very low resistance, but it does not adhere well to glass. It is usual to cushion the gold with a layer of Nichrome, which is an alloy of about 80 per cent nickel and 20 per cent chromium. Aluminium is also used as conductor material. Although it is cheaper than gold it has a higher resistance and reacts chemically with the gold wires which are sometimes used to connect semiconductor chips to the tracks. Nichrome is the most popular resistor material. It has excellent adhesion to glass and gives resistors with a low temperature coefficient. For high valued resistors materials consisting of a compound of a dielectric and a metal, known as cermets, are used. Capacitors can be added in chip form or built onto the substrate as layers of conductor and dielectric materials. Both tantalum oxide and aluminium oxide are used, and they are formed by first putting down a layer of the metal, oxidizing it to give the dielectric, and then adding the top conductor layer.

The film can be deposited on the thin film substrate by several techniques. Evaporation is the most direct and consists of placing the source material and the substrate in a partial vacuum. The source is heated to vaporize it and this material then settles on the substrate. Source heating can occur by resistance elements or by bombarding it with a beam of high energy electrons. The disadvantage of the evaporation system is that it gives a film which has low adhesion to the substrate and a low density. An alternative technique, known as sputtering, overcomes these disadvantages but is slower in forming a film of a given thickness. In sputtering a glow discharge is formed in argon at between 0.01 to 1.0 torr, by a high voltage applied between the source to be sputtered (cathode) and the substrate (anode). Argon ions are formed and these strike the source releasing molecules which are negatively charged. These molecules bombard the substrate, giving a film which has good adhesion and density. The track pattern can be formed on the substrate by either covering the whole surface with the track material and then using photolithography to etch out the areas not required, or by placing a mask in contact with the substrate, prior to film deposition, with cut outs in the regions where the film is to be placed on the substrate.

The materials used for thick film production are called *inks* or *pastes*. Many different ink compositions are available, depending on the supplier and whether conductor, resistor or dielectric material is required. Basically all inks are made from fine metal and glass powders which are mixed with an organic solvent. The substrate must be pure and flat with a high mechanical strength. It should have good electrical resistance and low thermal resistance. The thick film process involves very high temperatures, above 1000°C, and the substrate must remain chemically stable throughout. A good general purpose substrate material is a compound of 96 per cent alumina and 4 per cent glass, although 99.5 per cent alumina substrates are also used. Beryllia is preferred for power circuits since it has good thermal conductivity.

Thick film tracks are formed by printing the inks through a nylon or stainless steel mesh. The areas where the ink is not required are blocked off by emulsion. Fig. 1.7 illustrates the print operation. After the ink has been printed onto the substrate it is allowed to stand for a few minutes for the ink to coalesce. It is then dried in an oven or under infrared heaters to remove the volatile components of the paste. Following the drying process the circuit is fired in a zoned oven. This first removes the remaining volatile elements from the ink and carbonizes and oxidizes the organic binders. Then the glass content

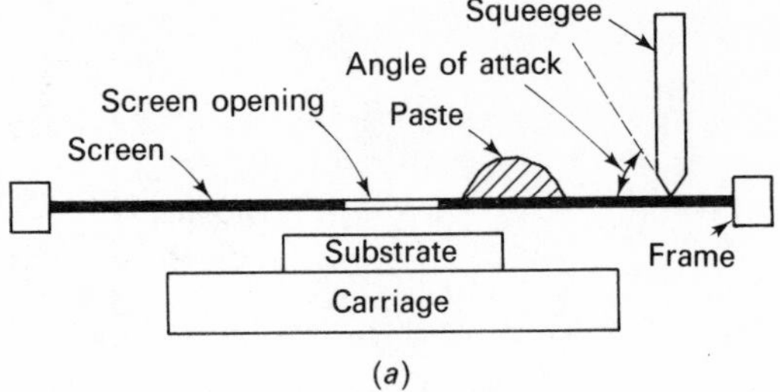

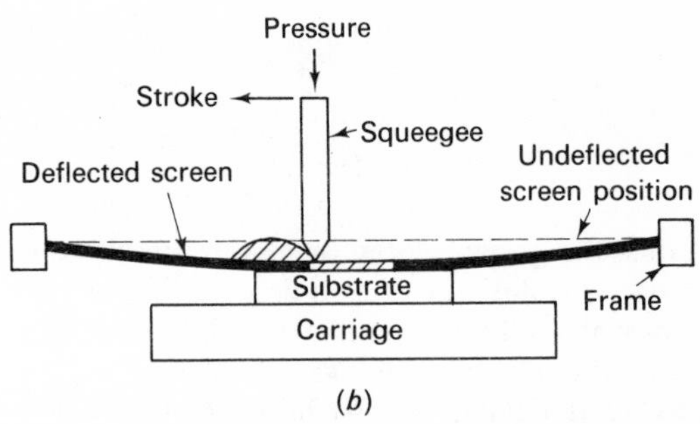

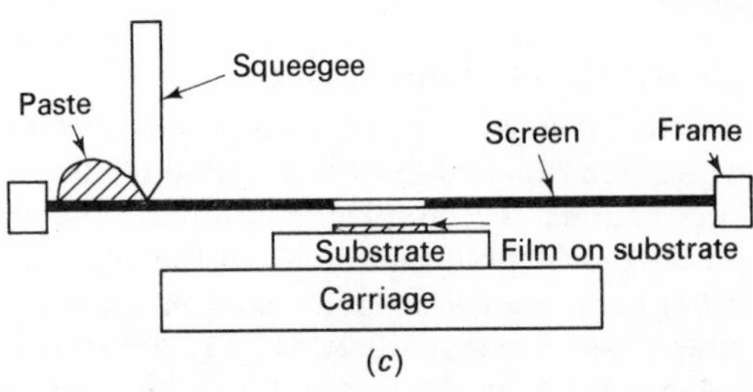

Fig. 1.7. Stages in thick film printing; (*a*) start of stroke, (*b*) during stroke, (*c*) end of stroke.

of the paste is melted, and this forms a seal around the metallic particles as well as fixing the track to the substrate.

Printed resistors have an accuracy between batches of less than 30 per cent. For tighter tolerance, trimming must be used which removes a portion of the track area and so adjusts the resistance value upwards. Two techniques exist for trimming. The first is *air abrasive trimming* and uses a high velocity stream of fine abrasive powder to wear away the film. It does not result in any appreciable temperature rise or shock or vibration, but it does give an 'overspray' action onto adjoining components. The trimming operation also removes the vitreous parts of the resistor paste and exposes the resistors to subsequent contamination. In laser trimming a high energy laser beam is directed onto the film, raising its temperature and vapourizing the required area. It uses more expensive equipment than air abrasive trimming and the laser needs to be re-tuned each time a different colour resistor is trimmed. However, there is no overspray and the high temperature results in a flow of the vitreous contents of the film so that the cuts are resealed.

1.4 Integrated circuit bonding

There are many techniques in use for connecting a semiconductor die onto a hybrid substrate or into its package. First the die must be connected to the substrate and then conducting bonds must be made to its terminals. Eutectic bonding is commonly used for connecting the chip to the substrate. It relies on the fact that an eutectic alloy such as gold–silicon has a lower melting point than both gold and silicon. The area to which the die is to be attached is gold plated and the chip is then pressed and scrubbed onto it to form the eutectic bond. Alternative techniques for die *bonding* include using silicon or epoxy adhesives, and soldering, which requires both the back of the die and the area on the substrate to be gold plated.

The terminals on the die can be wire bonded to the package pins, or hybrid tracks, using thermocompression or ultrasonic techniques. Thermocompression ball bonding uses gold wires and it relies on heat and pressure to form the bonds. Ultrasonic bonding does not require the substrate to be heated and it uses aluminium wire. The wire is pressed and scrubbed at high frequency against the metal contact area and this removes surface oxides and results in a strong molecular bond. Instead of using wires for bonding, flip chip or beam lead techniques can be used. In flip chip bonding contact bumps are formed on the die or the tracks, and the die is 'flipped' or inverted and then connected to the bumps by soldering, ultrasonic, or thermocompression bonding. In beam lead bonding metal leads or 'beams' overhang the semiconductor die and these are attached to the substrate contact areas. In both flip chip and beam lead techniques no separate die bonding is required since the electrical contact and mechanical support are both provided by the beams and balls.

Film carrier bonding is a relatively new system which is intended to greatly reduce the labour costs involved in wire bonding small chips. In general a trained operator should be capable of making a conventional wire bond in a second. Add to this the time involved in mounting and positioning a die under a microscope, and the average time for a 14 lead package (28 bonds in all) is about a minute. This can be expensive when compared to the cost of a small die mounted in a cheap plastic package. Semi-automatic wire bonders are currently available, but these generally do not perform much better.

The film carrier technique uses a continuous film of polyimide which has copper lead frames mounted on its surface. The whole system resembles a film spool. A window in the film allows the inner leads of the frame to protrude. It is to these fingers that the silicon die is bonded. After bonding the lead frame and die are wound onto a second spool for storage until required. The lead frames are then removed from the film and connected into the required circuit or package by thermocompression, ultrasonic, or solder bonding.

1.5 Integrated circuit packaging

The previous section has described some of the methods used for bonding semiconductor chips to substrates or to metal lead frames. The packages for these frames can be made from *hermetic* or non-hermetic (plastic) material. In addition the packages can have a variety of different configurations, such as TO-5, dual-in-line and flat pack.

The basic principle involved in a plastic package is to assemble the circuit on a substrate or metal lead frame and then to mould the entire structure, apart from the leads, in plastic to form the body of the device. The most commonly used plastic materials are epoxy, phenolic and silicone resins of which epoxy is the most popular. Several requirements are placed on the plastic material. It should adhere well to the lead frame so that it prevents moisture from creeping in along the package legs. The material must also have a thermal coefficient of expansion which is matched to the rest of the circuit in order to prevent stresses being set up in it which could damage the leads or the thin bonding wires. It is also important that the material does not contain or release any impurities which would contaminate the enclosed silicon.

Plastic packages are low cost, especially when compared to hermetically sealed devices. They are also ideally suited to volume production techniques. However their resistance to moisture and environmental contaminants is not as good as hermetically sealed packages and they can be damaged under thermal cycling conditions. Plastic packages also have lower heat dissipation compared to other types, especially metal packages. Heat dissipation properties can be considerably improved by wrapping the circuit in a metal surround before plastic encapsulation and in some cases bringing out part of this metal for bolting onto an external heatsink. Plastic packages are also usually characterized for operation over the industrial temperature range of 0°C to 70°C while hermetic devices can cover the full military range from −55°C to 125°C. However for normal industrial use plastic devices are very suitable, and with chips which have been passivated with a glass or silicon nitride layer these devices can be made to operate in fairly hostile environments.

A semiconductor chip is sensitive to the presence of contaminants such as sodium ions, hydrogen, oxygen and water vapour. These reduce the collector–base breakdown voltage of on-chip transistors, increase the leakage current and parasitic capacitance, reduce current gain, and attack the chip metallization and the wire bonds. Hermetic sealing aims to minimize these effects by first removing the contaminants from the package, usually by heating it to about 250°C, and then sealing it in the presence of an inert gas. During sealing the package temperature is kept to as low a value as is practical.

A hermetic package usually consists of a base, a body, leads and a cover. Obtaining a seal between metal surfaces is relatively easy and methods of glass to glass seals also present little problem since both surfaces can be fused at high temperatures, or cemented by means of an interconnecting lower melting point glass. The problem arises when glass to metal seals need to be made, such as when metal leads pass through a glass package. The difficulties generally arise due to the unequal heat conduction and thermal coefficient of expansion of the two materials, which make the seal unreliable under temperature cycling conditions. Two methods are generally used to produce glass to metal seals. In the first method a thin oxide layer is grown on the metal surface prior to coating with glass. The surface of this oxide is dissolved into the glass and results in a smooth transition from metal to oxide to glass. This results in a good seal since the oxide in effect acts as an intermediate buffer. The second sealing technique uses a solder glass seal. This is a special composition low melting point glass which is coated onto the two surfaces to be sealed. On heating the interconnecting glass melts and wets the two surfaces, forming a good seal on cooling.

Three types of materials are used for hermetic packages. These are metals, ceramics and glasses. Of the many metals the most popular is a nickel–iron–cobalt alloy called Kovar. It has a relatively low electrical and thermal conductivity and is compatible with standard sealing glass. Ceramics make excellent hermetic packages and are very commonly used. They have high thermal conduction and a thermal coefficient of expansion which is compatible with that of glass. Many different ceramic materials can be used, the most popular being alumina having a purity between 75 and 99.5 per cent. For high power dissipation beryllia is most common although other ceramics in production are stealite, forsterite, titanate and zircon. The thermal coefficient of all these materials is close to that of metal. Glass packages are cheap since they can give good inexpensive seals. They have poor conductivity and are generally not suitable for anything apart from very low power circuits. Hybrid packages using ceramic bases and glass walls have been used for higher power dissipations.

So many different sizes and shapes of packages have been used in the manufacture of integrated circuits that it would be folly to attempt a description of them all in this short section. Instead only a few of the more popular packages are considered here.

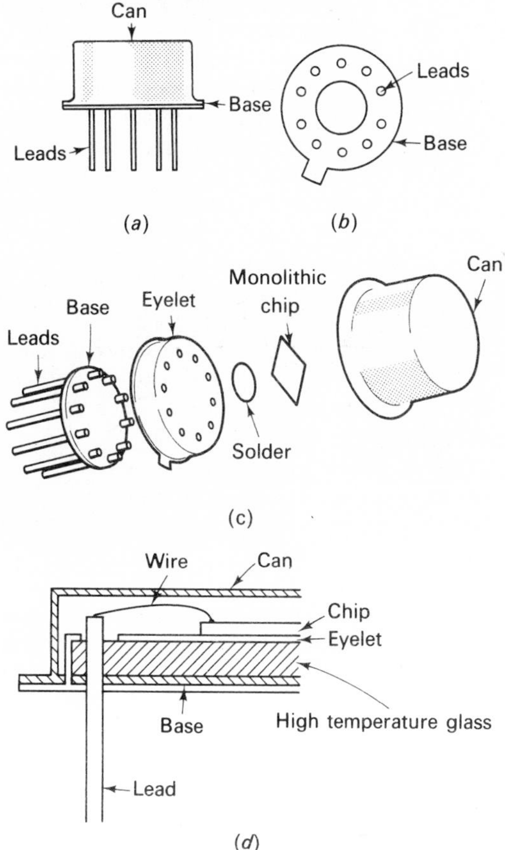

Fig. 1.8. TO-5 package; (a) side view, (b) underneath view, (c) exploded view, (d) cross sectional view.

The TO type of package was a natural development from the transistor case. Various sizes are in use such as TO-3, TO-5, TO-8, TO-99 and TO-100. The number of pins in these packages generally varies between 8 and 14. Fig. 1.8 shows the construction of one type of TO package. The base is a gold plated header made from Kovar. The leads pass into the

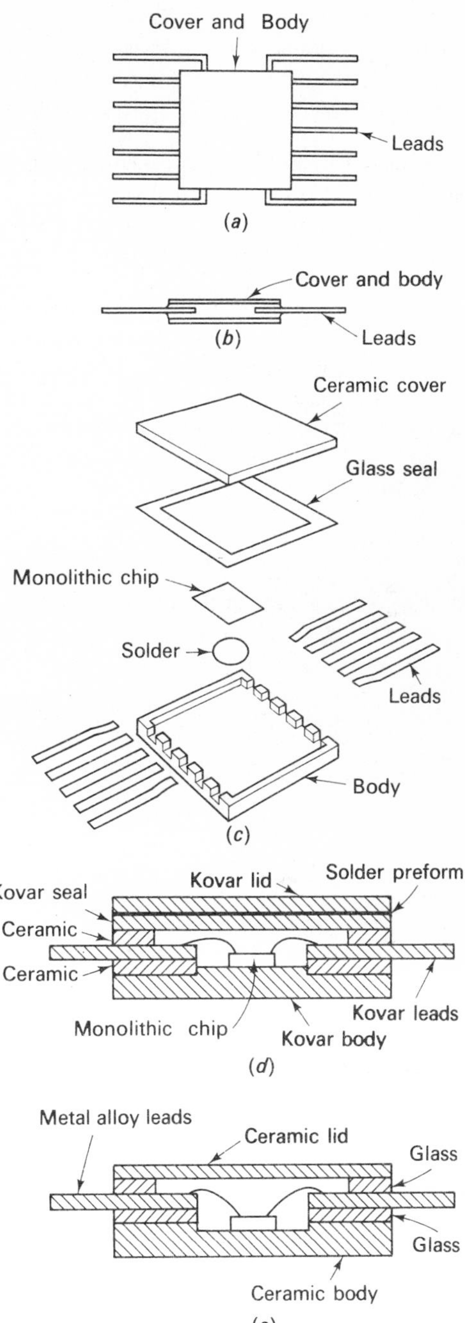

Fig. 1.9. Flat package; (a) top view of 14 lead package, (b) side view, (c) exploded view of 10 lead ceramic package, (d) cross section of metal package, (e) cross section of ceramic package.

header via glass to metal seals. The die is attached by means of a gold–silicon eutectic solder. The can or cover is usually also made from Kovar and is welded onto the header flange.

Flat pack encapsulations also come in many varieties. Fig. 1.9 shows typical structures. The dice is bonded to the base by eutectic solders. The gold plated leads are embedded in a glass frame and the lid is sealed with a low temperature glass frit. A metal flat pack employs Kovar for most parts except for the walls between the ring and base which is glass. The leads pass through this. A glass package is usually made from borosilicate glass and consists of a one-piece base and ring assembly in which the Kovar leads are sealed. This package has good thermal and mechanical properties. The lid can be made of glass, ceramic or metal and is attached by a low melting point sealing glass. Ceramic flat packs are similar to metal devices. The base, ring and lid are now usually made from alumina or, for higher power dissipation, from beryllia.

The dual-in-line package has become very popular especially for monolithic integrated circuits, since it is convenient to handle and can be readily adapted for automatic insertion into printed circuit boards. Fig. 1.10 shows a typical device, which can be made from metal or ceramic. The chip is placed in a cavity and bonded by glass frit. The leads also pass through glass frit seals in the package. The ceramic lid is initially metallized and then brazed or solder sealed to the body, or a glass frit can again be used to connect the two together. Plastic dual-in-line packages do not have a cavity for the chip. It is essential in these instances that the moulded package completely surrounds the die, to protect it from the hostile environment, and all seals between the body and the leads are reliable. The overall outlines of these packages are essentially as in Fig. 1.10.

Packages used to house *large scale integration* (LSI) devices can present a problem. These chips are relatively large and generally have many leads, up to about forty. The most popular configuration is at present still a dual-in-line, but due to the necessity of maintaining

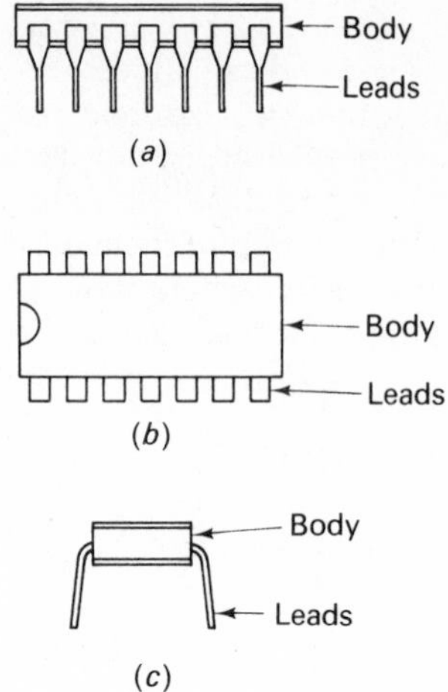

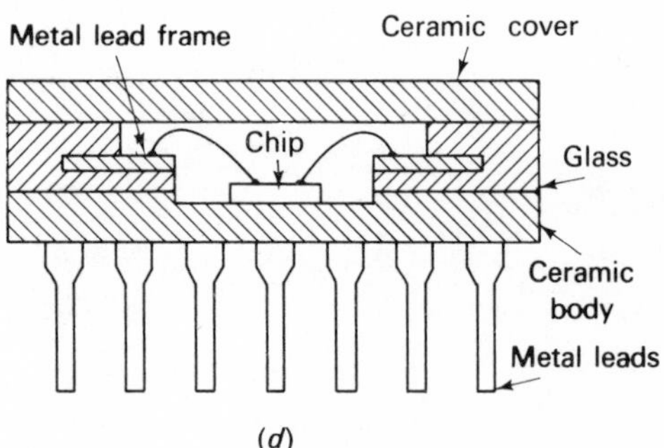

Fig. 1.10. Dual-in-line package; (*a*) side view, (*b*) top view, (*c*) end view, (*d*) cross section of ceramic package.

0.25 centimeters spacing between pin centres the package tends to be relatively bulky. It is difficult to align the pins and to keep the device in position during insertion into a board or socket. The pins also tend to be fragile and break, and unsoldering a board to remove a device for replacement is often a very tricky operation. To overcome these limitations several different types of packages are in use. The edge-mounted package, shown in Fig. 1.11 is similar to a miniature printed card. The chip is housed in a cavity which is sealed. Connections are made to the chip and to the thick film conductor tracks which are printed onto the ceramic substrate. The tracks have gold plated fingers which plug into an edge connector.

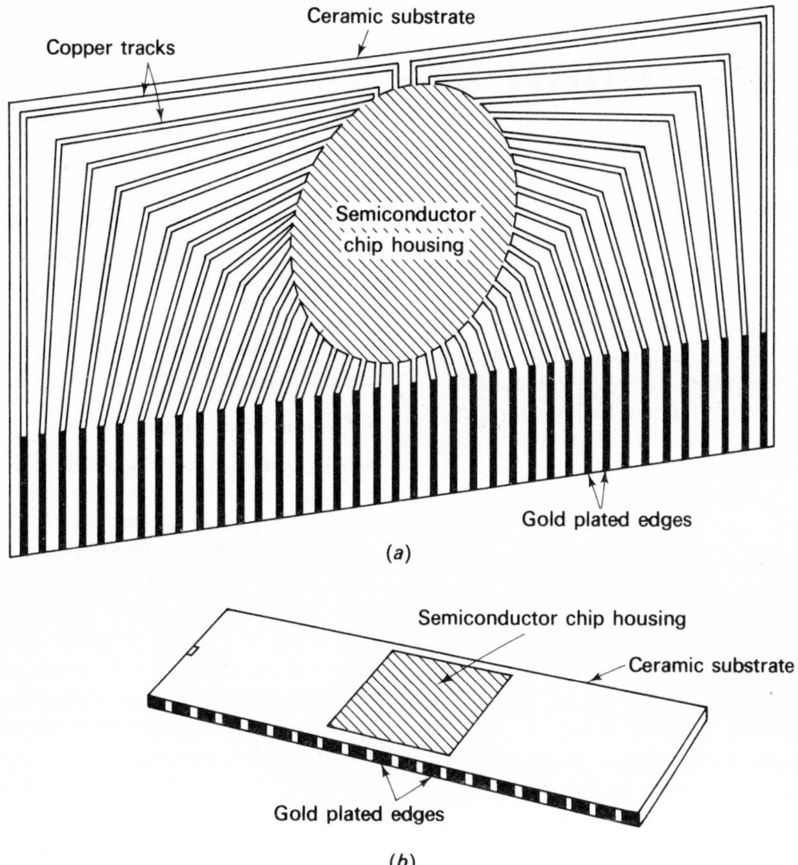

Fig. 1.11. LSI packages; (a) edge mount, (b) leadless.

Although such a package can be rapidly inserted and removed from its holder the fingers tend to wear with each operation. It is also unsuitable for many external connections since the long ceramic substrate has a tendency to bow. Furthermore the assembly is relatively expensive. An alternative approach is the leadless package. The chip is once again sealed in a central cavity and connected to printed tracks. The tracks are brought out to bumps (not pins) on either the side, or bottom of the package. When in use the package is placed in a receptacle which has conducting pads located near the bumps on the package. Once the cover of the receptacle is closed it squeezes the package bumps and pads into close contact, connecting the chip into circuit. This package is claimed to have several advantages such as low cost, freedom from pin damage and ease of replacement. It is however still relatively expensive and not widely used in industry.

Hybrid packages often resemble larger versions of those housing monolithic devices. These can be TO, dual-in-line or flat pack in construction. In addition the material used may be metal, ceramic or plastic.

2. Logic families

2.1 Introduction

Digital integrated circuits are available as functional building blocks to perform certain specialized routines such as addition, storage and so on. They can therefore be classified according to these functions, and this is done in chapter 4. However there is a much more fundamental difference between various integrated circuit types and this is related to the way a certain group or family of devices behaves regarding speed of operation, the ability to provide large output currents and so on. Each family has similar functional building blocks but the *characteristics* of these blocks are different. In this chapter the construction of the various logic families is described, with reference to a logic gate.

Fig. 2.1 illustrates how a gate works. Assume that the diodes are ideal in that they have zero forward volts drop and infinite reverse resistance. If all the inputs are capable of switching between zero and +V volts then the output will be at + V volts if, and only if, *A* AND *B* AND *C* are all at +*V* volts. Defining +*V* volts as a logic '1' signal and zero volts as logic '0' then all inputs must be at '1' to obtain a '1' output. This is an AND gate in positive logic. However if we define a logic '1' signal as zero volts and logic '0' as +*V* volts then if *A* OR *B* OR *C* goes to zero volts (logic 1) then its associated diode will conduct and the output will go to zero volts (logic 1). Therefore Fig. 2.1 also represents an OR gate in negative logic.

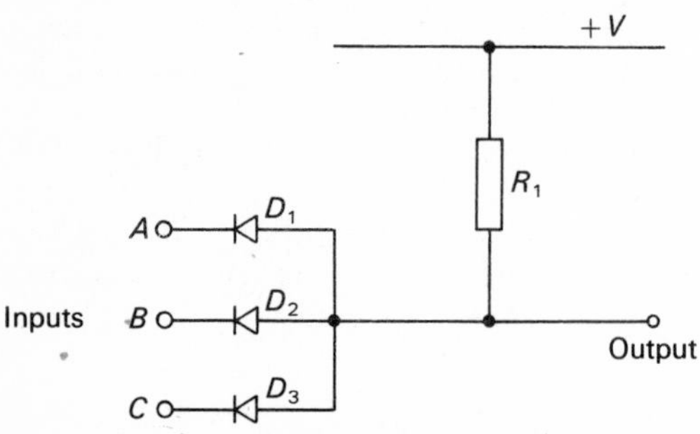

Fig. 2.1. Diode AND gate in positive logic.

2.2 Logic parameters

Before going on to look at the various types of gate structures which can be used it would be useful to consider some of the important parameters which may be used in comparing them.

(1) Speed. The speed at which a *logic family* can operate is obviously an important consideration in some systems. It is usually specified in terms of gate propagation delay, which is the time between equal events in the input and output waveforms. For an AND gate it is the delay between a point on the output waveform and an equal point on the waveform of the last changing input. For an OR gate the first changing input would be used.

(2) Loading. All logic systems are made up of many interconnecting functional blocks and it is often necessary to drive several blocks from a common output of a gate. The amount of output power which the gate can provide is therefore an important consideration. Usually families of the same type are interconnected and the drive capability is specified as the 'fan-out' which is defined as the number of similar gates which can be driven simultaneously from one output. The 'fan-in' of a gate is the number of parallel inputs. This is three for the gate shown in Fig. 2.1. However many logic families enable the number of inputs to be expanded by additional circuitry, provided a maximum value, determined by the gate design, is not exceeded.

(3) Noise immunity. *Noise* always exists in a logic system. It may be a slowly changing noise signal, such as the drift in the power supply rails, or a rapidly changing spike of noise. Noise

can cause unwanted changes in the logic states and result in faulty system operation. Some logic families are capable of withstanding higher levels of noise than others, without changing their state. In the case of fast changing noise, the wider the noise pulse the smaller the amplitude that the logic family can tolerate. This is shown in Fig. 2.2. Logic family *B* will be generally slower than family *A* so that it does not respond as quickly to the noise spikes. However other design considerations may also be used to increase the noise immunity of a logic family. This is especially the case when d.c. noises are considered.

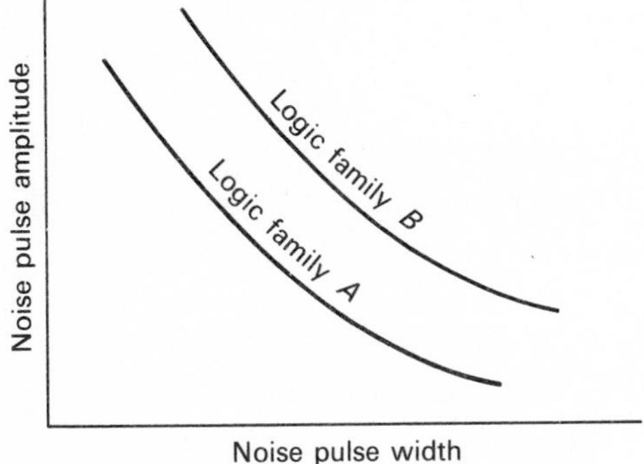

Fig. 2.2. Noise immunity of logic families.

(4) Noise generation. The amount of noise generated by a logic family when it switches between its various states is also an important consideration. This noise is fed through the power supply rail and affects other circuits connected to it. More noise is generated when current is rapidly switched between several levels. Logic families which do not vary the current appreciably during operation are prone to less noise generation.

(5) Power dissipation. The power which a logic circuit dissipates is equal to its supply voltage multiplied by the mean current it draws. This power varies with the loading on the gate and its operating speed. Therefore when making comparisons between families the operating condition must be clearly defined. Large logic systems can draw considerable power so that the cost of the power supply becomes appreciable. Furthermore the

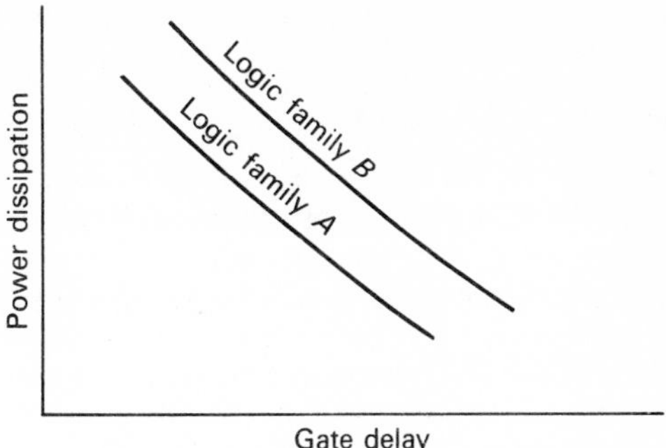

Fig. 2.3. Speed–power curves.

amount of logic one can usually build onto a single silicon chip is often determined by the power dissipating capabilities of the package. Therefore the lower the power dissipation of the logic family the higher its potential level of integration.

(6) Speed-power product. Generally logic gates can be designed to have a range of speeds. Unfortunately the power dissipation also increases with the operating speed. Fig. 2.3 shows typical speed–power curves in which the speed is measured in terms of gate delay. In the ideal gate both the power dissipation and gate delay should be small, therefore the lower the speed–power product the better the logic family. In Fig. 2.3 family *A* is preferable to family *B*.

(7) Flexibility. It is often possible to use logic circuits in relatively unconventional ways in order to derive a certain circuit function. For instance some logic families allow their outputs to be connected directly together. Others can drive various different types of families without the need for complex interface circuitry. These features enable minimum cost designs to be obtained in certain types of applications.

(8) There are many other parameters which are important when comparing logic families. These include cost, availability of different logic functions within a given family, and the environmental range over which the family can operate.

2.3 Saturating bipolar logic

The logic families which will be considered in

this section are all made from bipolar technology and their circuit transistors switch between off and saturated modes when the logic switches between its two states. There have been many different logic families. Some of these have proved to be more popular than others. Many of the families are not now available as general purpose small scale or medium scale devices. However, due to their unique advantages for certain applications they are still frequently used in LSI circuitry.

The diode gate shown in Fig. 2.1 is limited in its range of applications. It does not have any internal amplification so that fan-out is low. The first family to employ amplification was that using direct coupled transistor logic or DCTL. This was followed by resistor–transistor logic (RTL) and diode–transistor logic (DTL). However the logic family which first gained widespread acceptance was transistor–transistor logic or TTL. Many versions of this have been built, a few of these being shown in Fig. 2.4. Essentially it consists of a multi-emitter input transistor. This can be conveniently fabricated in integrated circuit technology. Referring to Fig. 2.4 (*a*), if any of the inputs are at logic 0 then current will flow down R_1 to that emitter and *TR*2 will be off so that the output is at logic 1. If all inputs go high then currrent will flow via the base–collector diode of *TR*1 and turn *TR*2 on. The output is now at logic 0. The circuit therefore performs the NAND function in positive logic. Transistor *TR*1 is designed to have a low inverse gain so that the current drawn from the inputs when they are at a high voltage is small.

Fig. 2.4 (*b*) shows an alternative TTL circuit in which a push–pull or totem-pole output stage is used. *TR*1 drives *TR*2 which acts as a phase splitter. It ensures that either *TR*3 or *TR*4 is on at any one time. Because of this R_4 can be made small and the TTL gate has a low output impedance in both the logic 0 and logic 1 states. This ensures rapid charging of output capacitors, and increases speed. However when *TR*2 switches on it rapidly turns *TR*4 on and attempts to turn *TR*3 off. Due to hole storage this is not accomplished quickly so that for a short time both *TR*3 and *TR*4 are conducting. This results in a spike of current from the supply which is limited by R_4. The current spike not only generates noise but it also increases the dissipation through the gate. This problem gets worse at high frequencies and can limit the maximum switching frequency of the TTL gate.

One of the disadvantages of the TTL gate shown in Fig. 2.4 (*b*) is that, due to its low impedance in both directions, two such gates cannot have their outputs connected together. This can however be accomplished using three state TTL gates, as shown in Fig. 2.4 (*c*). Essentially it consists of a control input which forces *TR*1 into a forward saturation state, so that *TR*2 is off and therefore *TR*5 is off. However the same input also pulls down the collector of *TR*2 so that the top transistor in the totem-pole arrangement is also off. The output is now in its third stage of high impedance. Three state gates enable some very useful multiplexing operations to be performed. Note that in Fig. 2.4 (*c*) a Darlington output stage is shown for the source transistor so that a greater output current is possible. No diode is now required in the output stage since the Darlington pair provides two base–emitter voltage drops.

The characteristics of a TTL gate can be readily changed by altering its circuit components. The ratio of R_2 to R_3 in Fig. 2.4 (*b*) and (*c*) is important as it determines the magnitude of the switching current spike. Resistors R_2 and R_4 determine the power dissipation and if these are made large the gate dissipation is considerably reduced. This is called a low power TTL gate (LPTTL). However the higher impedances give slower speeds, but generally the reduction of speed is not in proportion to the increase in resistor size. This is because the lower dissipation means that transistor geometries can be reduced so that their capacitance is also lower. Low power TTL has about one tenth the dissipation of a standard TTL gate and about three times its propagation delay.

2.4 Non-saturating bipolar logic

The logic families described in section 2.3 were referred to as saturating since their transistors were driven well into the saturation region when conducting. This results in hole storage within the transistors with the result that their operating speed is reduced. This can be

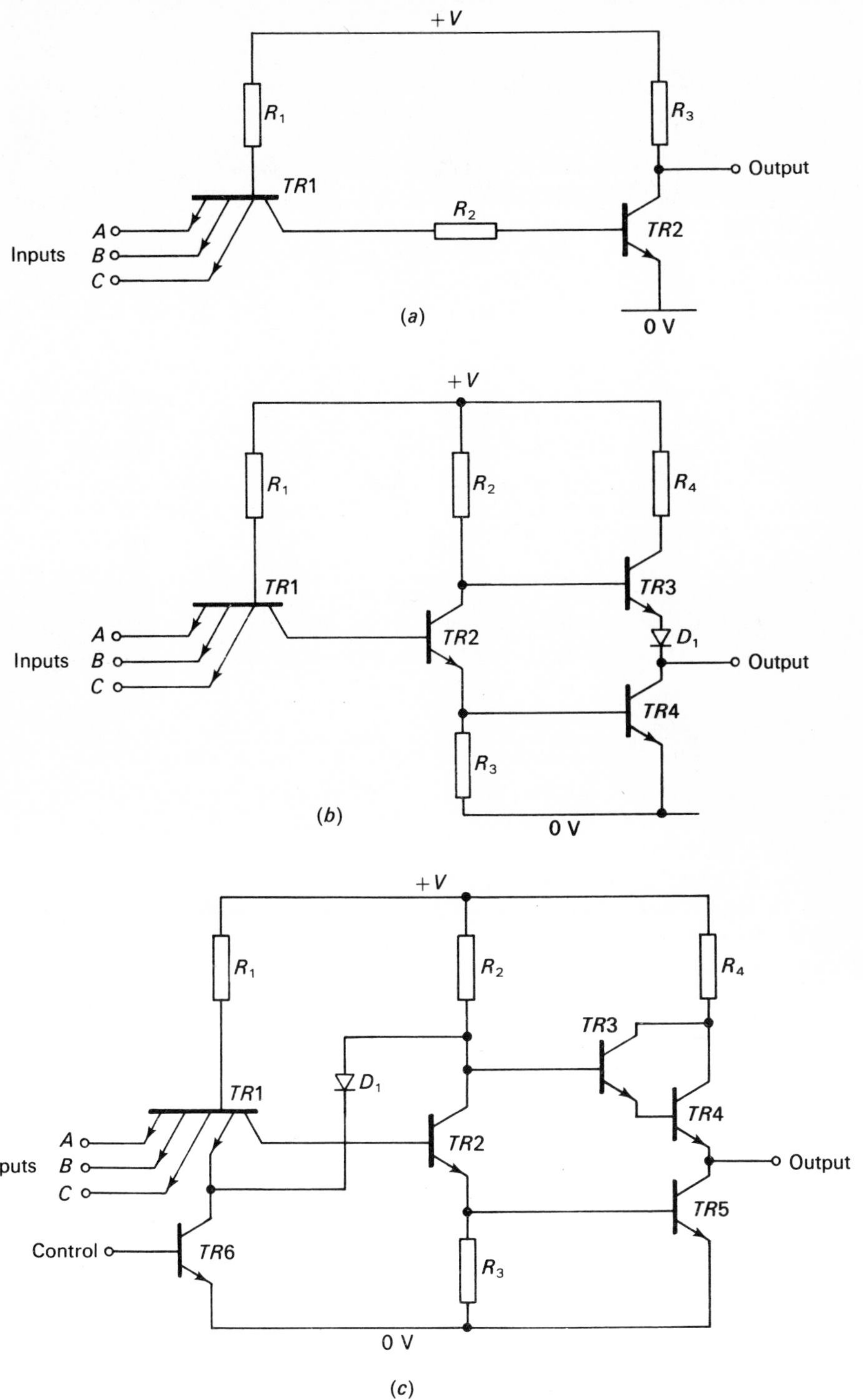

Fig. 2.4. Transistor–transistor logic gates (TTL), (*a*) simple gate, (*b*) improved gate, (*c*) gate with Darlington output and three state control.

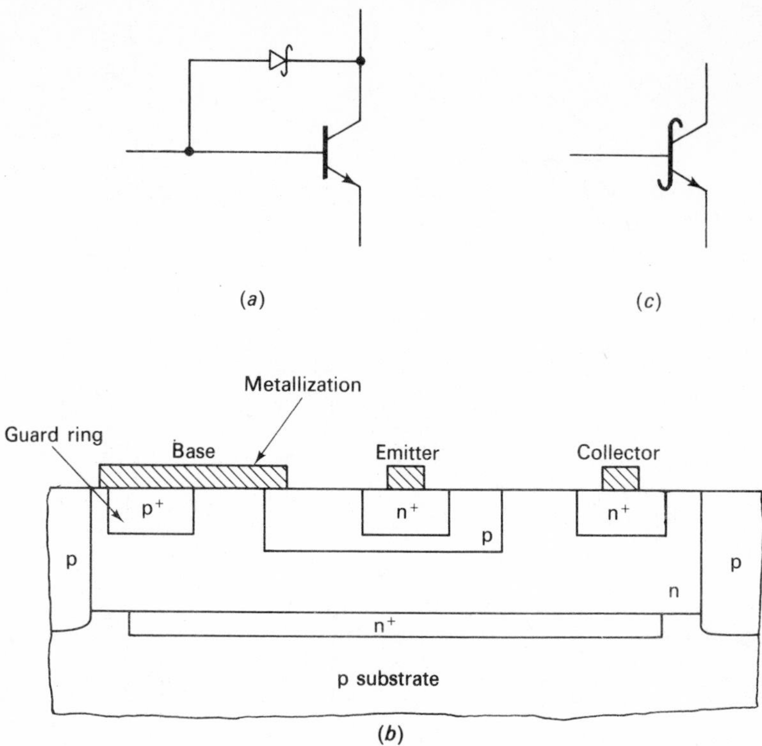

Fig. 2.5. Schottky transistor; (a) circuit arrangement, (b) construction, (c) symbol.

overcome by gold doping the devices but the process is expensive and can result in some yield loss. An alternative method of preventing the transistors from saturating is to connect a low voltage drop diode across their base–collector junction, as in Fig. 2.5 (a). Provided the diode voltage drop is less than that of the base–collector region, current will be diverted through it when the transistor is on so that excess charge build-up in the base region is avoided.

The diode used in Fig. 2.5 (a) is a Schottky device which has a voltage drop of about 0.4 volts. Fig. 2.5 (b) shows how the diode can be incorporated into the transistor structure remembering that the anode of the Schottky diode is the metal base connection and its cathode is the collector n diffusion. A p^+ guard ring is used to avoid edge effects. Fig. 2.5 (c) gives the symbol for a Schottky transistor and Fig. 2.6 shows a TTL–Schottky gate. It is similar in principle to the TTL gates illustrated in Fig. 2.4 except that Schottky diode clamped transistors are used. Note that *TR*4 is a conventional device since it is prevented from saturating by the clamping action of transistor *TR*3. Two additional features are illustrated in Fig. 2.6. These can also be used in conventional TTL circuits. The inputs have been clamped by diodes to protect against high speed oscillations, and *TR*5 ensures a rapid turn off of *TR*6. The gain of the transistors increases with temperature and this would cause the turn off time of *TR*6 to increase. A resistor across the transistor base–emitter would also increase in value with temperature so that its effectiveness would decrease. The current through *TR*5 which is called an active pull down circuit, increases with voltage so that it compensates for temperature changes.

Schottky transistors can be designed with higher gains than conventional circuits since

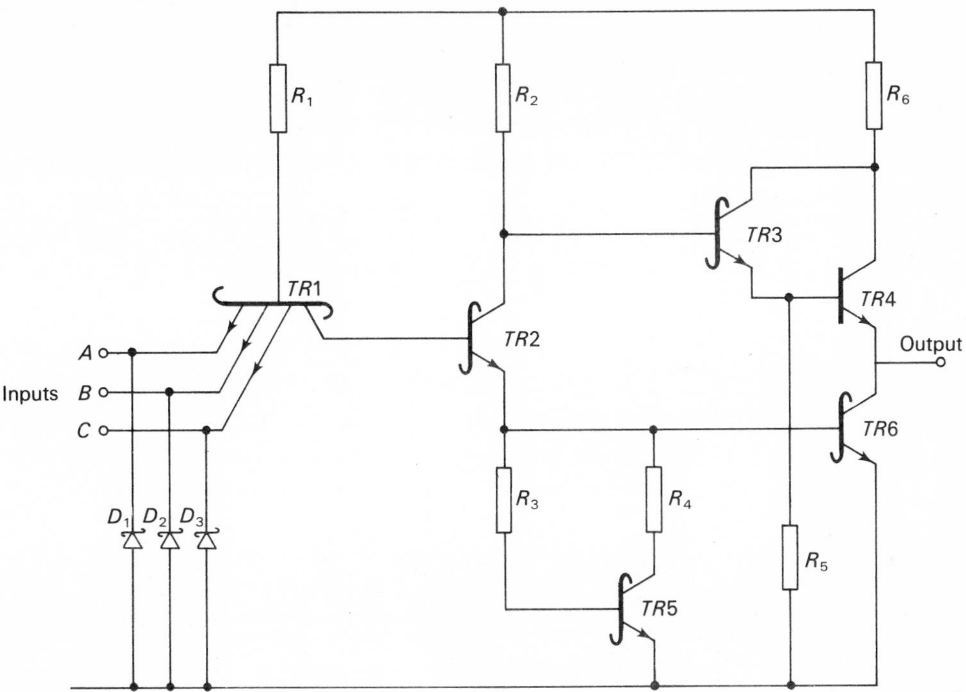

Fig. 2.6. A TTL–Schottky gate.

they have no hole storage. However the Schottky junction introduces capacitance into the circuit so that *RC* time constants need to be minimized by careful design.

Another form of logic family, which has inherently non-saturating properties, is called emitter coupled logic (ECL) or current mode logic (CML). It is currently the fastest logic family and is illustrated in Fig. 2.7. The inputs are compared against a stable reference voltage and, depending on their value, current is switched in the system. Transistors *TR*1, *TR*2, and *TR*3 form the inputs and are also part of a differential amplifier with transistor *TR*4. This gives the system its high input impedance. Transistor *TR*5 and its associated circuitry provide a stable reference voltage at point *P* and *TR*6 and *TR*7 give the low impedance outputs.

With all inputs at a low voltage transistor *TR*4 is on so that *TR*7 is off and *TR*6 is on. If any of the inputs go high its corresponding transistor comes on and diverts the current away from *TR*4. Therefore *TR*6 goes off and *TR*7 comes on. The gate is therefore capable of simultaneously producing both the OR and NOR functions and this adds considerably to its versatility in designs. It should also be noted that the gate draws almost constant current from the supply, if the load is not considered, irrespective of its state since the current is merely switched around the system. It does not therefore generate current spikes like TTL. The output transistors are high gain, low impedance devices capable of providing current drives of the order of 50 milliamperes. This gives the gate a high fan-out although the additional *loading* reduces its operating speed.

The speed of an ECL gate results from its low output impedance, which allows it to drive capacitive loads, and its non-saturating operation. Its logic voltage swings are in the region of only 800 millivolts and this makes it generally difficult to interface to other forms of logic. The low voltage levels also means that it is more susceptible to noise.

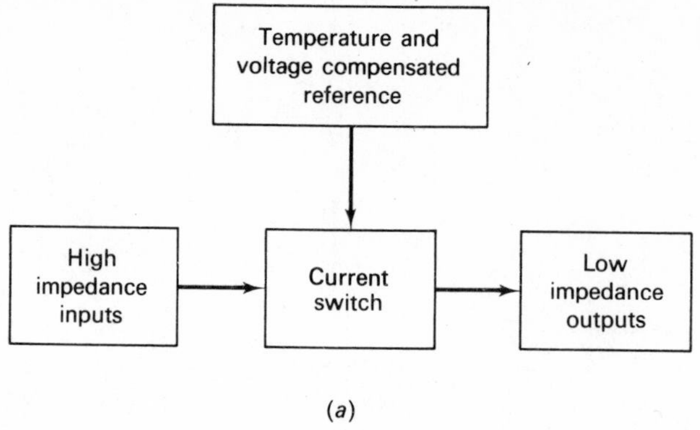

(a)

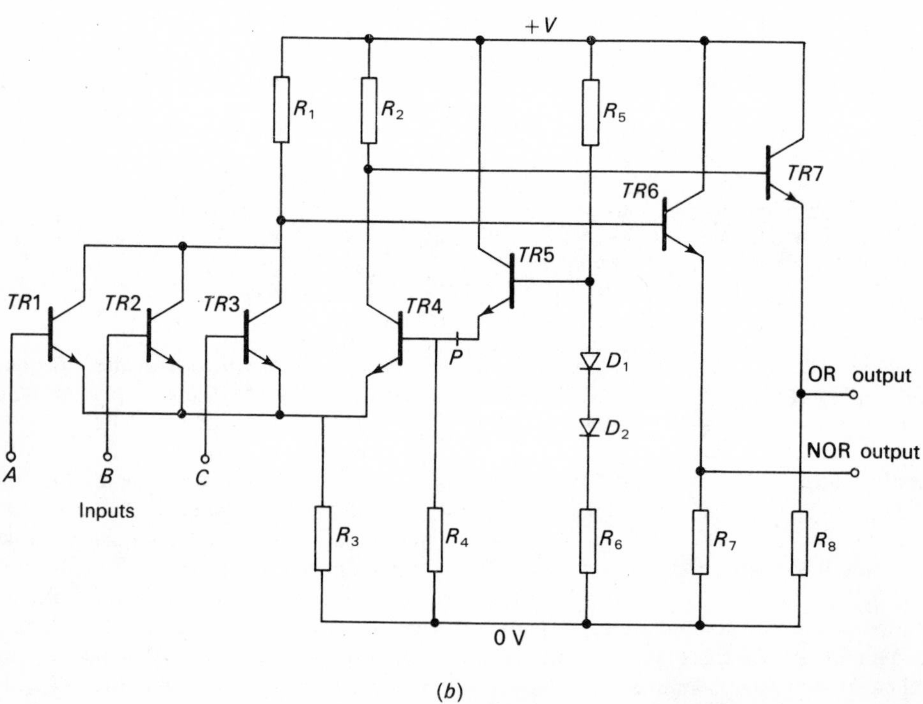

(b)

Fig. 2.7. An ECL gate; (a) block schematic, (b) circuit.

2.5 Bipolar LSI logic

The logic families described in previous sections of this chapter are used to make a variety of logic devices, such as gates, which are marketed as building blocks for larger systems. This section deals with logic families which have primarily been used for LSI systems.

Fig. 2.8 shows a logic family known as integrated injection logic (IIL), merged transistor logic (MTL), or multicollector logic (MCL). It is based on the old DCTL gate in which all the emitters go to a common point and so also do the bases of transistors which are driven from the same resistor. Therefore for these devices only the collectors are separate. It is therefore

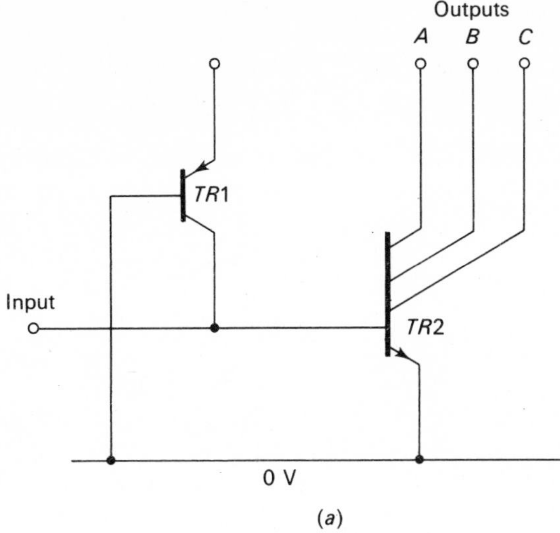

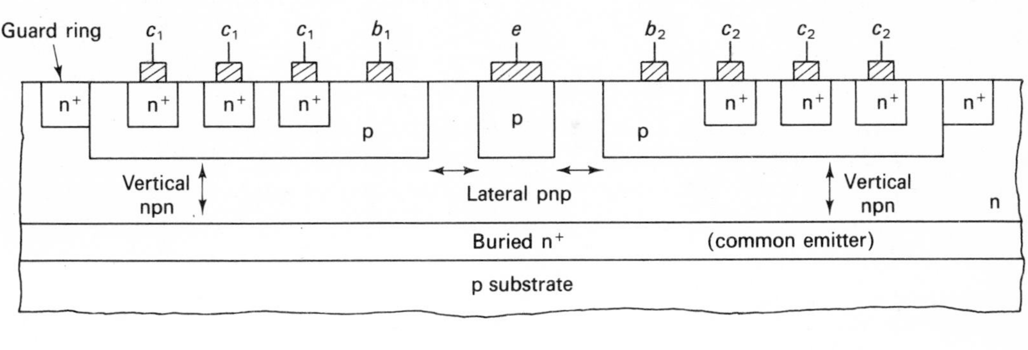

Fig. 2.8. Integrated injection logic (IIL); (a) electrical schematic, (b) construction.

possible to use inverted transistors with multicollectors to combine all devices with common emitters and bases into one structure. The driving base resistor can also be replaced by a pnp current source. Such an arrangement is shown in Fig. 2.8 (a). The pnp transistor provides the injection current into the base of the npn, hence the name injection logic. This transistor has its base and collector regions common to the emitter and base regions of the npn device. If it is made into a lateral transistor then it can be merged into the npn structure, hence the alternative name of merged transistor logic. Such an arrangement is shown in Fig. 2.8 (b). In this structure a single injector is used to provide current to two npn devices. It should be noticed that no isolation regions or resistors are required with this form of logic so that its circuit density is considerable. Guard rings are normally included to prevent crosstalk between adjacent devices. They can be brought right up against the base region, as shown, and therefore take up little extra space. A single injector can be used for several npn transistors provided they are all operating at similar injector current levels. Since this current varies exponentially as the voltage it is essential to use a low impedance injector rail or transistors at the far end of the injector will receive insufficient current.

In the basic IIL device shown in Fig. 2.8 (*a*) the output transistor is on when the input voltage is high and it goes off when the input is low since injection current from *TR*1 is diverted from the base of *TR*2. Fig. 2.9 shows how this basic structure can be interconnected to give

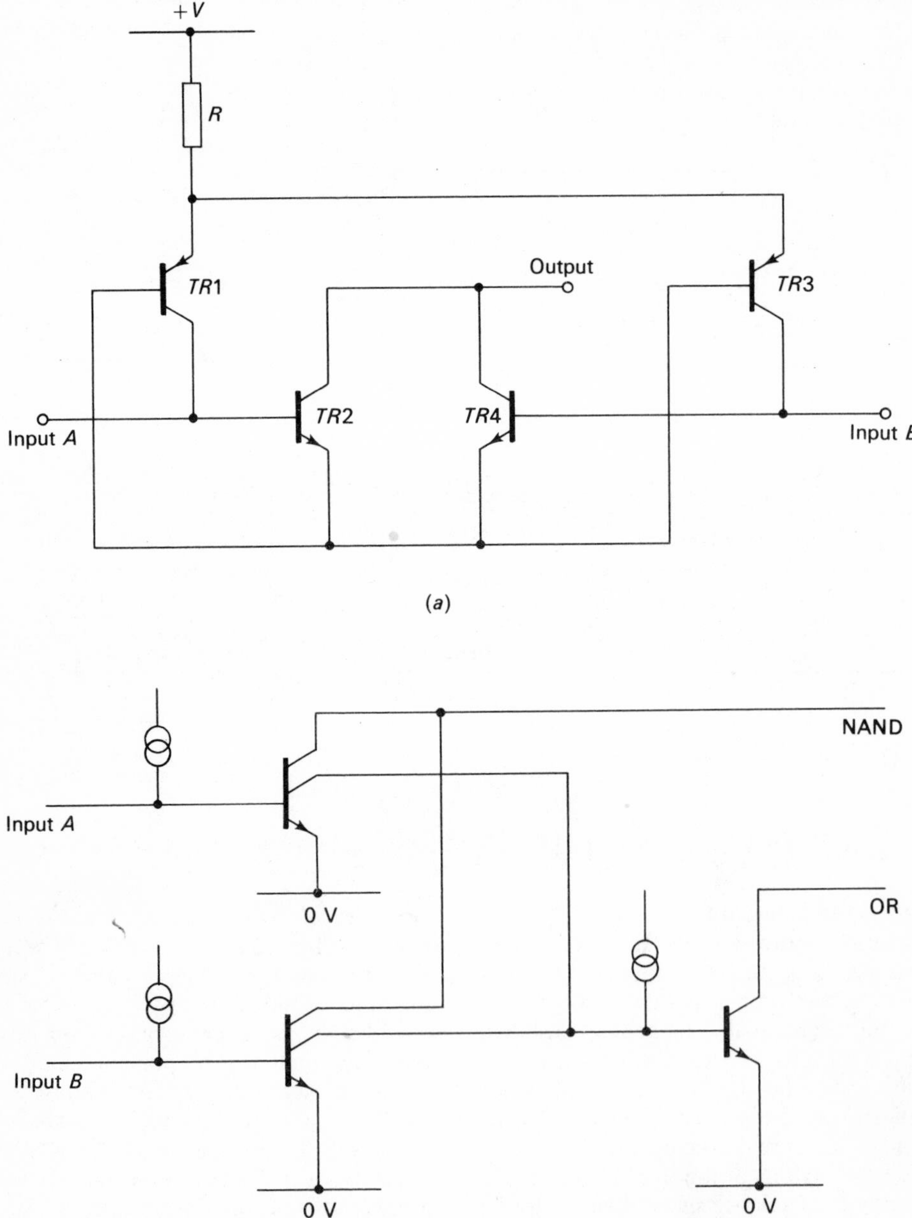

Fig. 2.9. Integrated injection logic gates; (*a*) NAND, (*b*) NAND and OR.

NAND and OR gates. For the NAND gate the output is a logic 1, defined as an open circuit in this case, only if both inputs A and B are at logic 0. When one considers that this NAND gate can be fabricated as in Fig. 2.8 (b) and two collectors from each npn are still spare, the advantages of IIL become evident. Also, as seen from 2.9 (b), the multicollector structure enables several logic combinations to be generated using a minimal number of devices.

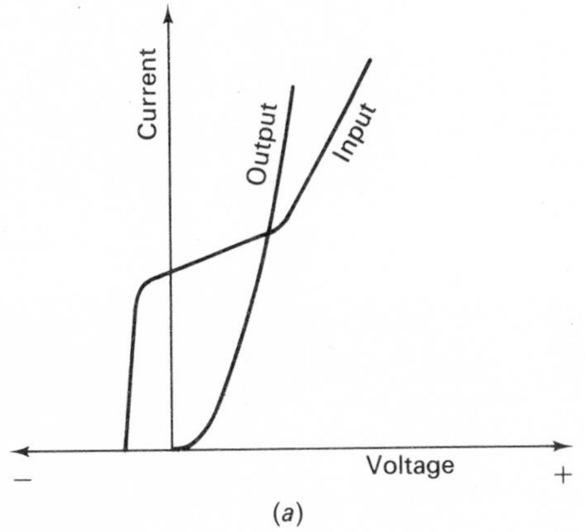

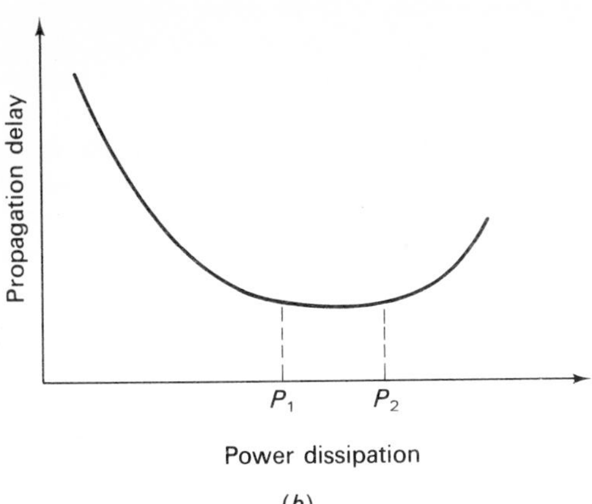

Fig. 2.10. Injection logic characteristics; (a) input and output, (b) speed–power.

The speed and noise immunity of an IIL gate is improved by increasing the injection current. However this also increases the power dissipation. The speed–power product of the logic family is very low, below 1 picojoule, and this means that if high speed is not required the dissipation can be made small enabling large single chip LSI devices to be fabricated. Fig. 2.10 (a) shows the input and output characteristics of the logic family. The input not only sinks current from the injector but it can also feed current into the base of the npn transistor. The output characteristic is basically that across the collector–emitter of the npn device. The speed–power curve shown in Fig. 2.10 (b) illustrates the increase in speed as injection current, and therefore power dissipation, is increased. If V is the power supply voltage, I the injection current, C the parasitic junction capacitances, then gate delay D and power dissipation P are given by

$$D = \frac{CV}{I} \tag{2.1}$$

$$P = VI \tag{2.2}$$

Therefore the speed–power, or more correctly the delay–power product is equal to

$$DP = CV^2 \tag{2.3}$$

It is therefore independent of the injection current. Fig. 2.10 (b) therefore shows this as a relatively straight line up to P_1. For power greater than P_1 the main delay in the gate is caused by the active charge in the transistors which is proportional to the injection current so that the delay remains constant, although the dissipation increases from P_1 to P_2. After P_2 the series resistance of the transistor base prevents rapid removal of accumulated charge and the delay starts to increase.

Several techniques have been used to improve the speed–power product of the IIL gate. Fig. 2.11 for instance shows a three-sided injector construction which reduces the parasitic capacitance of the side walls of the npn transistor and therefore reduces the speed–power product. However the packing density of this arrangement is also less. Fig. 2.12 shows an alternative configuration in which the npn transistor is vertically above the injector. The substrate forms the emitter of the injector hence this type of arrangement is called substrate fed logic (SFL).

The final form of logic to be described here is

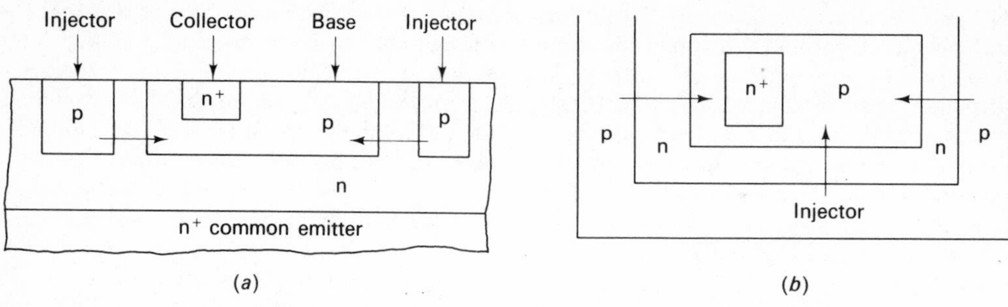

Fig. 2.11. Injection logic with current injection from three sides; (a) cross sectional view, (b) plan view.

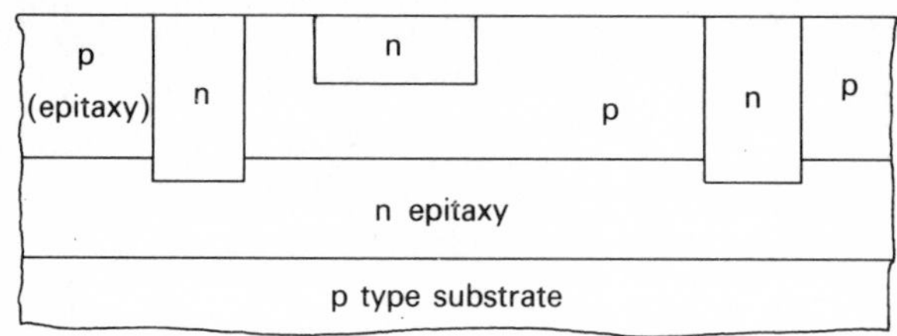

Fig. 2.12. Substrate fed logic (SFL).

called current hogging logic (CHL). It is based on the lateral pnp transistor as shown in Fig. 2.13. If the collector of this transistor is biased negative then charge from the emitter will be collected by the collector and a negligible amount will pass to the substrate. However if the collector is left floating then the charge from the emitter flows to the collector and then onto the substrate as shown by the dotted arrows. Therefore the substrate current can be regulated by the collector voltage. The CHL structure shown in Fig. 2.14 uses two collectors. When the inner control collector c_1 is biased negative no charge flows to the output collector c_0. When c_1 is left floating, charge flows to c_0. By using two control collectors it is possible to make NAND and NOR gates. The output collector feeds the base of a npn transistor in a logic system and this provides the output voltage signal.

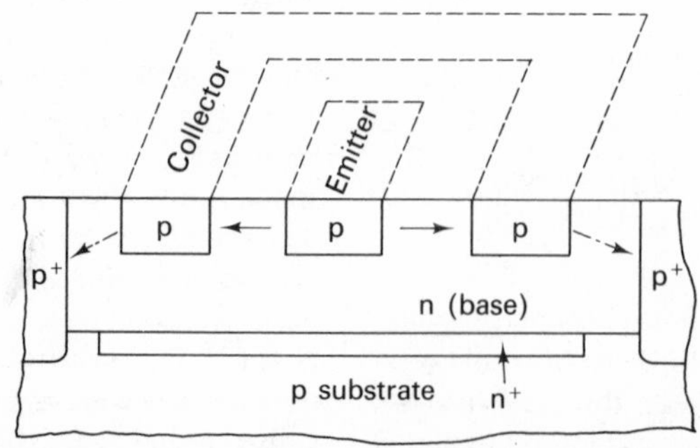

Fig. 2.13. A lateral pnp transistor.

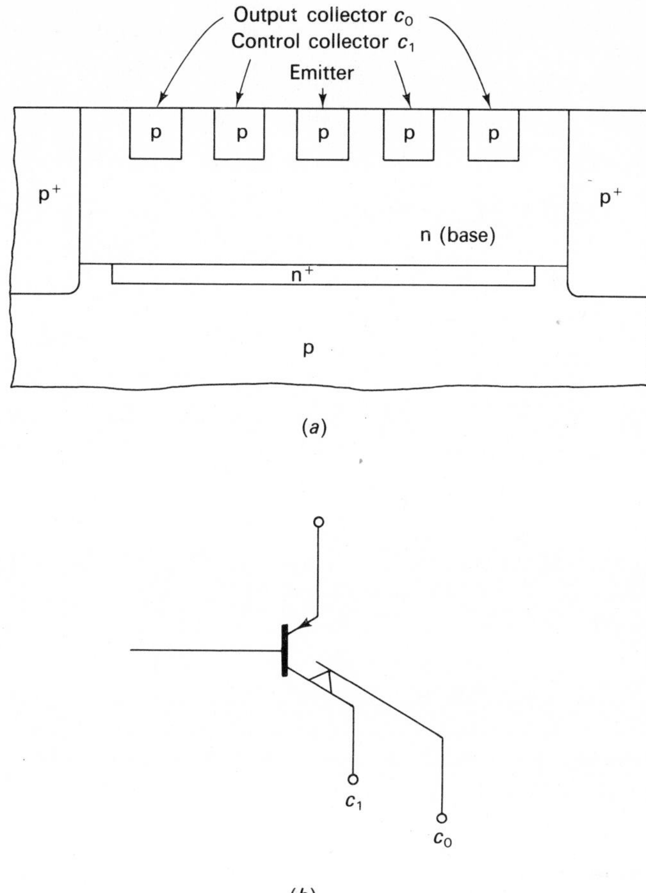

Fig. 2.14. Current hogging logic (CHL); (*a*) construction, (*b*) symbol.

2.6 Unipolar LSI logic

Most types of unipolar device fabrication processes are capable of producing a high density of circuits so that their potential is not fully utilized in *small scale integration* (SSI) and *medium scale integration* (MSI) where the chip size is often dominated by the bonding pad areas. Therefore the logic to be described here is primarily used in LSI systems.

Since a p-channel unipolar transistor turns on when a negative voltage is applied to its gate it is usual, in unipolar systems, to discuss the circuit in terms of *negative logic*. For instance Fig. 2.15 shows a unipolar *inverter*. V_{SS} is normally at ground potential and V_{DD} is a negative voltage. When the input is at zero volts (i.e. logic 0 in negative logic) *TR*1 is off and the output is at V_{DD} (logic 1). When the input goes to a negative voltage (logic 1) transistor *TR*1 turns on and the output goes to logic 0. A NAND gate using unipolar logic is illustrated in Fig. 2.16 (*a*). The diffused load resistor of Fig. 2.15 is replaced by transistor *TR*3 whose gate is connected to its drain. This biases the transistor on. However, since a unipolar transistor has a relatively high 'on' resistance it can be designed to provide the required value of load resistance. Using a transistor in place of a diffused resistor saves space on the silicon chip. When inputs *A*

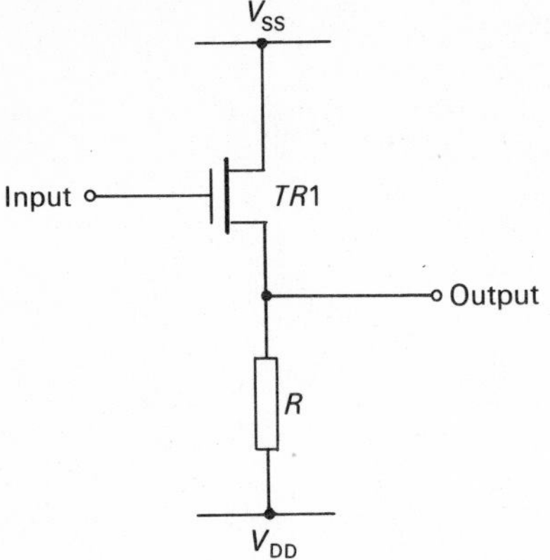

Fig. 2.15. A unipolar inverter using a diffused resistor.

and B go to logic 1 (negative) the output goes to logic 0 (positive). In order to ensure that the logic 0 state is close to V_{SS} the combined impedance of transistors $TR1$ and $TR2$ must be much less than that of $TR3$.

This impedance is determined by its aspect ratio, i.e. the ratio of width to length of the transistor channel. If the output is to swing to a value equal to 0.1 V_{DD} at logic 0 then $TR1$ and $TR2$ must each have an aspect ratio equal to 20 times that of $TR3$. Therefore logic transistors are short and fat and the load transistor is long and thin. Generally, in order to avoid excessively large aspect ratios, and hence to conserve space, a NAND gate is limited to about 2 inputs.

A further disadvantage of the NAND gate of Fig. 2.16 (*a*) is that the output can only swing negative to a value equal to $V_{DD} - V_T$ where V_T is the threshold voltage of $TR3$. This is because $TR3$ will come out of saturation once the difference between its gate and source voltage is less than V_T. To obtain the full logic swing a separate gate voltage is used as shown in Fig. 2.16 (*b*) where $|V_{GG}| = |V_{DD} + V_T|$. This figure also illustrates a NOR gate since the output is at logic 1 provided none of the inputs goes to 1. Since the transistors are in parallel each need now only have an aspect ratio of ten times that of the load device. Therefore NOR gates with many more inputs, usually up to ten, are economical. The upper limit is determined by transistor leakage when off, and parasitic capacitance which reduces the operating speed. Both these are additive since the transistors are in parallel.

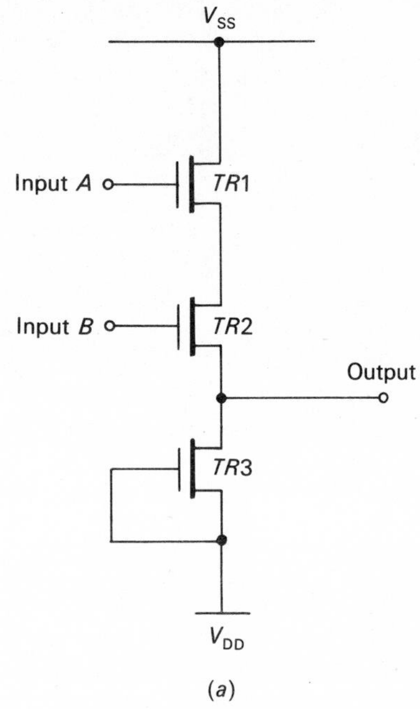

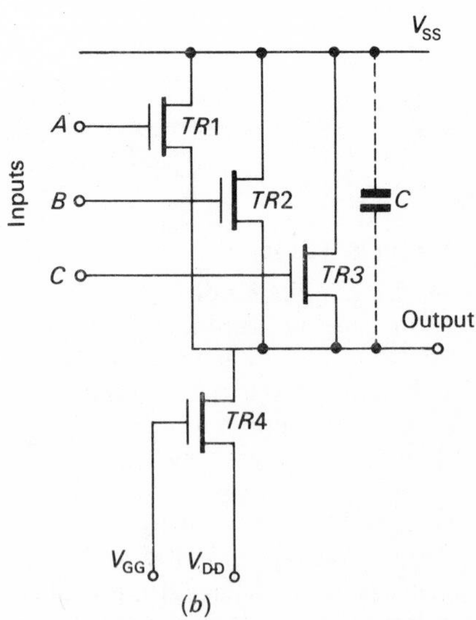

Fig. 2.16. Unipolar gates with transistor loads; (*a*) NAND, (*b*) NOR.

The logic gates described so far in this section are called *static* circuits since the load transistor is on all the time. This has the disadvantage that the dissipation in the system is relatively high. To reduce this dissipation V_{GG} can be periodically switched on and off by an external *clock*. Provided the clock rate is fast enough the existing output state will be stored in the parasitic capacitance C and will not leak away during the inter-clock periods. This assumes that logic transistors are off and the output feeds a high impedance load, such as the gate of the next logic stage. Such a system is called *dynamic* logic. A form of this is shown in Fig. 2.17. When clock φ goes negative transistors *TR*3 and *TR*4 are on. Depending on the state of the inputs the parasitic capacitors C_1 and C_2 are charged or discharged. When φ goes positive, load transistor *TR*3 is switched off but *TR*4 maintains the state on C_2 since it is also off.

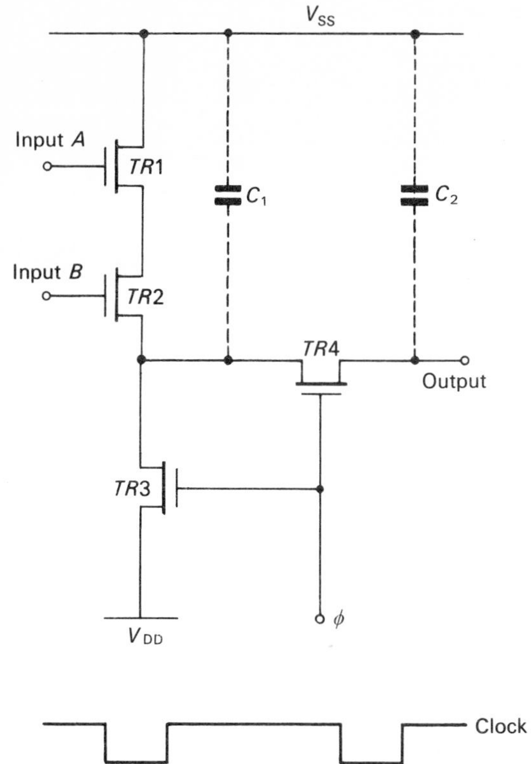

Fig. 2.17. Dynamic two phase NAND gate.

This device is called the transfer transistor and it needs to have a relatively large aspect ratio so that it does not add appreciably to the series resistance of *TR*3 or to the load capacitance, or else the switching speed will suffer.

The logic circuit illustrated in Fig. 2.17 is called two phase dynamic since two identical circuits, one after the other, normally form a complete stage of a device called a shift register. It should be noted that the output logic levels are determined when *TR*1, *TR*2 and *TR*3 are in series across the supply so that the aspect ratios of the various transistors must still be maintained in the correct proportions. Fig. 2.18 (*a*) illustrates a four phase circuit in which all transistors can have the same aspect ratio. Initially clock φ_1 charges parasitic capacitance C_2 to logic 1. When φ_1 goes off clock φ_2 turns on *TR*3. Now C_2 is either discharged or maintains its original charge depending on whether *TR*1 and *TR*2 are on or off, i.e. depending on the state of the logic inputs. Clearly *TR*3 and *TR*4 are never on at the same time so that no potential divider action is involved in determining the output voltage level. The ratios of the transistors are no longer important so this circuit is also called 'ratioless'.

An alternative form of four phase dynamic logic is shown in Fig. 2.18 (*b*). No external power supplies are used apart from the clock. At t_1 clocks φ_1 and φ_2 are negative but *TR*1 and *TR*2 cannot conduct since their source voltage also goes negative as they are supplied by φ_1. At t_2 *TR*4 goes off and φ_1 returns to zero so that the logic transistors may now turn on if their gate voltages are negative. At t_3 clock φ_2 goes off and prevents any further action. Note that the φ_1 and φ_2 clocks overlap and, as with all dynamic circuits, there is a minimum clock repetition period below which the charge on the parasitic capacitors will have leaked away to too low a value.

It was shown in section 1.2 that unipolar transistors can be designed to act in either a *depletion* or *enhancement mode*. All the circuits described so far have used enhancement mode load transistors. Fig. 2.19 shows an inverter using the three possible types of loads and Fig. 2.20 gives their salient characteristics. The depletion load transistor tends to act as a constant current source so that load current

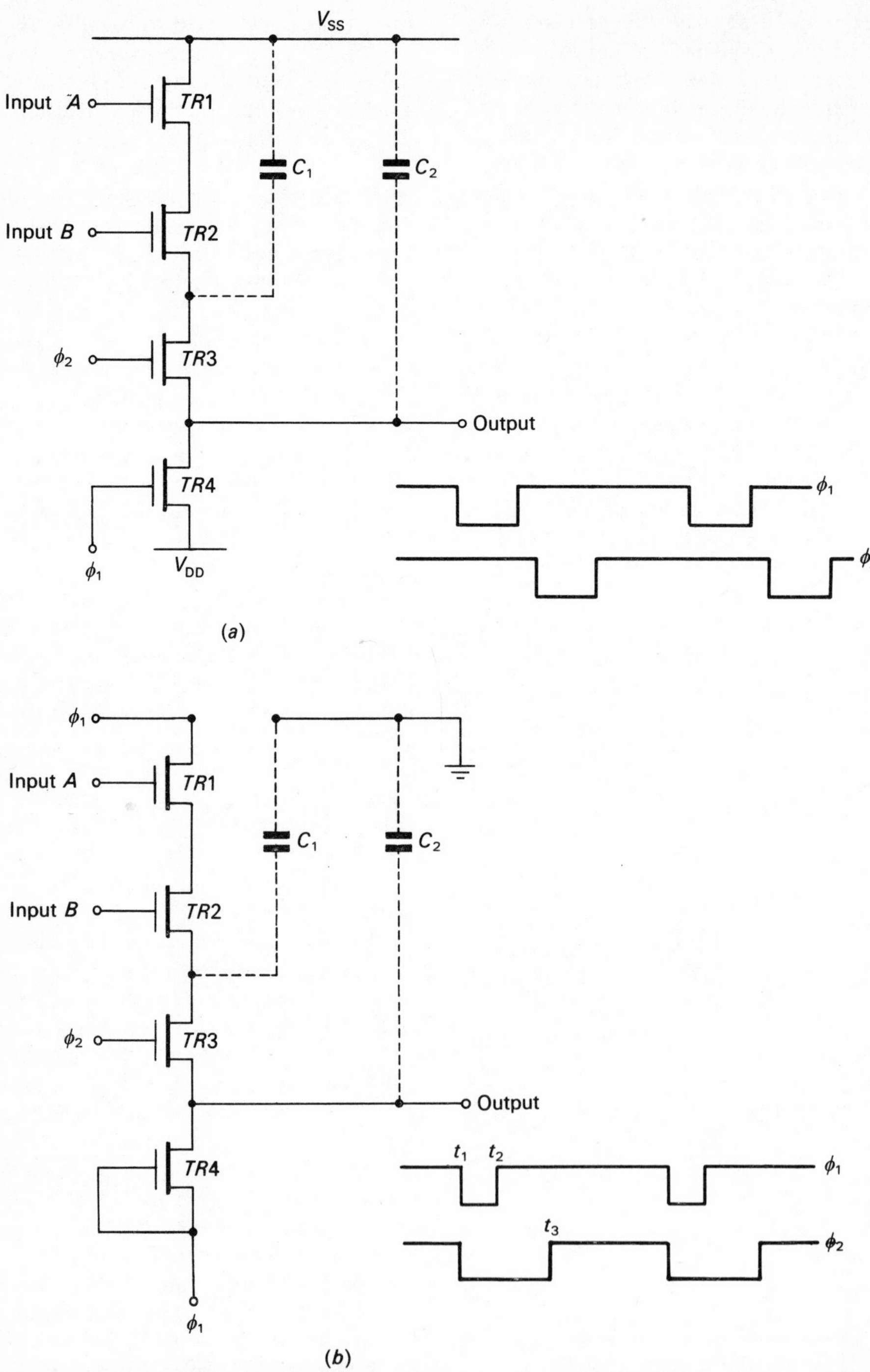

Fig. 2.18. Dynamic four phase (ratioless) NAND gate; (a) with separate clock and power supplies, (b) with no additional power supplies.

remains constant almost until the load voltage reaches V_{DD}. The enhancement load transistor cuts off at $V_{DD} - V_T$. The larger charging current of the depletion load gate gives shorter rise times as shown in Fig. 2.20 so that these devices can work at higher speeds. However the speed

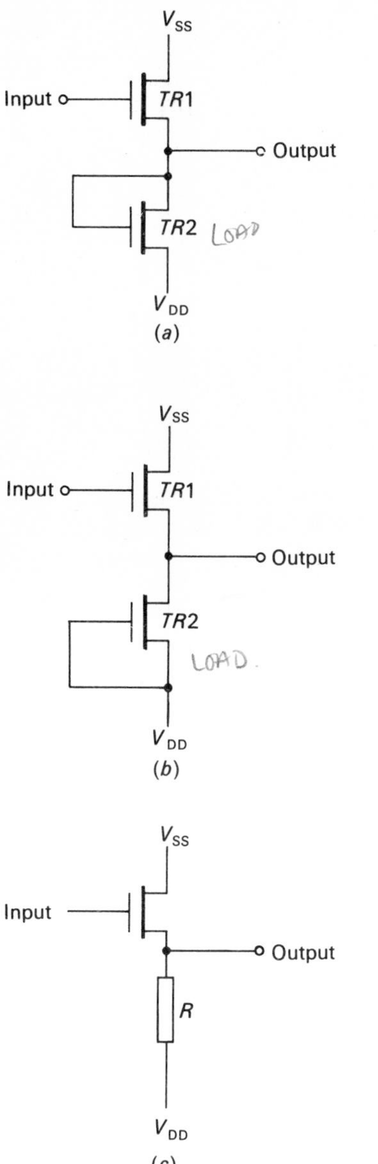

Fig. 2.19. Unipolar inverters with different types of loads; (a) depletion, (b) enhancement, (c) resistive.

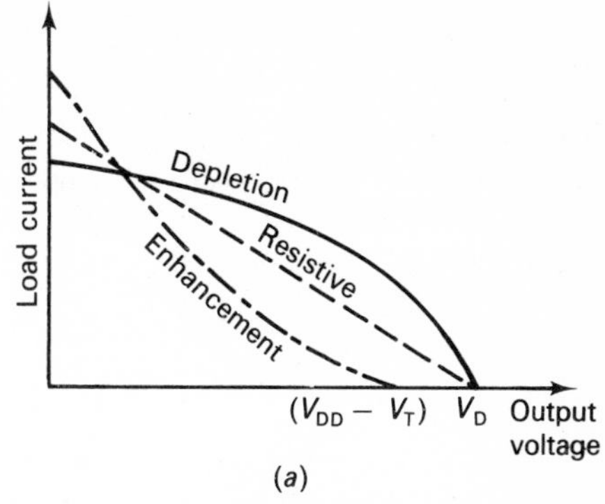

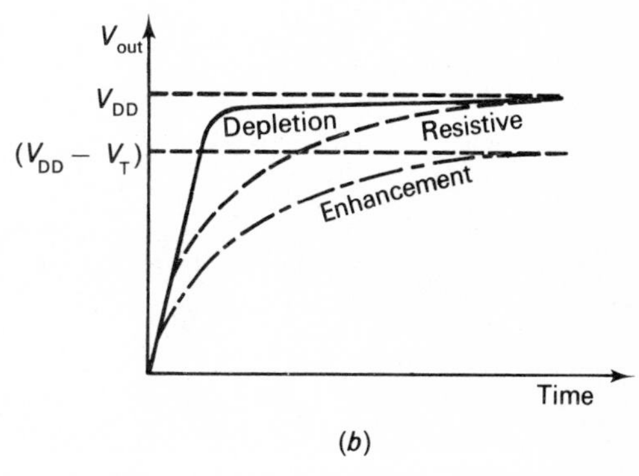

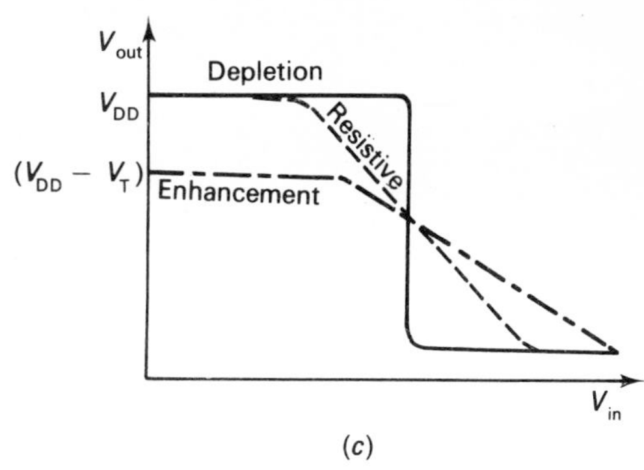

Fig. 2.20. Unipolar inverter characteristics with different types of loads; (a) load current, (b) output risetime, (c) transfer curve.

of unipolar circuits is generally poor due to high parasitic capacitance and relatively large saturation resistance. The depletion load circuit also has the squarest transfer characteristic, as shown in Fig. 2.20 (c). This gives the best d.c. noise immunity since it results in the largest differential between the logic 0 and logic 1 signals.

2.7 CMOS logic

Complementary MOS logic has been treated separately from the other unipolar logic families because it has several advantages for use in SSI and MSI systems and is presently extensively used in this area. Fig. 2.21(*a*) shows the arrangement of a CMOS inverter. Positive

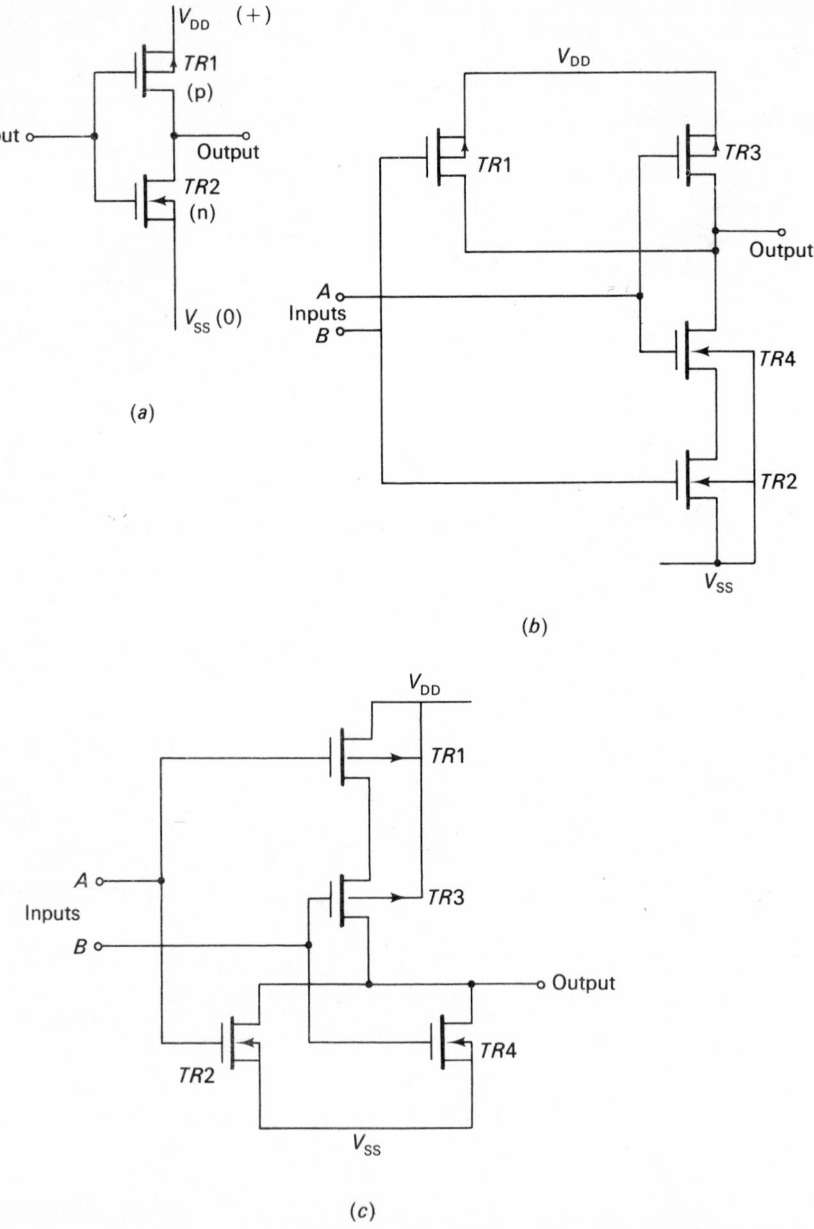

Fig. 2.21. CMOS logic circuits; (*a*) inverter, (*b*) NAND gate, (*c*) NOR gate.

logic conventions are now used so that V_{DD} is positive with respect to V_{SS} which is normally at zero potential. Remember that a p-channel device turns on when its gate potential goes negative with respect to its source, and a n-channel device turns on when its gate goes positive. Therefore for the inverter when the input goes positive (logic 1) *TR*1 goes off and *TR*2 comes on so that the output is at zero volts (logic 0). When the input goes negative *TR*2 is off and *TR*1 is on so that output goes to logic 1.

An extension to the inverter is the NAND gate shown in Fig. 2.21 (*b*). The output voltage is at a logic 0 only if both inputs *A* and *B* are at logic 1 such that *TR*2 and *TR*4 are conducting. Similarly for the NOR gate shown in Fig. 2.21 (*c*) the output is at logic 0 if *TR*2 or *TR*4 is on, that is if inputs *A* or *B* go to logic 1.

The CMOS logic circuit is seen to represent TTL systems with totem-pole outputs since no resistive loads are used. This means that the circuit is capable of providing a relatively low impedance drive when in the logic 0 and 1 state and this enables it to run at higher speeds. The low output impedance and high input impedance also give this family a very large fan-out, in the region of 1000. However, like TTL, the circuit passes through a stage during its switching cycle when both the p- and n-channel transistors are momentarily on. This gives rise to current spikes and circuit noise. Since unipolar devices have inherently higher impedance than bipolar no series resistance is necessary to limit the value of this current spike.

An advantage of CMOS is its ability to operate over a wide range of supply voltages. Fig. 2.22 (*a*) shows the transfer characteristic at three different supply voltages. The curve has a

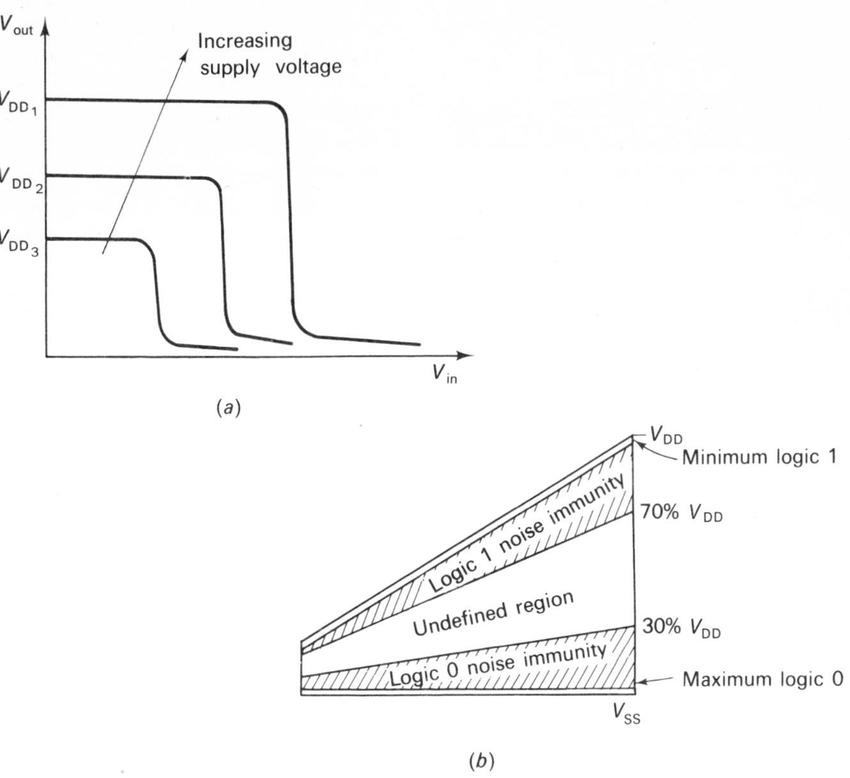

Fig. 2.22. CMOS noise immunity characteristic; (*a*) transfer curve, (*b*) guaranteed noise immunity bands.

very sharp knee which results in excellent voltage noise immunity, as illustrated in Fig. 2.22 (b). Since the load is usually very capacitive the minimum logic 1 level is close to V_{DD} and the maximum logic 0 level is close to V_{SS}. The logic 0 and logic 1 input levels are normally guaranteed as 30 per cent and 70 per cent of V_{DD} so that the noise immunity in both state approximates to a minimum of 30 per cent of V_{DD}. Clearly the higher the supply voltage the greater the ability of CMOS to withstand absolute levels of noise.

The d.c. power dissipation of a CMOS gate is very low since one or other of the transistors is always off and it presents a high impedance. However when the gate is switching it is charging and discharging its loading capacitance which is usually the input parasitic capacitance of the next gate. This power P is given by

$$P = CV^2f \tag{2.4}$$

where C is the load capacitance, V the voltage to which it is charged, usually that of the supply, and f is the switching frequency.

At high frequencies the power dissipation of CMOS equals that of most other logic families.

The speed of CMOS logic reduces as the load capacitance increases. There is, however, an improvement in speed at higher supply voltages. Clearly the value of this voltage needs careful selection for any application since it is seen to effect several of the CMOS logic's important parameters.

2.8 Comparisons

A table showing the comparative performance of the various logic families is given in Fig. 2.23. The table uses a grading system in which 1 represents best and 11 worst. Therefore, of all the families, ECL has the fastest speed, i.e. the lowest propagation delay, CMOS consumes the least power and has the highest fan-out and noise immunity whereas ECL generates the least internal noise.

Absolute values for the various parameters have been purposely left out from Fig. 2.23 since these can vary from manufacturer to manufacturer depending on his fabrication process. Furthermore it must be remembered that the logic designer has considerable freedom in trading off one parameter against another. For instance IIL circuits can be built to have a gate delay anywhere between 25 and 250 nanoseconds. The power dissipation at these two extremes is 5 nanowatts and 50 microwatts respectively. The table represents typical values with the systems designed for their most frequently used applications.

The most popular logic families for SSI and MSI work are presently low power Schottky TTL, CMOS, and ECL. For the highest speeds, with gate delays into the subnanosecond range, ECL is the obvious choice. This speed is however obtained at the expense of power dissipation. For more modest speeds CMOS or low power Schottky TTL may be used. CMOS has a gate delay of about 35 nanoseconds compared

Logic family	Speed	Power dissipation	Fan-out	Noise immunity	Noise generation
RTL	5	6	5	11	2
DTL	7	6	5	10	2
TTL	3	6	5	5	9
TTL–S	2	9	5	5	9
LPTTL–S	3	4	5	5	9
HNIL	10	10	5	1	2
ECL	1	11	2	5	1
PMOS	11	2	2	3	2
NMOS	8	2	2	4	2
CMOS	9	1	1	1	2
IIL	5	4	5	5	2

Fig. 2.23. Comparison of logic families; 1 = best, 11 = worst.

to low power Schottky TTL's delay of 10 nanoseconds. However CMOS has a much higher noise immunity and its d.c. power dissipation is over one thousand times less. It is therefore the obvious choice for battery operated systems. Note that the fan-outs indicated in Fig. 2.23 are those for a gate which is driving devices in its own family. CMOS has a fan-out of almost 100 times greater than its nearest rival, not because it can provide the largest current output, but because its input impedance is so high and it has a modest output current capability.

For LSI systems the choice is largely between NMOS and IIL or its variants. Generally speed is the determining factor and unless the shorter gate delays are required by the system the unipolar approach is likely to prove more economic.

3. The design of logic circuits

3.1 Introduction

Techniques for logic circuit design have been developed and refined over many decades. Logic systems are ideally suited to mathematical analysis and a wealth of literature exists on the subject. In this chapter only the more elementary concepts will be introduced so as to give the reader a familiarity with the subject. These concepts will be utilized and extended in subsequent chapters when the various types of logic circuits are discussed.

The basis of a logic design is the number system. Counting in *decimal* is not convenient since a logic circuit is essentially two state, either logic 0 or logic 1. A *binary* system is preferred in which each column, starting from the right, is weighted by a multiple of 2, i.e. 2^0, 2^1, 2^2, 2^3 etc. Therefore the binary number 10110 is equivalent to $1 \times 2^4 + 0 \times 2^3 + 1 \times 2^2 + 1 \times 2^1 + 0 \times 2^0$, or decimal 22. Long strings of binary numbers are sometimes difficult to use and octal coding can then be employed. A base of eight is now used and it can be obtained by splitting the binary numbers into groups of three, starting from the right, and then applying their weights. For example binary 110 011 101 is equivalent to octal $1 \times 2^2 + 1 \times 2^1 + 0 \times 2^0$ $0 \times 2^2 + 1 \times 2^1 + 1 \times 2^0$ $1 \times 2^2 + 0 \times 2^1 + 1 \times 2^0$ or 635.

Mathematical manipulation in various number systems is similar to that used in decimal except that the base is now different. For instance when adding two binary numbers a carry will be generated whenever the base exceeds two. For example

```
   1001
 + 1011
 ------
  10100
```

Subtraction can be accomplished either by borrowing the base number from the left, when required, or by the complement and add method. In binary, the 2's complement is found by taking the 1's complement and then adding 1. For example 10101 − 1110 can be found by taking the 2's complement of 1110 and adding. The 1's complement of 1110 is 0001 and adding 1 to this gives the 2's complement as 0010. Therefore 10101 − 1110 is the same as 10101 + 0010, i.e.

```
     10101
   +  0010
    ------
   [1] 0111
```

The carry shown in the box is ignored so that the result is 0111.

Multiplication of numbers is obtained by conventional shift and add techniques, and this is especially easy for binary numbers due to its simple base system. For example

```
      1101
   ×   101
   -------
      1101
    00000
    1101
   -------
   1000001
```

The concept of the logic gate was introduced in chapter 2. Fig. 3.1 summarizes the function of most of the more commonly used types. The AND gate gives a logic 1 output only when inputs A and B and C are at a logic 1. This can be written in mathematical form as $D = A \,.\, B \,.\, C$ where the dot represents a logical AND symbol. For the OR gate the output is at logic 1 when A or B or C are at logic 1. This can be written as $D = A + B + C$ where the cross symbol represent a logical OR symbol. The EXCLUSIVE–OR gate gives a logic 1 output when only one of the inputs is at logic 1, not if more than one input is

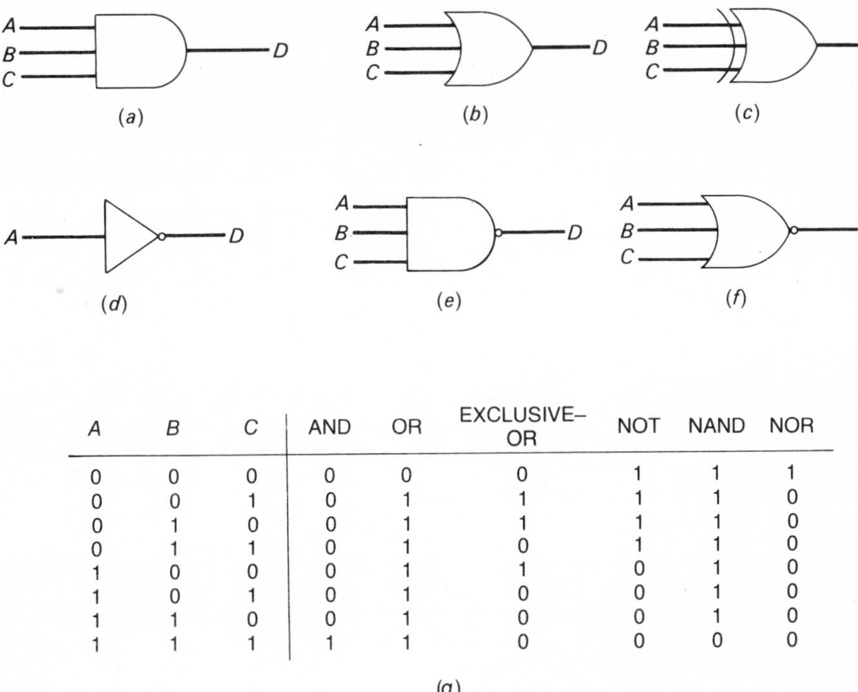

A	B	C	AND	OR	EXCLUSIVE–OR	NOT	NAND	NOR
0	0	0	0	0	0	1	1	1
0	0	1	0	1	1	1	1	0
0	1	0	0	1	1	1	1	0
0	1	1	0	1	0	1	1	0
1	0	0	0	1	1	0	1	0
1	0	1	0	1	0	0	1	0
1	1	0	0	1	0	0	1	0
1	1	1	1	1	0	0	0	0

Fig. 3.1. Logic gates; (*a*) AND, (*b*) OR, (*b*) EXCLUSIVE-OR, (*d*) NOT, (*e*) NAND, (*f*) NOR, (*g*) truth table for the gates.

at logic 1. It is written as $D = A \oplus B \oplus C$. The circle at the end of a gate symbol indicates inversion. For the NOT gate the output is an inverse of the input i.e. $D = \overline{A}$. For the NAND gate $D = \overline{A\,.\,B\,.\,C}$, and for the NOR gate $D = \overline{A + B + C}$.

It is extremely useful to be able to reduce logic systems to mathematical terms since they are then much easier to manipulate and simplify. The technique which is used in this manipulation is called Boolean algebra.

3.2 Boolean algebra

In the middle of the eighteenth century an English mathematician called Boole introduced a form of algebra in which only two types of statements were allowed, i.e. false and true. If one assigns the values 1 and 0 to these states then the form of algebra is ideally suited to the analysis of logic circuits. The concept of logical notations was introduced in section 3.1 where a truth table was used to represent logic states. In this section it will be assumed that if a normally open switch is labelled A then $\overline{A}$ is given by a normally closed switch. Furthermore a permanent open circuit is logic 0 and a continuous short circuit is logic 1.

With the above in mind several of the basic concepts of Boolean algebra can be described. These are given in Fig. 3.2 and Fig. 3.3. The illustrations show switches and from these the logic of the theorems can be readily verified. Remember that an AND function (dot) is given by series switches whereas an OR function (cross) is represented by parallel switches. Therefore the first theorem states that having a switch in series with a short circuit is equivalent to having the switch itself. This is logical enough. Theorem 2 states that if one has a switch in parallel with a short circuit then cur-

Theorem number	Postulate	Illustration
1	$A . 1 = A$	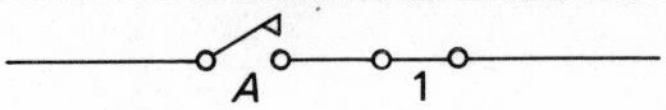
2	$A + 1 = 1$	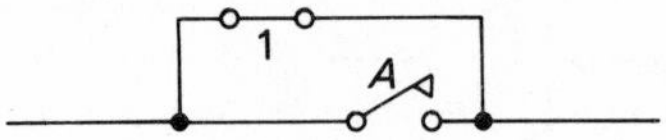
3	$A . 0 = 0$	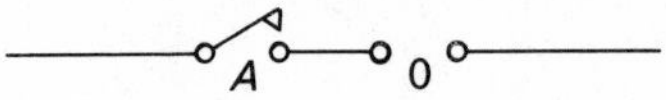
4	$A + 0 = A$	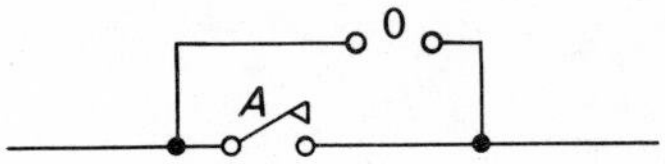
5	$A + \overline{A} = 1$	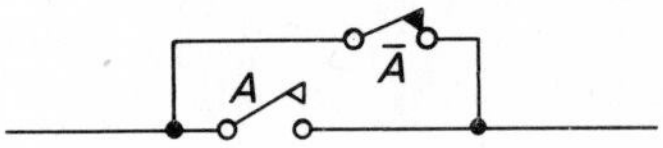
6	$A . A = 0$	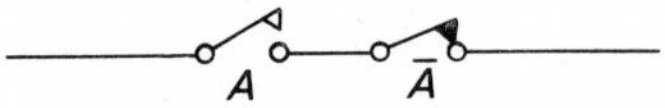
7	$A . A = A$	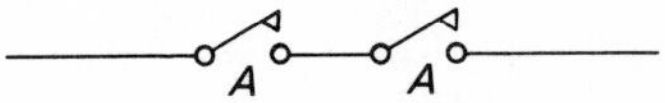
8	$A + A = A$	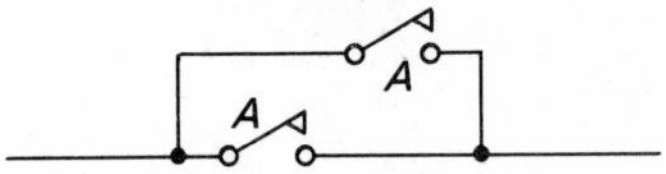
9	$A + B = B + A$	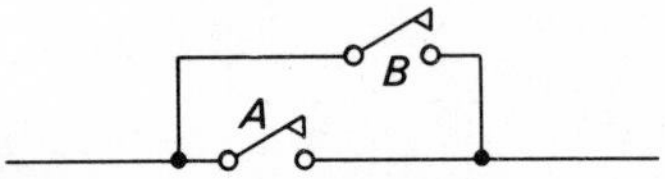
10	$A . B = B . A$	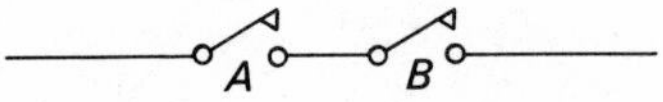
11	$A . (A + B) = A$	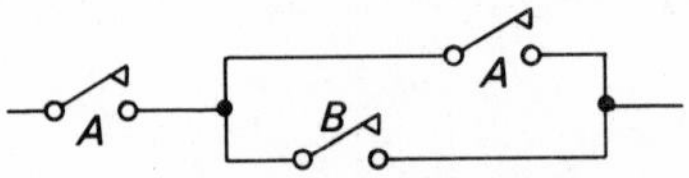
12	$A + (A . B) = A$	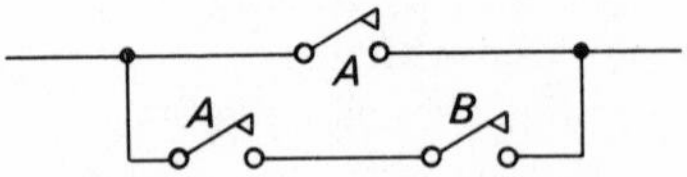
13	$(A + B) . (A + \overline{B}) = A$	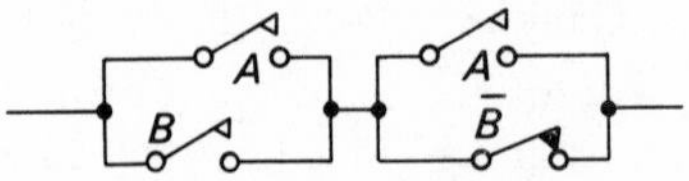
14	$A . B + A . \overline{B} = A$	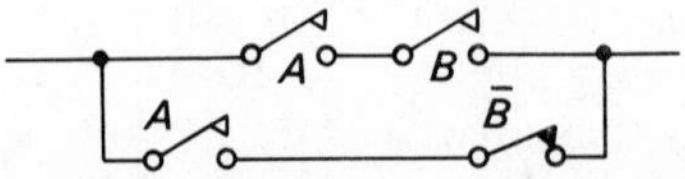

Fig. 3.2. Some theorems used in Boolean algebra.

Theorem number	Postulate
15	$\overline{A + B + C} = \overline{A} . \overline{B} . \overline{C}$
16	$\overline{A . B . C} = \overline{A} + \overline{B} + \overline{C}$
17	$A . B + A . C = A . (B + C)$
18	$(A + B) . (A + C) = A + B . C$
19	$A . B + B . C + C . \overline{A} = A . B + C . \overline{A}$
20	$(A + B) . (B + C) . (C + \overline{A}) = (A + B) . (C + \overline{A})$
21	$A . B + \overline{A} . C = (A + C) . (\overline{A} + B)$

Fig. 3.3. Further theorems used in Boolean algebra.

rent will continuously flow through the short circuit path irrespective of whether the switch is open or closed. Therefore the switch can be omitted. Similar considerations apply for theorems 3 and 4 in which the logic 0 state is represented by an open circuit.

In theorem 5 the states of the A and $\overline{A}$ switches are always opposite. Therefore if A is closed $\overline{A}$ is open and vice versa. Therefore there is always a current path through one or the other of the switches so that, as far as the rest of the system is concerned, the two switches behave like a short circuit. For theorem 6, current will never flow since one or other of the switches is always open so the two behave like an open circuit. In theorems 7 and 8 since the two switches always have identical states they may be replaced by one switch. For theorems 9 and 10 the position of the switches can be interchanged without affecting their overall logical operation. In theorems 11 and 12 switch A forms the controlling element since when A is closed a current path always exists, irrespective of the state of switch B, and when A is open no current can flow. Therefore switch B is redundant. Similar considerations apply for theorem 3 and 14 since the reverse states of switches B and $\overline{B}$ cause them to be ineffective.

The theorems given in Fig. 3.3 can be illustrated by switches if desired. Theorems 15 and 16 are commonly known as de Morgan's theorem. Theorem 15 states that if there is an OR function of three (or more) variables then the NOR function of these variables is obtained by replacing each function by its complement and each OR notation by an AND notation. Similarly to go from AND to NAND all the variables are complemented and all AND notations replaced by OR notations.

A useful concept for proving logic theorems is the Venn diagram, shown in Fig. 3.4. In Fig. 3.4 (a) if the shaded circle represents the variable A then everything outside this circle excludes A so that it is NOT A or $\overline{A}$. If there are now two circles each of which encompasses two variables A and B respectively then the region in which they overlap is equal to A and B. Therefore in Fig. 3.4 (b) the shaded area is A AND B or $A . B$. Now the unshaded area is clearly the opposite of the shaded area, i.e. $\overline{A.B}$. However the unshaded area is also obtained from the knowledge that it does not enclose the area A or area B, i.e. it is not A or not B $(\overline{A} + \overline{B})$. Therefore equating these two gives $\overline{A . B} = \overline{A} + \overline{B}$. This is equivalent to de Morgan's theorem, number 16, using two variables. Any number of circles can be drawn on the Venn diagram to represent different variables. However their logical manipulation becomes very difficult as the number of variables increase.

Another method for proving theorems is by truth tables. For example the table for theorem 18 is given in Fig. 3.5. For each possible combination of variables A, B and C the values of $(A + B) . (A + C)$ and $A + B . C$ are seen to be identical so that these two expressions are equivalent. Similarly the other theorems given in Fig. 3.3 can be proved, and this would be a useful exercise for the reader.

Boolean algebra is extensively used in the design of logic systems and it can result in considerable savings in circuitry. For instance suppose one wished to represent the system given by the truth table of Fig. 3.6 (a). From this table it is seen that C is a logic 1 when A and B

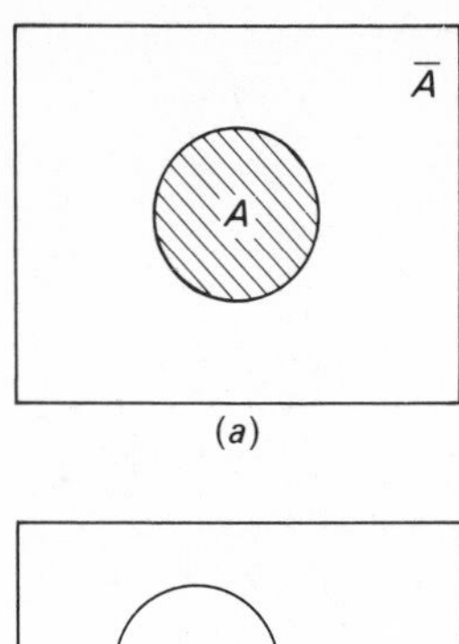

(a)

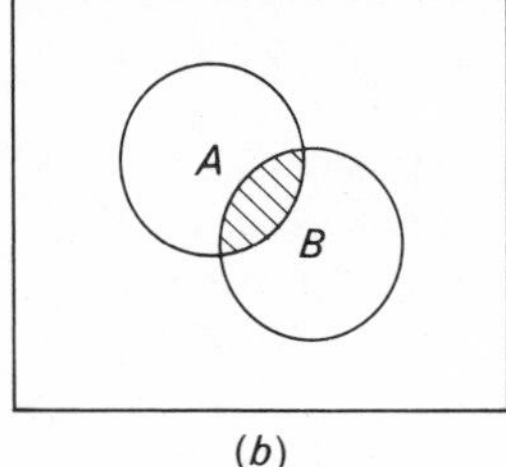

(b)

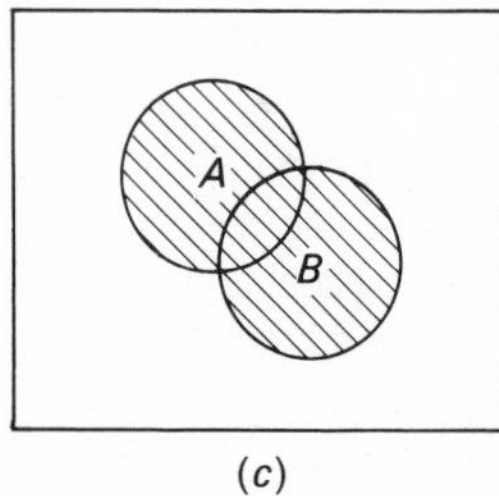

(c)

Fig. 3.4. Use of the Venn diagram to represent logic functions; (a) variable A, (b) AND of A and B, (c) OR of A and B.

are both 1 or when A is 1 and B is 0. Therefore the expression for C is given by

$$C = A \,.\, B + A \,.\, \overline{B}. \tag{3.1}$$

Function C can therefore be formed from A and B by using two AND gates, a NOT gate and an OR gate, as shown in Fig. 3.6 (b). However equation (3.1) can be manipulated by using algebra;

$$\begin{aligned} C &= A \,.\, B + A \,.\, \overline{B} \\ &= A\ (B + \overline{B}) \\ &= A \qquad \text{since } B + \overline{B} = 1 \text{ from} \end{aligned}$$

theorem 5.

Therefore to generate function C requires a direct link to input A and no additional circuitry so that the system shown in Fig. 3.6 (b) is redundant.

Using Boolean algebra on its own for logic minimization is often difficult since it is not at once clear which is the best way to proceed. In these instances several minimization techniques may be used as described in the next section.

3.3. Minimization techniques

These may be broadly classified into the mapping method and the tabulation method. Only one example from each of these methods is described in this chapter.

Karnaugh maps are extensively used in the minimization of logic circuits and a map for two variables A and B is illustrated in Fig. 3.7 (a). Each variable can have two states so that for two variables there is a total of four states represented by four squares. These are labelled $A \,.\, B, \overline{A} \,.\, B, A.\,\overline{B}$, and $\overline{A} \,.\, \overline{B}$. An alternative representation which is preferred in this book, is shown in Fig. 3.7 (b). The true and false states are now represented by 1 and 0. Fig. 3.7 (c) shows the map for the function $A \,.\, B + \overline{A} \,.\, \overline{B}$. Logic 1 in the squares which represent $A \,.\, B$ and $\overline{A} \,.\, \overline{B}$ indicates that these AND functions are present, and all squares of the Karnaugh map are ORed together.

The Karnaugh map for three variables is illustrated in Fig. 3.8. Note that the actual position of the groups $A, \overline{A}, B, \overline{B}, C, \overline{C}$ is not important. It is essential, however, that each square differs from those next to it by only one variable. Therefore the top left hand square in Fig. 3.8 (a) represents $A \,.\, B \,.\, \overline{C}$ and it differs from the square to the right by the variable A being changed to $\overline{A}$, and from the square below it by $\overline{C}$ being changed to C. It is important that in Fig. 3.8 (b) the labelling for BC does not go from 10 to 01 or from 11 to 00 in a single step since both these represent changes of two variables. It is left as an exercise for the reader to construct a Karnaugh map for four variables. Once again note that there is only a single variable change between adjacent squares and provided this is maintained the actual labelling sequence of the variables is not important.

The significance of having adjacent squares which differ by only one variable is illustrated in Fig. 3.9. The ones in the two squares would normally mean that there are two product terms

A	B	C	A + B	A + C	(A + B) . (A + C)	B . C	A + B . C
0	0	0	0	0	0	0	0
0	0	1	0	1	0	0	0
0	1	0	1	0	0	0	0
0	1	1	1	1	1	1	1
1	0	0	1	1	1	0	1
1	0	1	1	1	1	0	1
1	1	0	1	1	1	0	1
1	1	1	1	1	1	1	1

Fig. 3.5. Truth table for theorem 18.

in the function which are ORed together. However since these terms contain a single different variable, and this is equal to the variable and its complement, the resultant is a single product term in three variables. Therefore two adjacent squares on a Karnaugh map can be grouped together to eliminate a variable. Inspection of the map will show immediately which variable is eliminated. For instance in Fig. 3.9 (*a*) variable C is present both as 0 and 1 so it is removed, and in Fig. 3.9 (*b*) B appears as a 0 and 1 and so this is eliminated in the final expression. The variable which remains must appear either as a 0 or a 1 in all the squares which are grouped together.

Squares can be grouped in larger units than two so long as they are even numbers and the

A	B	C
0	0	0
0	1	0
1	0	1
1	1	1

(*a*)

A, B → $A . B$; A, $\overline{B}$ → $A . \overline{B}$; $C = A . B + A . \overline{B}$

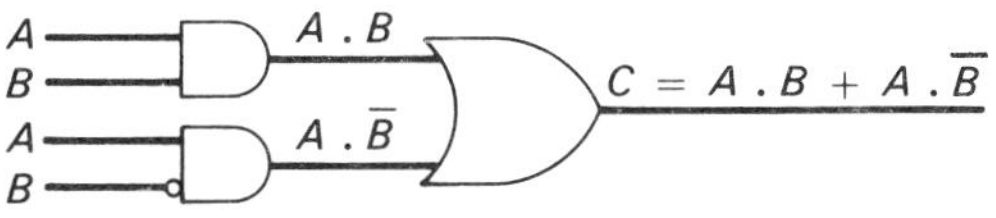

(*b*)

$C = A$

(*c*)

Fig. 3.6. Implementation of $C = A . B + A . \overline{B}$; (*a*) truth table, (*b*) logic circuit, (*c*) simplified version ($C = A$).

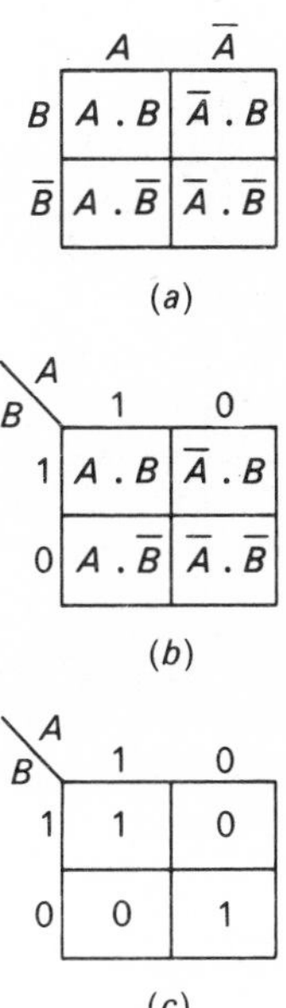

Fig. 3.7. Two variable Karnaugh map; (*a*) map construction, (*b*) alternative representation, (*c*) map for $A . B + \overline{A} . \overline{B}$.

squares are adjacent to each other. The general rule is that to remove *n* variables the number of squares which have to be grouped are 2^n. Therefore to remove one variable needs two squares. For two variables four squares are needed, for three variables eight squares are required, and so on. It would be a useful exercise for the reader to draw maps for four and eight groups, and to verify the result using Boolean algebra. Remember that in each case those variables which are present in both 1 and 0 forms for the group are eliminated.

Although the Karnaugh map has been drawn on a two dimensional plane for convenience, it is important to realize that it is really toroidal in shape. The top and bottom join together and so also do the two sides. This is so because these squares differ by only one variable from each other and therefore, in the context of the Karnaugh map, they are adjacent to each other. This means that squares at these extremes can be grouped together and three examples of this are illustrated in Fig. 3.10. In Fig. 3.10 (b) two separate groups of four squares each are formed so there are two individual product terms. Remember that squares which differ by one variable may only be grouped together. Therefore one cannot group squares which are diagonally opposite on their own although the grouping shown in Fig. 3.10 (c) is valid.

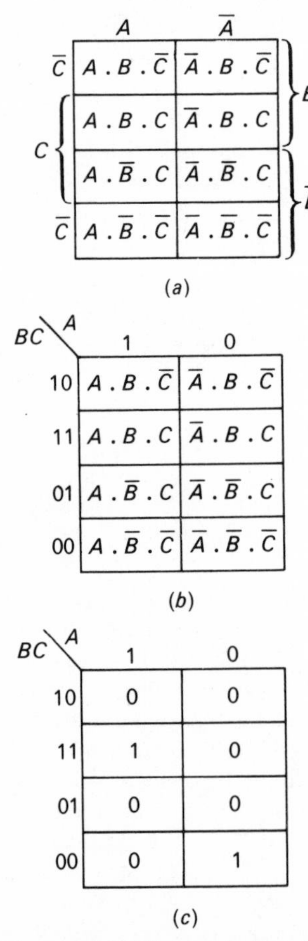

Fig. 3.8. Three variable Karnaugh map; (*a*) map construction, (*b*) alternative representation, (*c*) map for $A.B.C.+\overline{A}.\overline{B}.\overline{C}$.

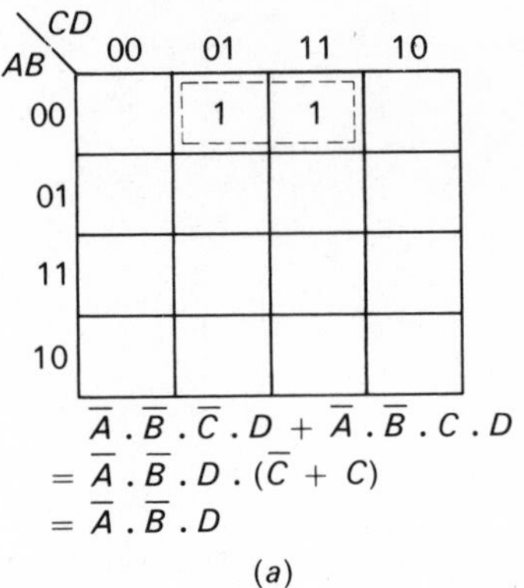

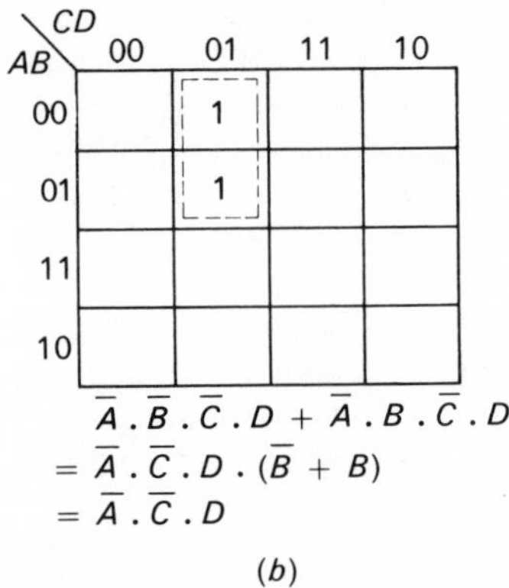

Fig. 3.9. Grouping of two squares; (*a*) map for $\overline{A}.\overline{B}.D$, (*b*) map for $\overline{A}.\overline{C}.D$.

It is also permissible to have several groups of overlapping squares. This is illustrated in Fig. 3.11. The object in all cases when forming groups is to encompass as many squares as possible so long as they are adjacent and the total number is even. The larger the group size the fewer the number of variables in the product term. All the logic 1 states on the map must be covered by groups and the groups are finally ORed together to give the function for the map. Fig. 3.12 illustrates how Karnaugh maps may be used for minimization in instances where 'redundant' or 'don't care' terms exist. For instance suppose a function Y is defined by the truth table shown in Fig. 3.12 (*a*). It is assumed that only the first ten terms can ever be attained in this circuit although sixteen states are possible. Since the last six will not be used they represent redundant terms and are denoted by

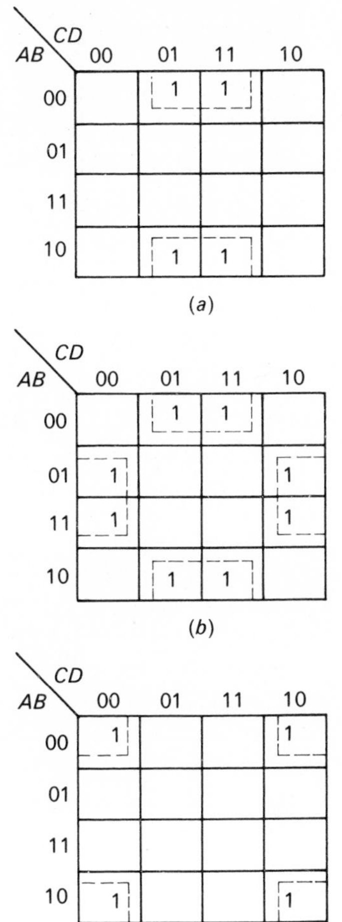

Fig. 3.10. Continuity of Karnaugh maps; (a) map for $\overline{B} . D$, (b) map for $\overline{B} . D + B . \overline{D}$, (c) map for $\overline{B} . \overline{D}$.

X. The Karnaugh map is plotted with these redundant terms. Each X can represent a 0 or 1 state so that all these squares need not be covered by the groupings. However a X may be considered as a 1 if it will enable a larger group to be obtained. This is shown in Fig. 3.12 (b). The final function for the truth table is $C + B$. If the redundant terms are not used then the function would have been $\overline{A} . B + \overline{A} . C$ which is a more complex expression. It is left to the student to verify this last function.

If the number of variables used exceeds six it becomes very unwieldy to represent them on a Karnaugh map. In these instances the tabulation method of minimization is preferred. Essentially this is similar to the truth table and is shown in Fig. 3.13. An extension of this, which allows a rigorous treatment by computer, is known as the Quine–McCluskey technique. This is not described in this book.

Suppose it is required to find the expression for the function shown in Fig. 3.13 (a). This table could have been derived from an expression such as $Y = \overline{A} . B . \overline{C} + \overline{A} . B . C + A . \overline{B} . C + A . B . C$ with a redundant term at $A . B . \overline{C}$. The first step is to draw a second table, as in Fig. 3.13 (b), with all the 1 and X terms together. Groups are now formed from this table in such a way that for differences of n variables in each group there are 2^n numbers per group. For differences in three variables (A, B and C) eight groups are needed. Since only five groups exist in Fig. 3.13 (b) this is not possible. For differences in two variables, groups of four are needed. The group made up of numbers 3, 4, 7 and 8 vary in variables A and C. For differences in one variable, groups of two are required. There are several of these, 3 and 4 vary in C. 7 and 8 vary in C. 3 and 7 vary in A. 4 and 8 vary in

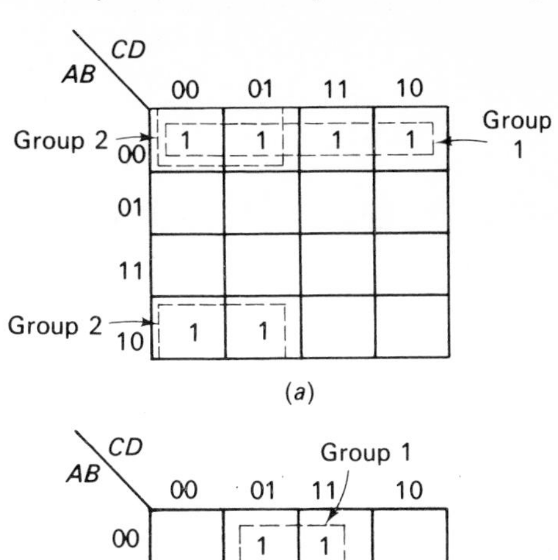

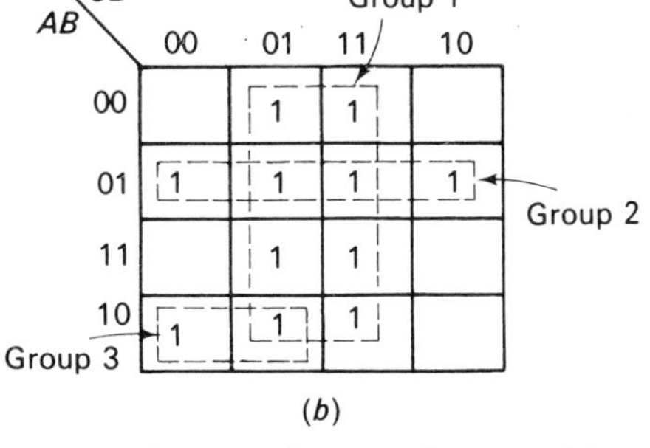

Fig. 3.11. Overlapping of groups of squares; (a) map for $\overline{A} . \overline{B} + \overline{B} . \overline{C}$, (b) map for $D + \overline{A} . B + A . \overline{B} . \overline{C}$.

A	B	C	D	Y
0	0	0	0	0
0	0	0	1	0
0	0	1	0	1
0	0	1	1	1
0	1	0	0	1
0	1	0	1	1
0	1	1	0	1
0	1	1	1	1
1	0	0	0	0
1	0	0	1	0
1	0	1	0	X
1	0	1	1	X
1	1	0	0	X
1	1	0	1	X
1	1	1	0	X
1	1	1	1	X

(a)

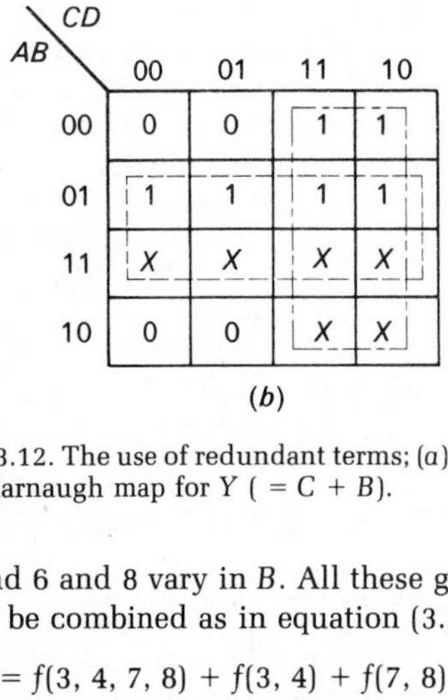

AB \ CD	00	01	11	10
00	0	0	1	1
01	1	1	1	1
11	X	X	X	X
10	0	0	X	X

(b)

Fig. 3.12. The use of redundant terms; (a) truth table, (b) Karnaugh map for Y (= C + B).

A and 6 and 8 vary in B. All these groups can now be combined as in equation (3.2),

$$Y = f(3, 4, 7, 8) + f(3, 4) + f(7, 8) + f(3, 7) + f(4, 8) + f(6, 8) \quad (3.2)$$

It is important that each number from Fig. 3.13 (b) is covered at least once. If not, then groups of single numbers are used. For each of the groups in equation (3.2) variables which appear as a 0 and 1 are eliminated so that the expression for Y is given by

$$\begin{aligned} Y &= B + \overline{A} . B + A . B + B . \overline{C} + B . C + A . C \\ &= B + B(\overline{A} + A) + B(\overline{C} + C) + A . C \\ &= B + B + B + A . C \\ &= B + A . C \end{aligned}$$

This expression is verified by drawing a Karnaugh map for the truth table of Fig. 3.13 (a) as in Fig. 3.13 (c).

Number	A	B	C	Y
1	0	0	0	0
2	0	0	1	0
3	0	1	0	1
4	0	1	1	1
5	1	0	0	0
6	1	0	1	1
7	1	1	0	X
8	1	1	1	1

(a)

Number	A	B	C	Y
3	0	1	0	1
4	0	1	1	1
6	1	0	1	1
7	1	1	0	X
8	1	1	1	1

(b)

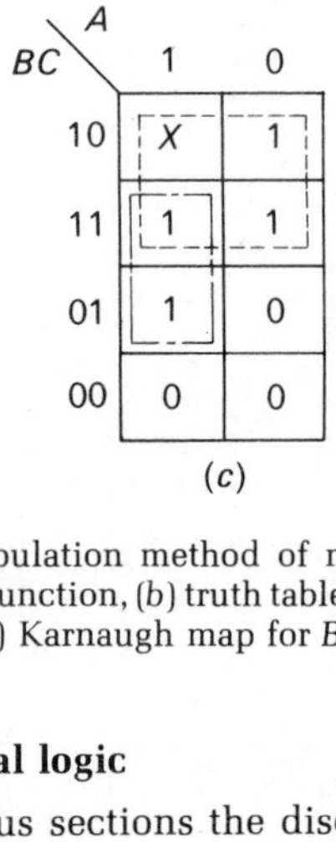

BC \ A	1	0
10	X	1
11	1	1
01	1	0
00	0	0

(c)

Fig. 3.13. Tabulation method of minimization; (a) truth table of function, (b) truth table of '1' and redundant states, (c) Karnaugh map for B + A . C.

3.4 Sequential logic

In the previous sections the discussions have concentrated on *combinational logic* in which the output state at any instant in time is determined by the state of the input at that time. In a *sequential logic* system the output state is determined not only by the state of the inputs at the time under consideration but also by the state of the output at a previous time interval. Therefore a sequential circuit implies the use of a memory to store the previous state as well as feedback from output to input.

A simple sequential circuit is shown in Fig. 3.14. The output Y is given by the NAND function of external or control inputs A and B together with the internal or feedback input y. This input y is equal to output Y after time interval T. The truth table for this system is

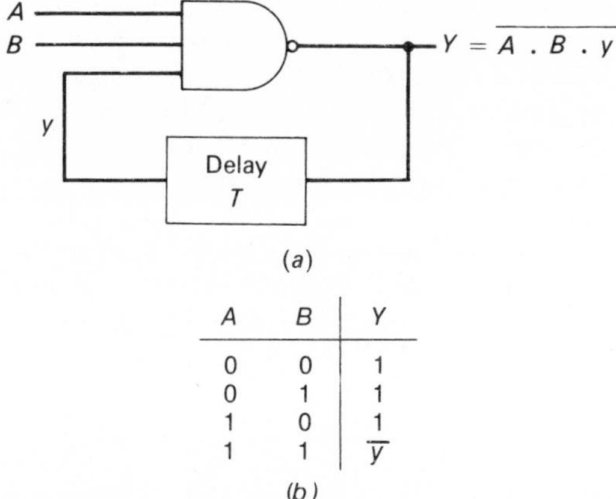

A	B	Y
0	0	1
0	1	1
1	0	1
1	1	$\bar{y}$

Fig. 3.14. Simple sequential circuit; (a) circuit arrangement, (b) truth table.

similar to that of a combinational circuit (without feedback y) so long as either A or B is 0. When both A and B are 1 a combinational circuit would give a steady state output of 0. However for the sequential circuit this 0 is transferred to input y after delay T. This causes the output Y to change to a 1. This is fed back to y after time interval T and so the output swings back to 0. Therefore the output oscillates between 0 and 1 with a time period equal to $2T$. Since the output Y is always equal to the input $\bar{y}$ this is illustrated as such in the truth table.

There are three maps which are much more useful than the truth table in representing and analysing sequential circuits since they indicate the path between various time intervals. These are shown in Fig. 3.15 for the system given in Fig. 3.14. The excitation map shows the state of the output for each of the input states. Inputs A and B are externally controlled whereas y will change itself to suit the system's operation. For instance if A and B are both made 0 then if y was momentarily at 0 it would give a 1 output which, after a delay, would change y to 1 and move the state in Fig. 3.15 (a) from the top left hand square to the bottom square. Similarly if A is 0 and B is 1 or B is 0 and A is 1 the value of y will change to the stable state of 1 which equals that of Y. In the case of A and B both being 1, y will oscillate between 0 and 1 as output Y changes between 1 and 0. The stars in the squares indicate stable states and it is important to note that these correspond to those instances when the feedback input y equals the output Y, since no further change can then occur after the time delay has expired. The arrows indicate the movement between states. An arrow going both ways indicates an unstable state.

The transition map indicates more clearly the stable and unstable states. A dot is used for an unstable state and a circled cross for a stable state. Finally the flow map shows the position which the output can occupy for any combination of inputs. Numbers in circles indicate stable states. The system can only move between the same numbers, that is 1 to ①, 2 to ② and so on.

Fig. 3.16 shows a second example of a sequential circuit in which two gates are cross coupled to provide the delay and feedback. Remembering that $y = Y$ and $x = X$ the truth table can be obtained. For A and B both equal to

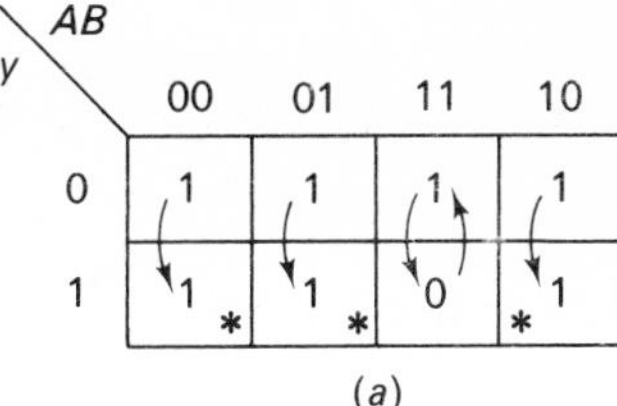

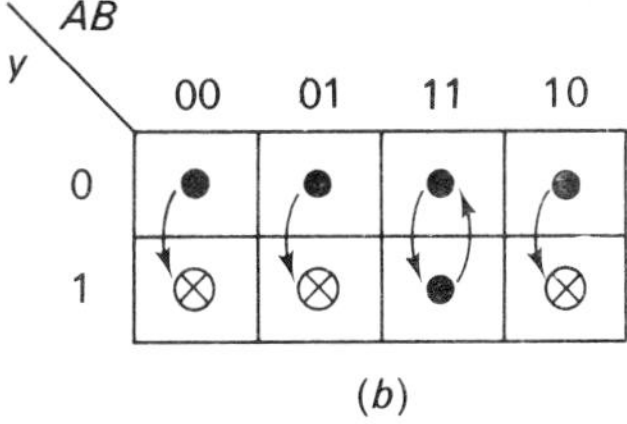

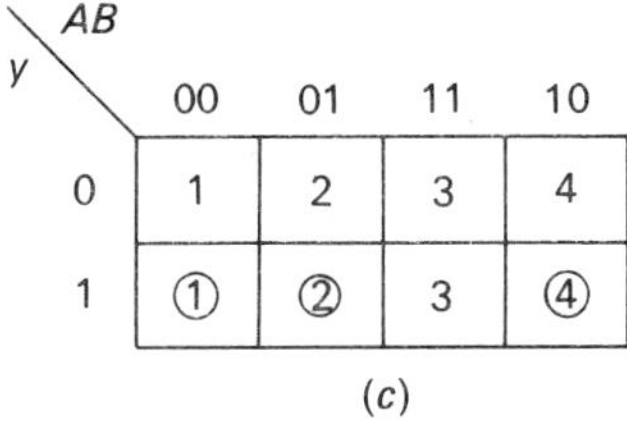

Fig. 3.15. The use of maps in sequential circuits; (a) excitation map, (b) transition map, (c) flow map.

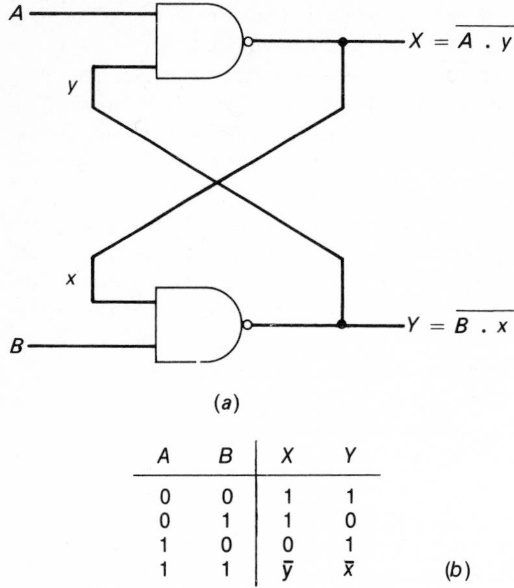

A	B	X	Y
0	0	1	1
0	1	1	0
1	0	0	1
1	1	$\bar{y}$	$\bar{x}$

Fig. 3.16. Cross coupled gate sequential circuit; (*a*) circuit diagram, (*b*) truth table.

0 outputs X and Y are both 1 since they are given by

$$X = \overline{A \,.\, y} = \overline{0} = 1 \tag{3.3}$$

$$Y = \overline{B \,.\, x} = \overline{0} = 1 \tag{3.4}$$

When A equals 0 and $B = 1$ the value of X is 1 as before. Therefore the value of Y is given by

$$Y = \overline{B \,.\, x} = \overline{1.1} = 0. \tag{3.5}$$

Similarly for $A = 1$ and $B = 0$ the value of Y is 1 and this gives X as 0. For A and B both equal to 1 the outputs are $X = \overline{Y}$ and $Y = \overline{X}$. This means that the system can be stable with $X = 0$ and $Y = 1$ or $X = 1$ and $Y = 0$ since both these satisfy the output conditions.

Fig. 3.17 (*a*) shows the excitation map. This is obtained by putting the corresponding values of x, y, A and B into equations (3.3) and (3.4). Therefore for the first square $x = y = A = B = 0$ so that X and Y are both 1. The first figure inside the square is the value of X and the second is that of Y. The stable states in this map are marked by a star and are obtained as those instances when the outputs XY equals the feedback inputs xy. The transition map and flow map are derived from the excitation map and these can be used to describe the system. For instance suppose the input $A = 0, B = 1$ so that the circuit is in its stable state with X equal to 1 and $y = 0$. If the value of B is changed to 0 the

xy \ AB	00	01	11	10
00	11	11	11	11
01	11	11	01 *	01 *
11	11 *	10	00	01
10	11	10 *	10 *	11

(*a*)

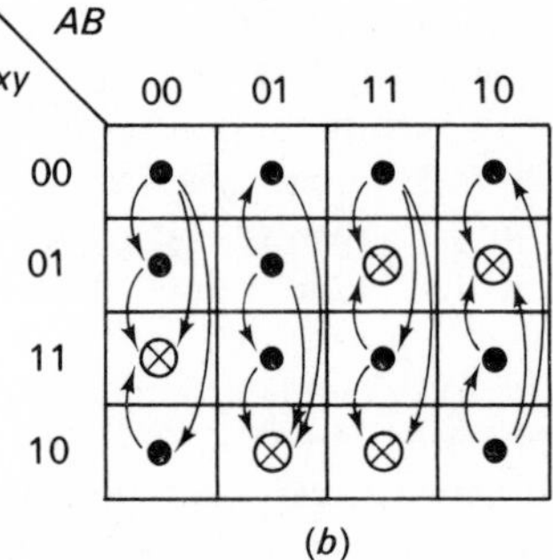

(*b*)

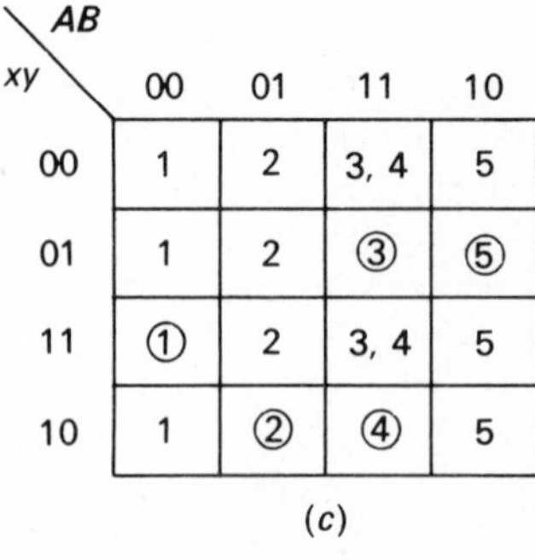

xy \ AB	00	01	11	10
00	1	2	3, 4	5
01	1	2	③	⑤
11	①	2	3, 4	5
10	1	②	④	5

(*c*)

Fig. 3.17. Maps for the circuit given in Fig. 3.16; (*a*) excitation map for X and Y, (*b*) transition map (*c*) flow map.

operation of the system will move to column AB equal to 00. The value of xy will remain at 10 so that the movement is sideways to the left. After a short delay, however, y will change from 0 to 1 so that stable state ① is achieved. Suppose now that the system was in stable state ⑤ and that the values of A and B are interchanged so that they become $A = 0$ and $B = 1$. This would result in a horizontal shift to state 2. Now both x and y need to change to move to the stable state given by ② for this combination of A and B inputs. If they both change simultaneously then a jump will be made directly to the stable state. However it is possible to first move to some of the other unstable 2 states depending on whether x or y changes its state first. The end result is a move to the square marked ②.

Suppose now that the system is in stable state ① and AB is changed to 11. The circuit will move horizontally to the square corresponding to $xy = 11$ and $AB = 11$. Now both input x and y will attempt to change to 0. If x changes first the system will move to stable state ③. If y changes first the move is to stable state ④. Therefore for a unique input change there exist two possible output states. This represents a race condition since both the feedback inputs are racing to change their state first. It is clearly undesirable to be unable to predict the output state accurately in all conditions so that $A = B = 1$ represents an illegal state and must not be used for the circuit shown.

3.5 Threshold logic

A form of logic which has not been widely used in industry in spite of its excellent potential is threshold logic. It is similar to conventional logic in having binary inputs and outputs. However in a threshold circuit the inputs are all weighted and the output switches when the accumulation of these weights exceeds some threshold value. The value of the weights and the threshold can be altered to provide a variety of functions from the same basic gate structure.

Fig. 3.18 (a) shows the symbol for a threshold gate. The inputs x_1 to x_n may achieve a logic 1 or logic 0 state. Each input is weighted by ω_1 to ω_n.

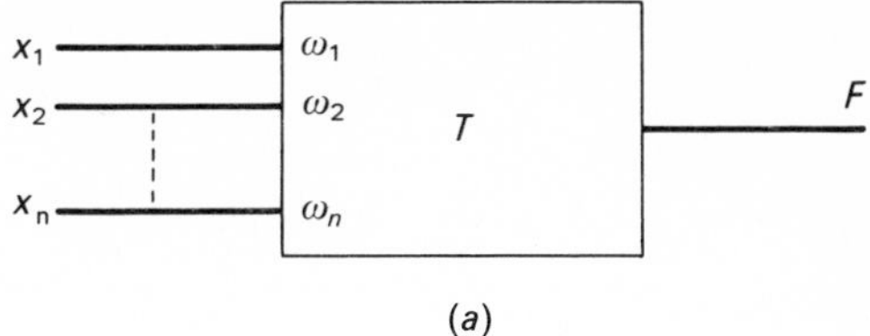

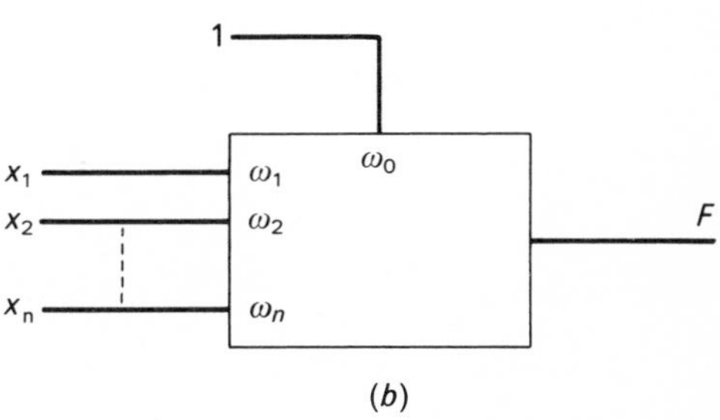

Fig. 3.18. Representation of threshold logic gates; (a) with threshold T, (b) with weighted threshold ω_0.

The threshold of the gate is T. The output F will now be 0 if

$$\sum_{i=1}^{n} \omega_i x_i < T.$$

The output F will be 1 if:

$$\sum_{i=1}^{n} \omega_i x_i \geqslant T.$$

The threshold gate may also be represented as in Fig. 3.18 (b) in which the input weight ω_0 is always at a 1. This now represents the gate threshold. For this gate

$$F = 0 \quad \text{if} \quad \omega_0 + \sum_{i=1}^{n} \omega_i x_i < 0. \tag{3.6}$$

$$F = 1 \quad \text{if} \quad \omega_0 + \sum_{i=1}^{n} \omega_i x_i \geqslant 0. \tag{3.7}$$

Fig. 3.19 represents a gate with three inputs and two control inputs y_1 and y_2. Fig. 3.20 shows the operation table for this gate under three conditions when y_1 and y_2 are changed between 0 and 1 so as to give thresholds of -1, -2 and -3. It is usual in these circuits to write

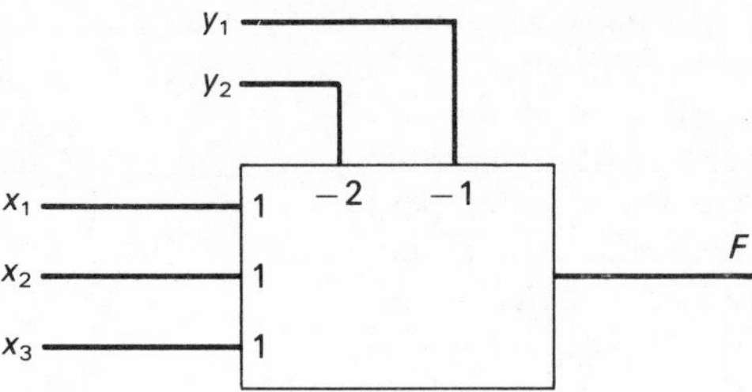

Fig. 3.19. Variable threshold gate.

x_1	x_2	x_3	$y_1 = 1$ $y_2 = 0$ −1, 1, 1, 1 $\omega_0 + \Sigma\omega_i x_i$	F_1	$y_1 = 0$ $y_2 = 1$ −2, 1, 1, 1 $\omega_0 + \Sigma\omega_i x_i$	F_2	$y_1 = 1$ $y_2 = 1$ −3, 1, 1, 1 $\omega_0 + \Sigma\omega_i x_i$	F_3
0	0	0	−1	0	−2	0	−3	0
0	0	1	0	1	−1	0	−2	0
0	1	0	0	1	−1	0	−2	0
0	1	1	+1	1	0	1	−1	0
1	0	0	0	1	−1	0	−2	0
1	0	1	+1	1	0	1	−1	0
1	1	0	+1	1	0	1	−1	0
1	1	1	+2	1	+1	1	0	1

Fig. 3.20. Function variation by threshold switching in Fig. 3.19.

the weights, starting from the control input, in a straight line. Therefore a gate −3, 1, 1, 1 has three inputs each of unity weighting and a control input with a weight of −3.

In Fig. 3.20 consider the case −2, 1, 1, 1. This is obtained when y_1 is at 0 and y_2 is at 1 giving $\omega_0 = -2$. When x_1, x_2 and x_3 are at 0 the value of $\omega_0 + x_1 + x_2 + x_3$ is −2. For the next input $x_1 = x_2 = 0$ and $x_3 = 1$. This gives a weight of −1. In a similar manner the rest of the table can be obtained. The value of F_2 is 1 if the total weight is zero or positive. Fig. 3.20 shows that for −1, 1, 1, 1 the output corresponds to $F_1 = x_1 + x_2 + x_3$, that is, to an OR gate. For weights −2, 1, 1, 1 the output is given by F_2 which can be reduced by a Karnaugh map to $F_2 = x_1 . x_2 + x_2 . x_3 + x_1 . x_3$. This is called a majority gate. For weights −3, 1, 1, 1 the value of $F_3 = x_1 . x_2 . x_3$ and the circuit is an AND gate. Note that one gate has been used to obtain three different functions, hence the versatility of threshold logic. It is also clear from the discussions that the threshold weighting cannot exceed that of all the combined inputs or the gate will never attain the logic 1 output stage.

Threshold logic gates are usually not much more complex in construction than conventional gates. Fig. 3.21 shows the principle behind a two input gate, and from this it is seen to be very similar to an ECL gate. *TR*1 and *TR*2 provide low impedance emitter follow outputs and go to 0 or 1 depending on the voltage developed across their base resistors R_2 and R_3. Resistors R_5 and R_6 represent the weightings of the inputs. With x_1 and x_2 at zero volts the reference voltage V_{REF} keeps *TR*4 and *TR*6 con-

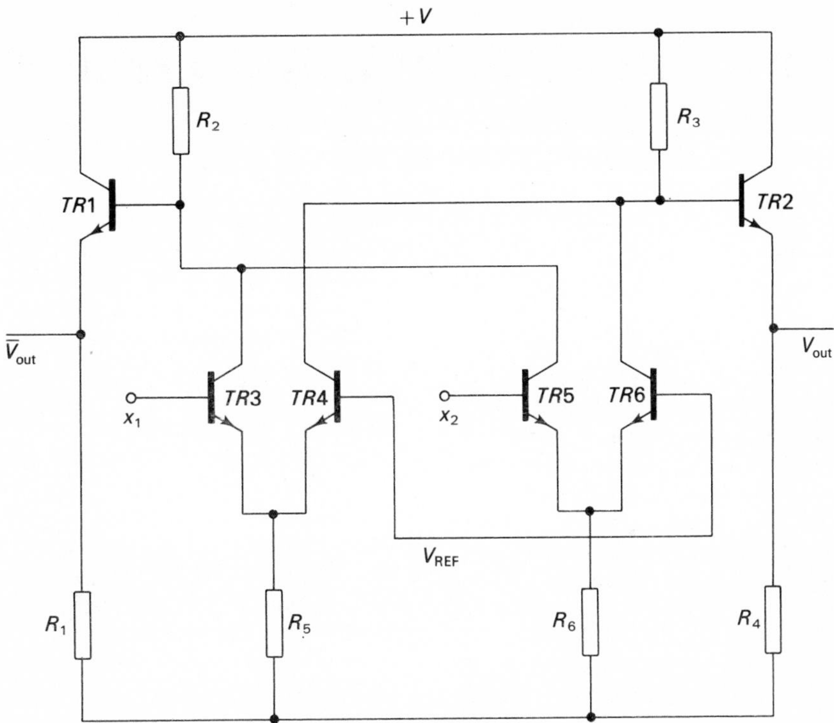

Fig. 3.21. Construction of a simple threshold gate.

ducting such that current is drawn through R_3. When an input exceeds V_{REF}, which represents the logic 1 state, its corresponding transistor conducts and current switches from R_3 to R_2. The magnitude of this current is determined by its emitter or weight resistor. Currents from all inputs are summed in resistor R_2 such that the output from *TR*1 falls and that of *TR*2 rises. When the voltage passes beyond V_{REF} the circuit represents a switch over from logic 0 to 1. The threshold voltage of this gate is determined by the magnitude of resistors R_2 and R_3.

4. Functional integrated circuits

4.1 Introduction

The development of integrated circuits was a boost to the field of systems engineering. With circuits of complexity ranging up to many thousands of transistors available in a single package all the engineer needed to do was to treat this part of the design as a *black box*. Provided the functional performance of this black box was known and certain basic rules were followed the circuit could be used in a total system without any detailed knowledge of its internal construction. These circuits are called functional integrated circuits. They fall into several groups and the actual fabrication technology in which they are made does not affect their functional characteristics. Therefore, for example, an AND gate with inputs A, B and C will produce an output given by $A \cdot B \cdot C$ irrespective of whether it is made in TTL or in CMOS. The difference between the two will be in the performance characteristics such as speed and power output. The present chapter will largely ignore the differences between logic families, as described in chapter 2, and will concentrate on the functional properties of small and medium scale integrated circuits. Large scale integration systems, such as *memories*, are described in chapter 5.

4.2 Gates

Gates are commercially available in a variety of combinations in a package. Generally, 14 pin packages are used and the number of gates and inputs are determined by the availability of pins. Two pins are required for the power supply which leaves 12 pins which can be used for gate inputs and outputs. If only single inputs and outputs are needed, as in a NOT or inverter circuit, it is possible to have six identical circuits. Such a package is called a hex inverter. If two input gates are needed only four circuits are possible. The package is a quad two input NAND. For three inputs only, three gates are possible within the limitation of the 12 pins. The system is a treble three input gate package. Also available are dual four input and single eight input gate packages.

Many of the gate families, such as DTL, present a relatively high impedance when off so that two gates can have their outputs strapped together. Some types of gates, such as TTL with totem-pole outputs and CMOS, cannot have their outputs connected together since they present a very low impedance in both states and this would cause a large current to flow through them. In these instances it is necessary to use an AND–OR–INVERT gate. These gates are described as 'm' wide, 'n' input where 'm' represents the number of AND gates and 'n' each of their inputs. For instance a gate with four AND gates having inputs 2, 3, 3 and 2 would be called a 4 wide 2–3–3–2 AND–OR–INVERT gate.

CMOS logic uses both p- and n-channel transistors and this gives rise to a gate which is unique to this family. It is called a transmission gate and is shown in Fig. 4.1 (*a*). $TR1$ is an n-channel transistor and it turns on when its gate voltage goes positive. $TR2$ is p-channel and it needs a negative gate voltage for conduction. Therefore the two gate voltages are complementary and this can be readily achieved as in Fig. 4.1 (*b*). When C is at a positive voltage both $TR1$ and $TR2$ are on so that an a.c. signal path exists between input and output. When C is at a low voltage both transistors are in a high impedance mode and the input and output are effectively isolated.

The last type of gate to be considered here is the Schmitt trigger. The symbol for a Schmitt trigger inverter is shown in Fig. 4.2 (*a*). The

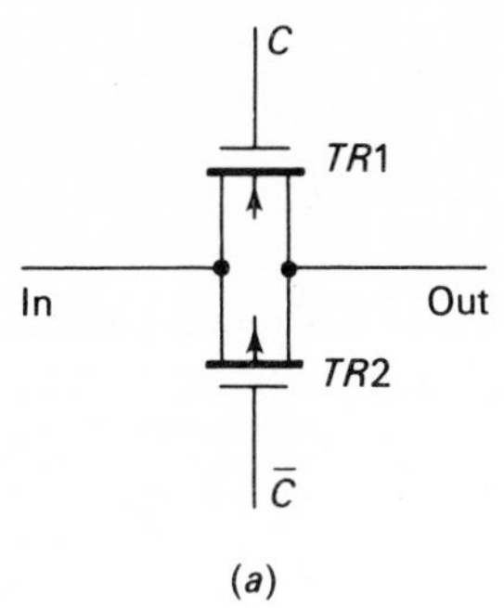

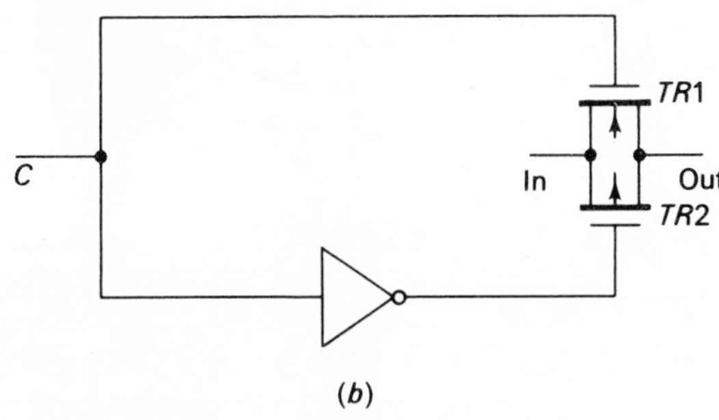

Fig. 4.1. Transmission gate; (a) basic gate, (b) gate with inverter control.

Schmitt trigger gate is also available in combinations such as quad two input, and so on. The integrated circuit version of the Schmitt trigger gate is similar in principle to the well known discrete component circuit. The transfer curve is given in Fig. 4.2 (b). For low voltage inputs the output is at a high voltage for the inverter configuration. As the input is increased it reaches a voltage V_1 at which the output changes from a high to a low voltage. Positive feedback is used in the Schmitt trigger gate so that all transitions are very fast. This feedback also introduces a hysteresis effect whereby the output will remain at a low voltage even when the input falls below V_1 and will return to its high state only after V_2 is reached. The difference between V_1 and V_2 is called the hysteresis voltage of the gate.

4.3 Flip-flops and latches

A flip-flop or bistable is primarily used as a storage element for a single bit of information. Latches perform the same function and are usually similar to the simpler types of flip-flops. There are several forms of flip-flops and although their operation is described with reference to gates in this section, in practice they are made from conventional circuitry and not whole gate structures.

A set–reset flip-flop is shown in Fig. 4.3. It can be made up of two cross-coupled NOR gates or NAND gates. The NAND gate version requires an additional pair of inverters. Similarly all the other systems in this section can be explained with reference to NOR or NAND gates although only the NAND version will be described. In the truth table given in Fig. 4.3 (d) Q_n and $\overline{Q}_n$ represent the state of the outputs prior to the S and R conditions shown, and Q_{n+1} and $\overline{Q}_{n+1}$ are the states following these conditions. Therefore if S and R are both at a low voltage (logic 0) then the state of the Q and $\overline{Q}$ outputs is unchanged. For instance, suppose these states were Q = 1 (high voltage) and $\overline{Q} = 0$ (low voltage). If R and S are both at 0 then the NOR and NAND gates both act as NOT gates so that the $\overline{Q} = 0$ input is inverted to Q = 1 and Q = 1 input is changed to $\overline{Q} = 0$. Therefore there is

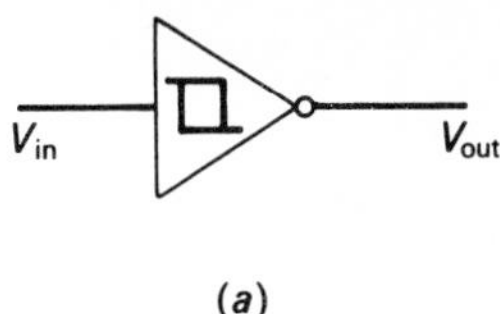

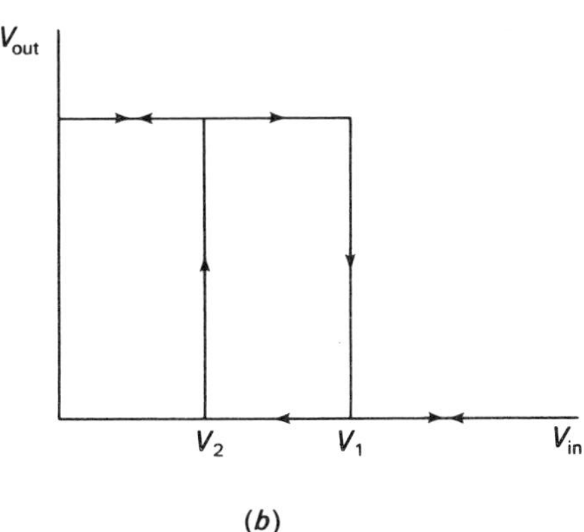

Fig. 4.2. Schmitt trigger gate; (a) symbol, (b) transfer characteristic.

no change over the previous flip-flop state. If now S = 0 and R = 1 then in Fig. 4.3 (*a*) the R = 1 input will force the output of gate G_1 to 0 and the two logic 0 levels at the input of G_2 will give an output of 1 so that Q = 0 and $\overline{Q}$ = 1. Similarly for Fig. 4.3 (*b*) the R = 1 state is inverted by G_2 to give a logic 0 input at G_4. This forces $\overline{Q}$ to 1 and the two logic 1 inputs at G_3 give Q = 0. A similar reasoning will show that for both the NOR and NAND versions the flip-flop will be forced into Q = 1 and $\overline{Q}$ = 0 for inputs S = 1 and R = 0. The S input of the flip-flop is called the set input since it sets the Q output to a 1 and the R input is the reset input since it resets Q to a 0 state. Clearly once the outputs have been set to their 0 or 1 state the inputs can both be removed, that is returned to 0. The output will remain at the previous state so that the circuit performs a memory function.

When the S and R inputs are both taken to a logic 1 the outputs in Fig. 4.3 (*a*) are both forced to a 0 and in Fig. 4.3 (*b*) they are 1. However when S and R return to a 0 the inputs to the NOR gates are all logic 0 and those to the NAND gate are all logic 1. A race condition will now be set up as each gate races to change its state. One of these will get there first and this will latch the flip-flop and prevent further changes. However since the final state is dependent on the delays involved through the gates it is very difficult to predict. Therefore the S = 1, R = 1 input states are prohibited in many circuits.

A more complex flip-flop, but one which is capable of higher operating speeds than the set–reset flip-flop, is shown in Fig. 4.4. It is a master-slave flip-flop and consists of two conventional flip-flops in series. When the clock is at logic 1 gates G_1 and G_2 are operative while G_5 and G_6 are inoperative due to the inverting action of G_9. This means that the output values of the master flip-flop, given by Q_1 and $\overline{Q}_1$, can be changed while the outputs of the slave flip-flop, given by Q and $\overline{Q}$ are unchanged. When the clock falls to logic 0 gates G_1 and G_2 are inoperative and G_5 and G_6 are operative. The S and R inputs are now disconnected while the information stored in the master flip-flop is transferred to the slave. Therefore now Q = Q_1 and $\overline{Q}$ = $\overline{Q}_1$. The higher speed of the master slave flip-flop is as a result of this two step operation. It means that the inputs do not need to remain steady during the period of the clock pulse, as in a clocked set–reset flip-flop, but can change ready for the next cycle while information is being transferred from the master to the slave flip-flop. Therefore during the clock period information enters the master at t_1 and transfers to the slave at t_2. The inputs can now change although the outputs have not stabilized.

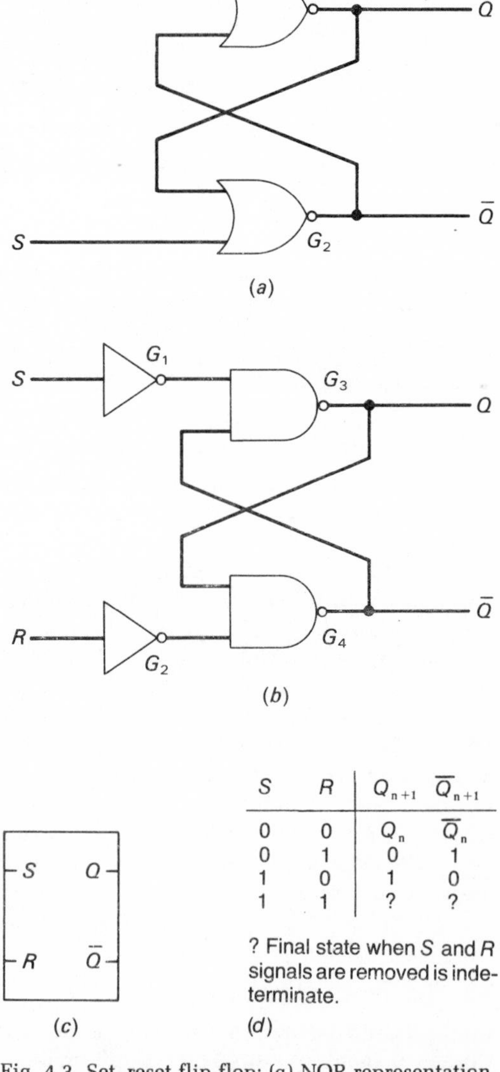

S	R	Q_{n+1}	$\overline{Q}_{n+1}$
0	0	Q_n	$\overline{Q}_n$
0	1	0	1
1	0	1	0
1	1	?	?

Fig. 4.3. Set–reset flip-flop; (*a*) NOR representation, (*b*) NAND representation, (*c*) symbol, (*d*) truth table.

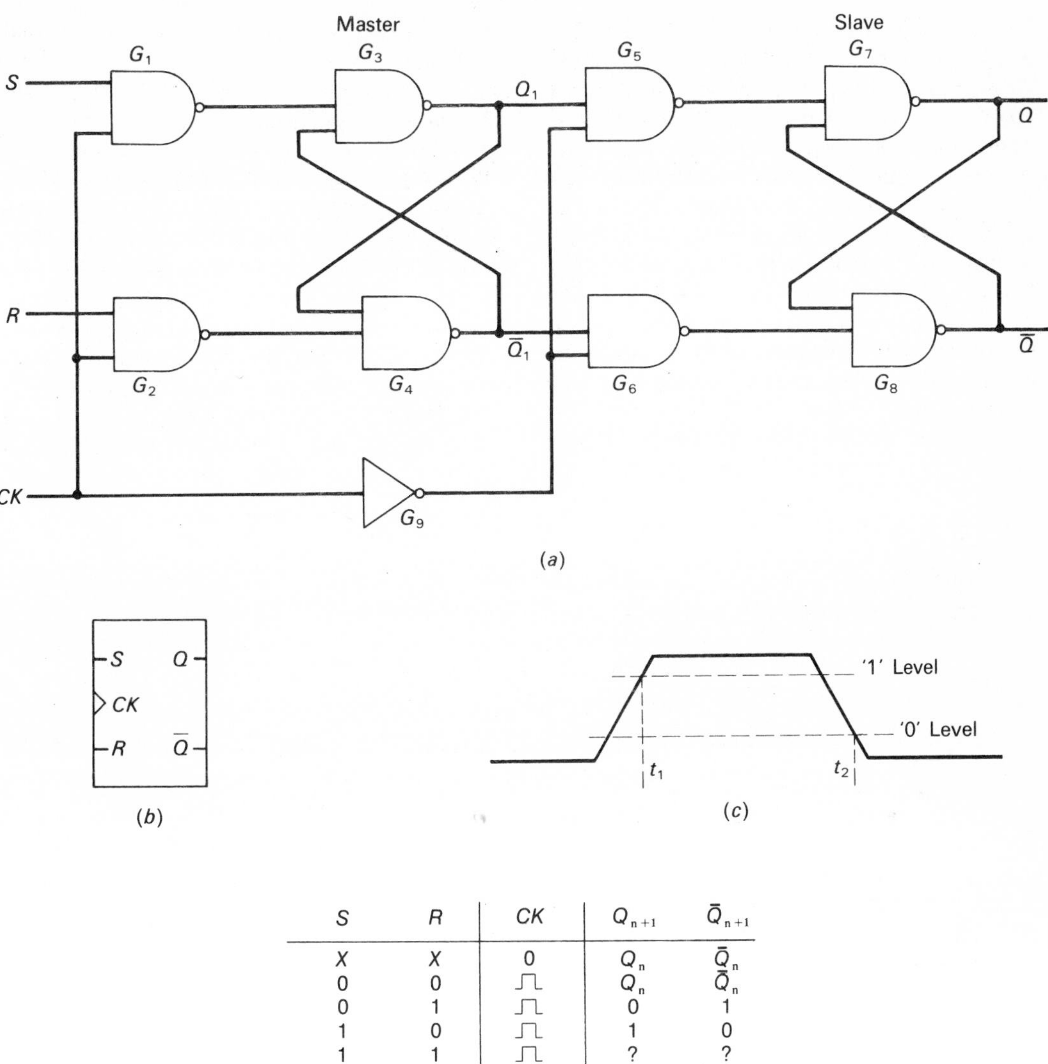

S	R	CK	Q_{n+1}	$\bar{Q}_{n+1}$
X	X	0	Q_n	$\bar{Q}_n$
0	0	⎍	Q_n	$\bar{Q}_n$
0	1	⎍	0	1
1	0	⎍	1	0
1	1	⎍	?	?

⎍ = A high level pulse. Data inputs are held constant while pulse is high. Data is transferred to output on falling edge of the pulse.

(d)

Fig. 4.4. Master–slave set–reset flip-flop; (a) NAND representation, (b) symbol, (c) clock waveform, (d) truth table.

In all the flip-flops described up to now the output is indeterminate if both the inputs are taken simultaneously to logic 1 during a clock pulse. A type of flip-flop in which this does not occur is shown in Fig. 4.5 and is called a *J–K* flip-flop. Gates G_3 to G_6 are similar to those of a set–reset flip-flop. The difference lies in gates G_1 and G_2 which are controlled by the feedback of Q and $\overline{Q}$. Suppose that $Q = 1$ and $\overline{Q} = 0$. This means that if *J* and K are both at logic 1 the value of S = 0 and R = 1. Therefore a clock pulse will result in Q = 0 and $\overline{Q}$ = 1. Now G_1 is closed and G_2 is open so that S = 1 and R = 0. The next clock pulse will change Q to 1 and $\overline{Q}$ to 0. Therefore when *J* and K are both at logic 1 the outputs complement at each clock pulse. This is sometimes called a toggle state. Fig. 4.5 (c) shows that apart from the *J* = 1, K = 1 state the *J–K* flip-flop is similar to the set–reset flip-flop. Master-slave *J–K* flip-flops, which are very similar to their set–reset counterparts, can also be commercially obtained as functional integrated circuit packages. These are not described here.

A variation of the *J–K* or S–*R* flip-flop is the *D* flip-flop shown in Fig. 4.6. It has only one control input, called the *D* input, and a clock. The *D* input is converted to a direct and a complementary input by an internal inverter G_1 and these feed the *J* and K or the S and *R* lines. Therefore a logic 0 at *D* will cause Q = 0, $\overline{Q}$ = 1 on the next clock pulse and a logic 1 at *D* will reverse this state. The *D* flip-flop operates in only two of the *J–K* and S–*R* modes, i.e. *J*(or S) = 1, K(or *R*) = 0 and *J*(or S) = 0, K(or *R*) = 1. The advantage of the *D* flip-flop is that since only one input is required it needs fewer interconnections when used in a larger system.

The S–*R*, *J–K* and *D* flip-flops described here are each available with several modifications.

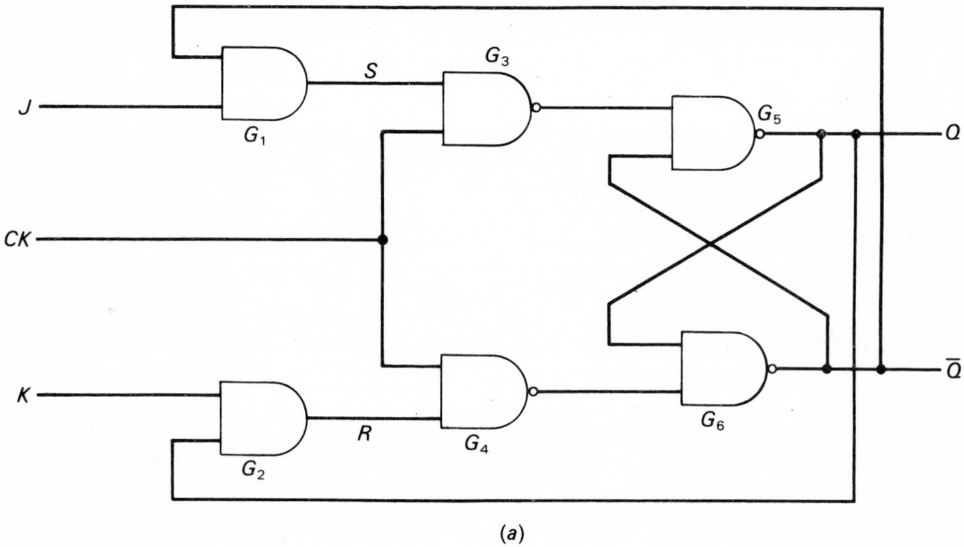

(*a*)

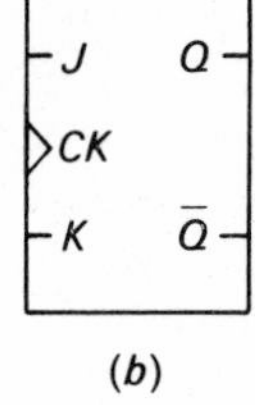

(*b*)

J	*K*	*CK*	Q_{n+1}	$\bar{Q}_{n+1}$
X	X	0	Q_n	$\bar{Q}_n$
0	0	↑	Q_n	$\bar{Q}_n$
0	1	↑	0	1
1	0	↑	1	0
1	1	↑	$\bar{Q}_n$*	Q_n*

↑= Clock transition from low to high level.
* This state also called toggle.

(*c*)

Fig. 4.5. *J–K* flip-flop; (*a*) AND–NAND representation, (*b*) symbol, (*c*) truth table.

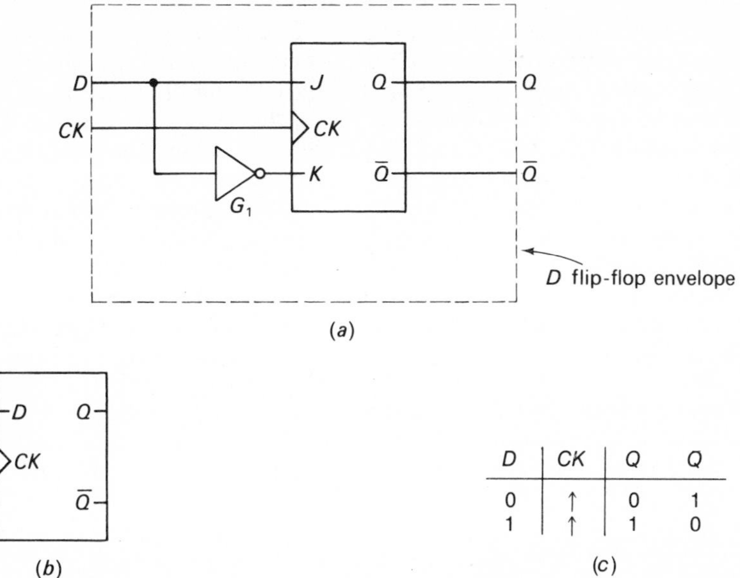

D	CK	Q	Q
0	↑	0	1
1	↑	1	0

Fig. 4.6. *D* flip-flop; (*a*) *J–K* representation, (*b*) symbol, (*c*) truth table.

Apart from the master–slave form they can also be given a preset and clear line. Irrespective of the *J* and *K* states a logic 0 on the preset line will force the flip-flop into $Q = 1, \overline{Q} = 0$, i.e. it will preset the flip-flop. Alternatively a logic 0 on the clear line will give $Q = 0, \overline{Q} = 1$, i.e. it will clear the flip-flop. Making the preset and clear lines simultaneously 0 results in an unstable state since the value of the outputs after this condition is removed is indeterminate.

A further variation is the edge triggered flip-flop. On-chip pulse shapers are used to generate a short clock pulse which is independent of the width of the input clock. Gated flip-flops are also available. These have added versatility since the inputs may be gated in various ways before the flip-flop stage. Several flip-flops are also often available within a single integrated circuit package. In these instances it is common practice to bring out only the Q output, or to use the same pins for the clock or reset and clear lines of all the flip-flops within the package in order to minimize the total number of package pins required.

4.4 Counters

The flip-flops described in section 4.3 can be interconnected to form a variety of circuits. In this section one of these, the counter, will be described. Although in general any of the flip-flop types may be used for counters the *J–K* arrangement is the most popular. The counters described here are almost all commercially available in a single package so that connection of individual packages is not required. There are basically two types of counters called asynchronous and synchronous. In addition the synchronous counters may be subdivided into those which use ripple enable and those with parallel, or look ahead enable. These are all described in this section.

Fig. 4.7 (*a*) shows four flip-flops in which the *J–K* inputs are all connected to a logic 1 level and the clock input is fed from the output of the preceding stage. An input line feeds the clock of the first flip-flop stage. Since the *J–K* inputs are connected to logic 1 the flip-flops will complement or toggle when their clock input goes from a 1 to a 0 state, if they are negative edge triggered devices. Assuming that all the outputs are initially at logic 0 and that the input line is supplied with a series of clock pulses then flip-flop *FA* will complement every time the input goes from 1 to 0, flip-flop *FB* will complement when the output of *FA* goes from 1 to 0, and so on. The truth table is given in Fig. 4.7 (*b*). Note that each stage divides the output frequency of the previous stage by two so that the counter also acts as a frequency divider. The change command also ripples through the chain of flip-flops. Therefore, for instance, for *FD* to go from 0 to 1, flip-flop *FA* must first change from 1 to 0 and when this is finished *FB* must change from 1 to 0 followed by *FC* changing from 1 to 0. Each change causes a delay due to the internal structure of the flip-flops so that the asynchronous counter is relatively slow in operation, specially for large counts. The advantage of an asynchronous counter is that it needs no additional gating so that it is relatively simple. Furthermore the halving of frequency between stages means that later stages can be designed for slower operating speeds than earlier ones giving a good overall speed–power product for the counter.

The circuit shown in Fig. 4.7 is called a binary counter since it counts in binary code. It is possible to modify this coding by additional gating. One such device is the BCD counter, also called a decade counter since it has ten steps. From the truth table of Fig. 4.7 (*b*) it is seen that for this counter the 1001 state must be followed by 0000, and internal gating is used to obtain this. Counters can also count up as well as down. To count down the flip-flops must toggle when the *A*, *B*, *C* outputs change from 0 to 1. However the flip-flops only respond to a 1 to 0 clock transition. Counting down can be

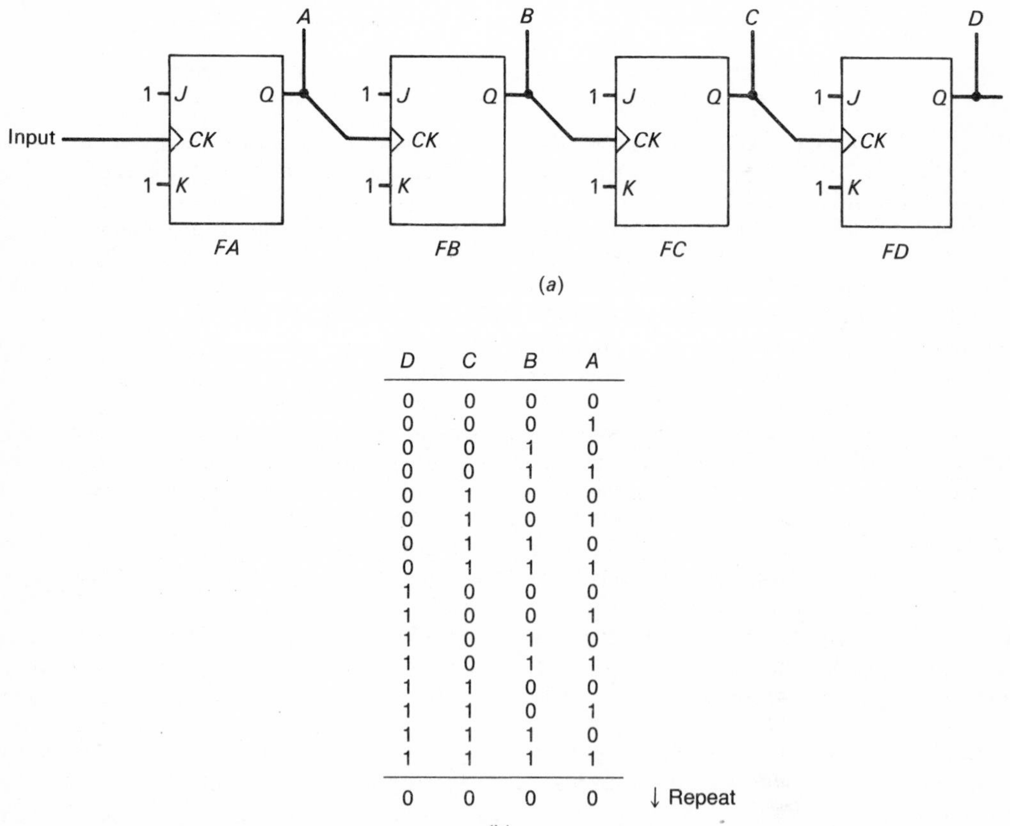

D	C	B	A
0	0	0	0
0	0	0	1
0	0	1	0
0	0	1	1
0	1	0	0
0	1	0	1
0	1	1	0
0	1	1	1
1	0	0	0
1	0	0	1
1	0	1	0
1	0	1	1
1	1	0	0
1	1	0	1
1	1	1	0
1	1	1	1
0	0	0	0 ↓ Repeat

Fig. 4.7. Asynchronous counter; (*a*) circuit arrangement, (*b*) truth table.

Present state				Next state				State of J–K inputs			
D	C	B	A	D	C	B	A	FD	FC	FB	FA
0	0	0	0	0	0	0	1	0	0	0	1
0	0	0	1	0	0	1	0	0	0	1	1
0	0	1	0	0	0	1	1	0	0	0	1
0	0	1	1	0	1	0	0	0	1	1	1
0	1	0	0	0	1	0	1	0	0	0	1
0	1	0	1	0	1	1	0	0	0	1	1
0	1	1	0	0	1	1	1	0	0	0	1
0	1	1	1	1	0	0	0	1	1	1	1
1	0	0	0	1	0	0	1	0	0	0	1
1	0	0	1	1	0	1	0	0	0	1	1
1	0	1	0	1	0	1	1	0	0	0	1
1	0	1	1	1	1	0	0	0	1	1	1
1	1	0	0	1	1	0	1	0	0	0	1
1	1	0	1	1	1	1	0	0	0	1	1
1	1	1	0	1	1	1	1	0	0	0	1
1	1	1	1	0	0	0	0	1	1	1	1

(a)

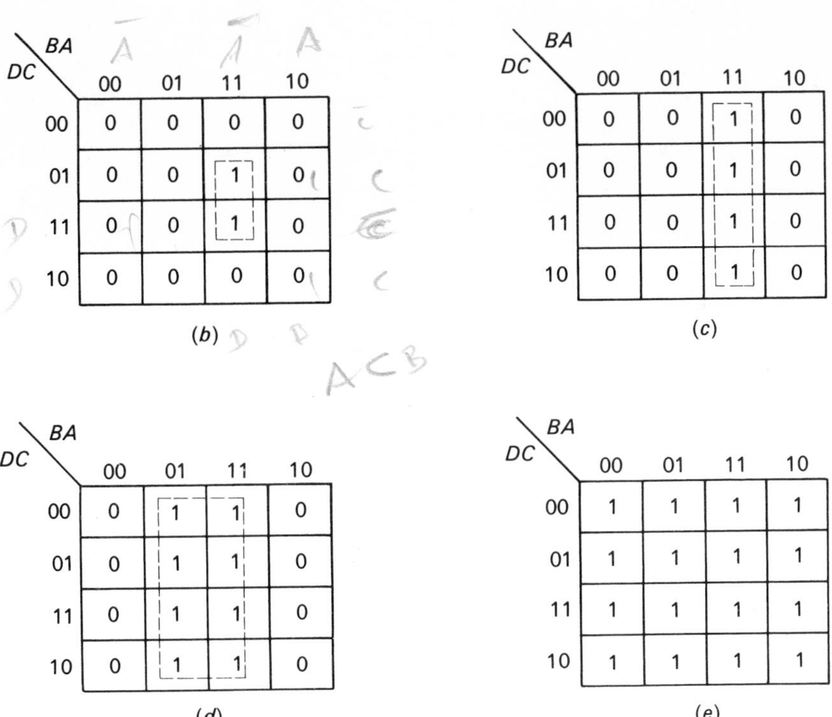

Fig. 4.8. Design of a synchronous binary counter; (a) truth table, (b) map for the J–K inputs of flip-flop FD, (c) map for the J–K inputs of flip-flop FC, (d) map for the J–K inputs of flip-flop FB, (e) map for the J–K inputs of flip-flop FA.

accomplished by using the complement of the outputs, obtained from the $\overline{Q}$ outputs of the flip-flops, to trigger the next stages.

In a synchronous counter the input line provides clock pulses to all the flip-flops simultaneously. They therefore operate synchronously with the input and there is no ripple action of the clock signal from one stage to the next. Fig. 4.8 shows the present and next state for each flip-flop of a 4 bit counter and indicates the state of their *J–K* inputs. If the flip-flop is to change state or toggle its *J–K* inputs must be at 1, otherwise they are at 0. The Karnaugh maps for these inputs can now be drawn as in Fig. 4.8. From these maps it is seen that *FA* must have its inputs at 1, *FB* at *A*, *FC* at *A* . *B*, and *FD* at *A* . *B* . *C*. Fig. 4.9 shows two methods by which these conditions can be obtained. The parallel or look ahead counter has the fastest operating speed since the enable *signal* is applied in parallel to all flip-flops. However the loading on the flip-flops increases for each additional stage, since it has to drive several gate inputs in parallel, and it can become very large for long counters.

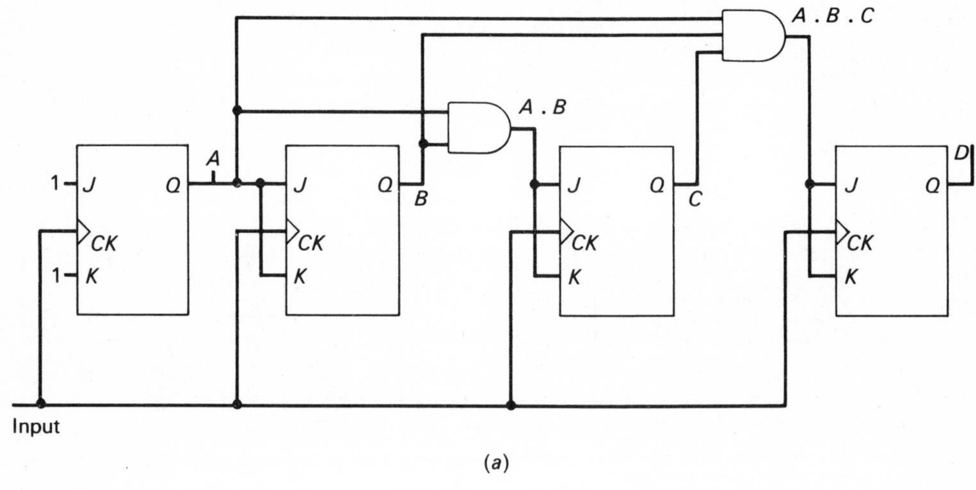

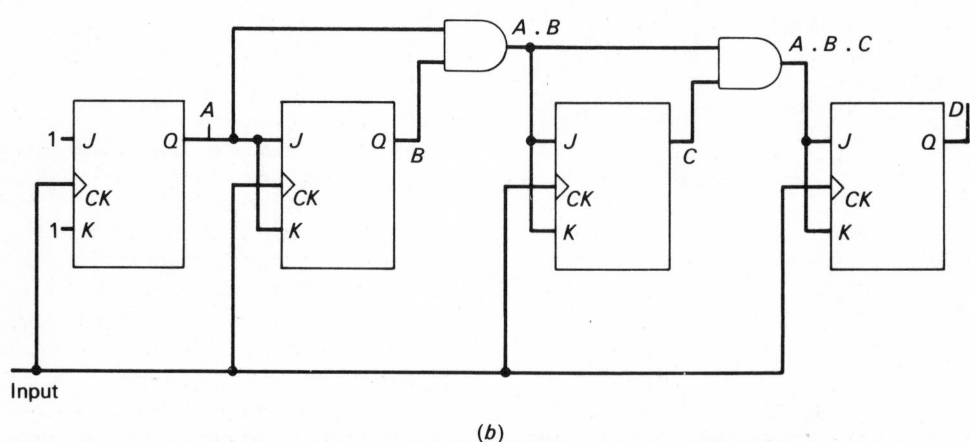

Fig. 4.9. Synchronous binary counter; (*a*) with parallel carry, (*b*) with ripple carry.

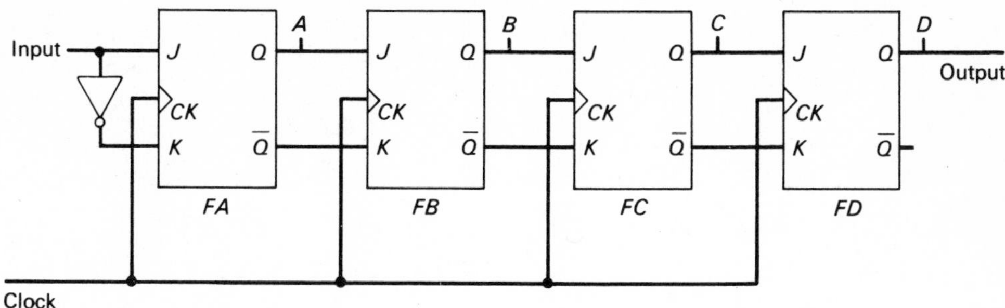

Fig. 4.10. A four bit serial-in serial-out shift register.

The number of inputs required per gate can also become very large. These disadvantages are overcome in the ripple carry counter. This is slower since although the clock is synchronous the enable signal between stages operates in a ripple mode.

4.5 Shift registers

The flip-flop represents a mechanism for storing one bit of information. Several flip-flops may be interconnected to store larger amounts of information. When these are arranged in such a way that all bits of the storage locations can be accessed in an equal time the device is called a *random access memory*. These devices are described in chapter 5. If the store is organized such that the information is transferred from one location to the next in serial fashion, it is called a serial memory or shift register. These registers are available commercially in sizes from four bits to many thousand bits. The general operating principles in all cases is the same. Shift registers are available for storing analogue or digital information. In this chapter digital shift registers are considered only, analogue devices being described in chapter 7.

Fig. 4.10 shows how four *J–K* flip-flops are connected together to form a shift register. All the flip-flops are clocked synchronously and their *J–K* inputs are fed from the *Q* and $\overline{Q}$ outputs of the previous stage. Therefore after a clock pulse the information moves along one bit to the right and flip-flop *FA* contains the information which was on the input line.

The shift register shown in Fig. 4.10 is serial-input serial-output. The *Q* lines from the individual bits can be brought out as well as the serial input and outputs. These parallel outputs are often required in a system to give faster data transfer. The disadvantage of parallel output is that it requires a much larger number of package pins. Similarly each bit can be set individually to a logic 1 or 0 either by gating the inputs or by utilizing the preset and clear inputs of the flip-flops. Therefore shift registers are commercially available in a combination of serial–parallel input–outputs.

Shift registers can operate in a left or right shift mode. In Fig. 4.10 for left shift the circuit needs to be modified such that the input feeds the *J* and *K* terminals of *FD* and the *Q* and $\overline{Q}$ outputs of this flip-flop feed the *J–K* lines of *FC*, and so on. For very large shift registers dynamic logic is often used to obtain lower power consumption and higher circuit density.

It is seen from Fig. 4.10 that data which enters the first bit of the shift register is not available at the output until it has rippled all the way through the bits. This ripple effect can only occur synchronously with the clock pulse which feeds in new data. It would be advantageous for many applications to be able to operate the output and input independently of each other. This would enable frequency buffering to be obtained between two systems. Such a device is available commercially as a first-in first-out memory or FIFO. Functionally it can be considered to be analogous to a tube, with access covers at both ends, as in Fig. 4.11 (*a*). If a ball is dropped into the tube it will roll to the other end and stop with its surface against the output cover. The next ball will similarly roll through the tube and reach the first. It is now possible to carry on putting balls into the tube and at the same time to remove balls from the other end. The rates at which the balls enter and leave the tube can be different so long as the tube is never completely full or completely empty. It should also be noticed that the first ball which enters the tube is also the first to leave it. Fig. 4.11 (b) shows the logic arrangement for a FIFO memory. It is very similar to the conventional shift register of Fig. 4.10 with the exception of the two external clocks, an internal clock and control logic, and the status bits associated with each storage bit. When a storage bit contains useful data its associated status bit is set to a logic 1. Otherwise it is at logic 0. The internal clocking system automatically moves data along from one bit to the next until it encounters a data bit which has its status bit at 1. It then stops. Suppose, for example, that the system in Fig. 4.11 (b) is initially reset so that all status and data bits are at 0. On the first input clock pulse the

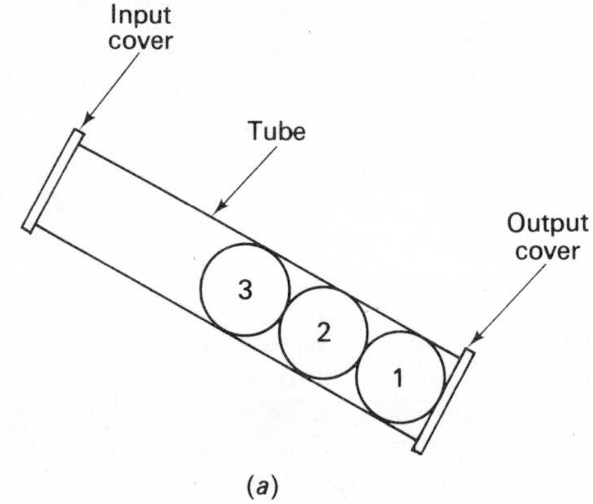

(a)

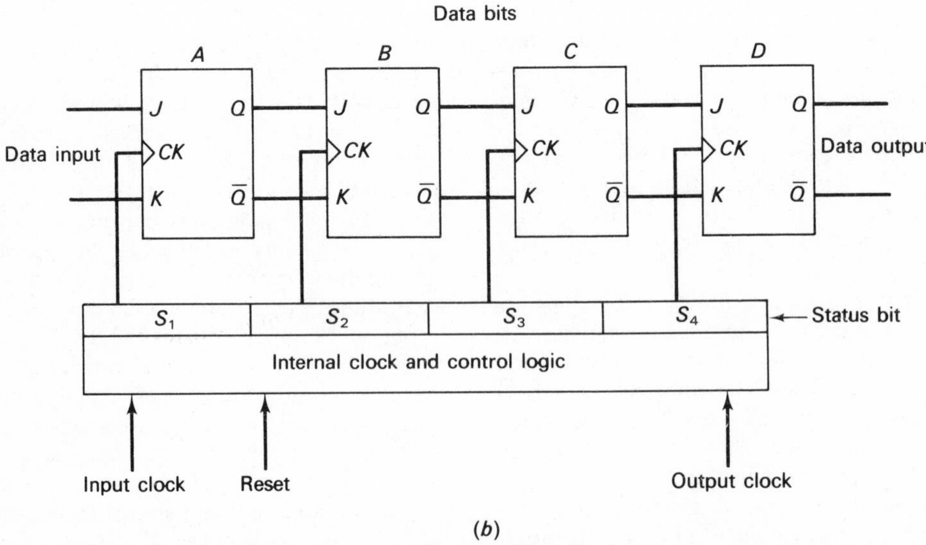

(b)

Fig. 4.11. A first-in first-out shift register; (a) mechanical analogy, (b) logic diagram.

input data, which is logic 1 say, is written into data bit A. The status bit S_1 associated with this data bit is also set to a logic 1. This enables the internal logic and clock system. The logic looks at bit data B and if its status bit S_2 is at 0 it transfers the information from A to B, sets the status bit S_1 to 0 and S_2 to 1. However, both data bits A and B are now at 1 although only bit B contains the stored information. Similarly the internal logic will examine the status bits S_3 and S_4 associated with data bits C and D and since these are both at 0 the information is shifted along to bit D so that eventually all the data bits are at logic 1 while all status bits are at logic 0 except for status bit S_4 which is at logic 1. The system now stops. If the next data input bit is logic 0 and this is clocked into bit A then this data bit will be set to 0 while S_1 will be set to 1. Now as before the internal logic and clock system will move this stored logic 0 from A to B to C. In each case status bits S_1, S_2 and S_3 are set to 1 when the information is stored in its corresponding data bit and then reset to 0 as it is moved along. When the 0 is stored in C the logic looks at status bit S_4 and since this is a logic 1 due to a previous operation, the internal system is disabled and no further operations occur until the next input or output clock pulse. On an output pulse the right shift operation is again commenced by the internal clock until a status bit with logic 0 is encountered when the operation stops. It is clear from this description that there are three clock speeds associated with a FIFO. These are the input and output clocks which are independent of each other provided the FIFO is not completely full or empty, and the internal clock which determines the ripple through, or bubble through time of the information from the first bit to the last empty bit. This bubble through rate is often the factor which limits the maximum operating speed of a FIFO.

4.6 Data conversion and routing

Several integrated circuits are available for converting data from one code to another. Applications for a few of these are described in this section. In addition multiplexer and demultiplexer circuits are introduced along with special integrated circuits which provide specialist functions such as data transmission and keyboard entry.

Coding principles were introduced in chapter 3 and, based on these, Fig. 4.12 (*a*) shows the truth table for a BCD to decimal code converter. A logic 1 on a decimal output indicates that this has been selected. The expressions for the decoding logic have been obtained with the help of the Karnaugh map shown in Fig. 4.12 (*b*). The logic shows that code conversion is obtained by feeding selected inputs and their complements into AND gates. The symbol for the code converter is shown in Fig. 4.12 (*c*) the total number of pins required being such that the device can be fitted into a 16 pin package.

Errors often arise during data manipulation or transfer since a spurious signal can change the state of a bit. Several special codes exist which make it easy to detect the error when it arises and some even enable limited self correction. The simplest way to detect whether an error has occurred is to look for the presence of redundant states in the transmitted data. Therefore for the BCD code shown in Fig. 4.12 the last six states transmit no useful information and are therefore redundant. When information is transmitted these six states will not be used. However if their presence is detected at the receiving end then it is clear that an error has occurred. The logic which would detect these states is seen from Fig. 4.12 to be
$A.\overline{B}.C.\overline{D} + A.\overline{B}.C.D + A.B.\overline{C}.\overline{D} + A.B.\overline{C}.D + A.B.C.\overline{D} + A.B.C.D.$
This can be simplified to $A.(B + C)$ so that the presence of error can be detected by a relatively simple circuit.

An alternative method of detecting errors is by the use of the parity bit. This is an extra bit which is added to the data at the transmission stage to modify its code to odd or even parity. Fig. 4.13 (*a*) illustrates the difference. The even parity bit makes the word contain an even number of logic 1 bits while for odd parity the word has an odd number of logic 1 bits. At the receiving end a check is made to see if the words contain odd or even bits, depending on the transmission mode adopted. A simple circuit for detecting the transmitted error is shown in Fig. 4.13 (*b*). The data is assumed to be transmitted serially and a flip-flop is inserted in its path. The flip-flop is cleared at the start of each word. Every time a 1 appears in the data stream the flip-flop is toggled. Therefore if even parity transmission is used the output must be 0 at the end of each word and for odd parity it must be 1. If these states are not present then an error has occurred. Clearly two errors in a word will result in the odd or even parity state being unchanged and it will go undetected.

A system which enables data from a number of input lines to be selected and transmitted to a smaller number of output lines is called a multiplexer. Conversely a system which converts data from a small number of lines to many lines is a de-multiplexer. Fig. 4.14 (*a*) shows a 5 to 1 multiplexer in which the circuit has been represented as a rotary switch such that it can connect any one of the five inputs to the output.

BCD Code				Outputs										Decoding
A	B	C	D	0	1	2	3	4	5	6	7	8	9	logic
0	0	0	0	1	0	0	0	0	0	0	0	0	0	$\bar{A}.\bar{B}.\bar{C}.\bar{D}$
0	0	0	1	0	1	0	0	0	0	0	0	0	0	$\bar{A}.\bar{B}.\bar{C}.D$
0	0	1	0	0	0	1	0	0	0	0	0	0	0	$\bar{B}.C.\bar{D}$
0	0	1	1	0	0	0	1	0	0	0	0	0	0	$\bar{B}.C.D$
0	1	0	0	0	0	0	0	1	0	0	0	0	0	$B.\bar{C}.\bar{D}$
0	1	0	1	0	0	0	0	0	1	0	0	0	0	$B.\bar{C}.D$
0	1	1	0	0	0	0	0	0	0	1	0	0	0	$B.C.\bar{D}$
0	1	1	1	0	0	0	0	0	0	0	1	0	0	$B.C.D$
1	0	0	0	0	0	0	0	0	0	0	0	1	0	$A.\bar{D}$
1	0	0	1	0	0	0	0	0	0	0	0	0	1	$A.D$
1	0	1	0	0	0	0	0	0	0	0	0	0	0	—
1	0	1	1	0	0	0	0	0	0	0	0	0	0	—
1	1	0	0	0	0	0	0	0	0	0	0	0	0	—
1	1	0	1	0	0	0	0	0	0	0	0	0	0	—
1	1	1	0	0	0	0	0	0	0	0	0	0	0	—
1	1	1	1	0	0	0	0	0	0	0	0	0	0	—

(*a*)

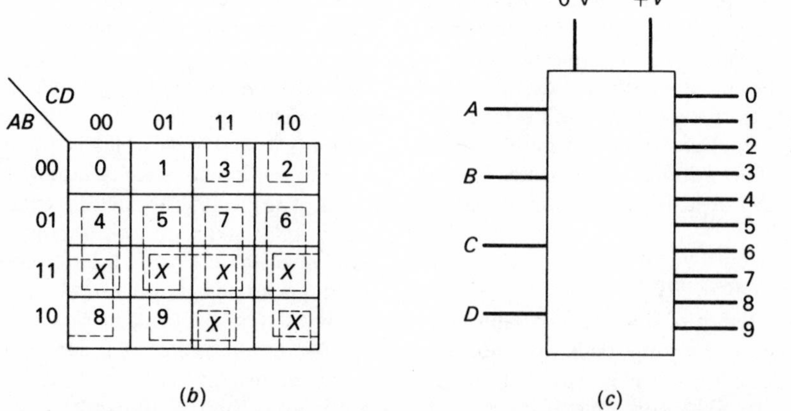

Fig. 4.12. BCD to decimal converter; (*a*) truth table, (*b*) Karnaugh map, (*c*) package symbol.

A	B	C	Even parity bit	Odd parity bit
0	0	0	0	1
0	0	1	1	0
0	1	0	1	0
0	1	1	0	1
1	0	0	1	0
1	0	1	0	1
1	1	0	0	1
1	1	1	1	0

(a)

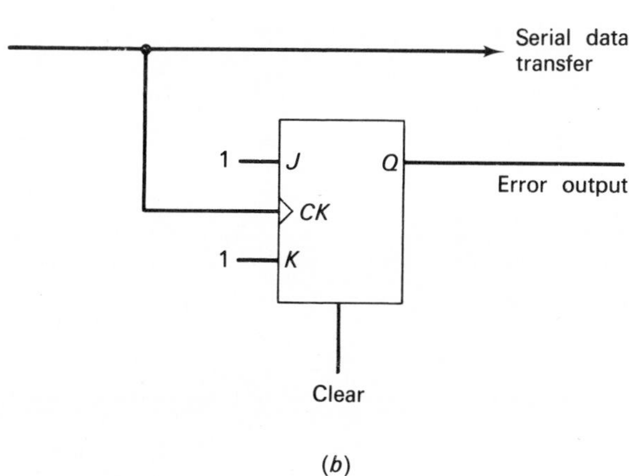

(b)

Fig. 4.13. Error detection using a parity bit; (a) truth table, (b) parity checker for serial data transfer.

The demultiplexer of Fig. 4.14 (b) can connect the input to any one of five output lines. The arrangement of a commercial 8 channel to 1 multiplexer is shown in Fig. 4.15 (a). The channel select lines A, B, C enable the eight input lines to be addressed in binary code and this addressed line is connected to the output. Fig. 4.15 (b) shows the functional logic diagram for the multiplexer. The binary code on the channel select lines will operate one of the AND gates and connect the relevant input line to the output. The truth table of Fig. 4.15 (c) illustrates this since the data output corresponds to the state of the selected input channel and is independent of the state of the other channels.

The code converter shown in Fig. 4.12 can also be used as a demultiplexer. Inputs B, C and D are used as the channel select lines and A is the data input. Output lines 0 to 7 are only used now. It will be seen from the truth table that for each state of B, C, and D the selected output line is a 0 or 1 depending on whether A is a 1 or 0.

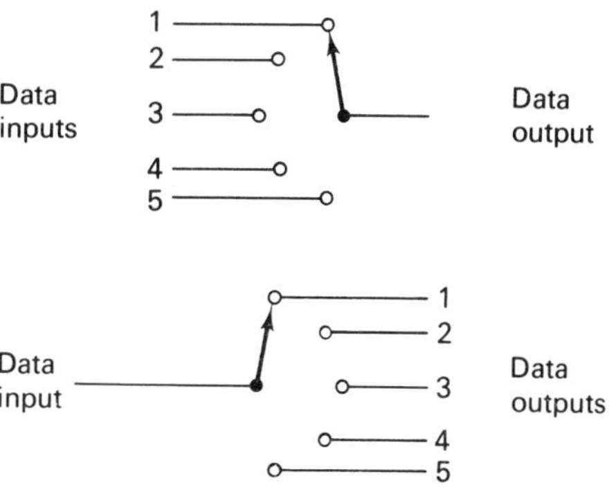

Fig. 4.14. Data routing; (a) multiplexer, (b) demultiplexer.

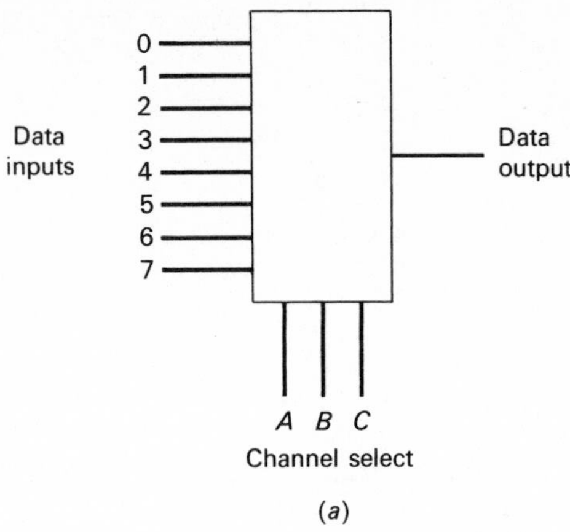
0
1
2
3
4
5
6
7
Data
inputs
Data
output
A B C
Channel select

(a)

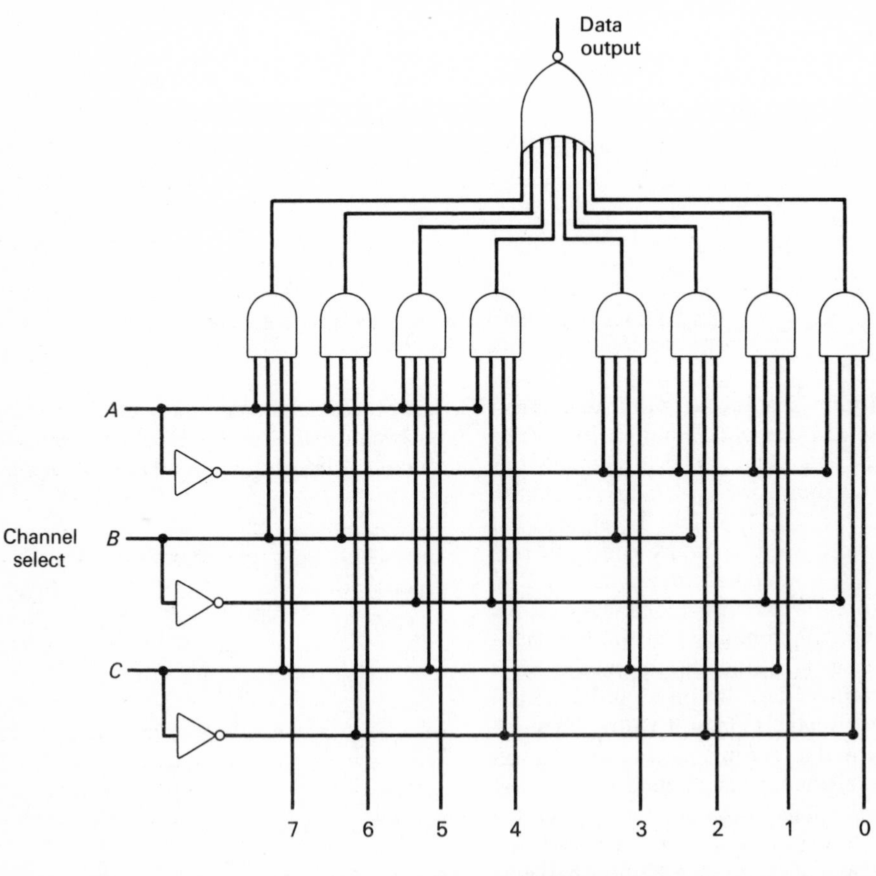
Data
output
A
B
C
Channel
select
7
6
5
4
3
2
1
0
Data inputs

(b)

Channel select A	B	C	Data inputs 0	1	2	3	4	5	6	7	Data outputs
0	0	0	0	×	×	×	×	×	×	×	0
0	0	0	1	×	×	×	×	×	×	×	1
0	0	1	×	0	×	×	×	×	×	×	0
0	0	1	×	1	×	×	×	×	×	×	1
0	1	0	×	×	0	×	×	×	×	×	0
0	1	0	×	×	1	×	×	×	×	×	1
0	1	1	×	×	×	0	×	×	×	×	0
0	1	1	×	×	×	1	×	×	×	×	1
1	0	0	×	×	×	×	0	×	×	×	0
1	0	0	×	×	×	×	1	×	×	×	1
1	0	1	×	×	×	×	×	0	×	×	0
1	0	1	×	×	×	×	×	1	×	×	1
1	1	0	×	×	×	×	×	×	0	×	0
1	1	0	×	×	×	×	×	×	1	×	1
1	1	1	×	×	×	×	×	×	×	0	0
1	1	1	×	×	×	×	×	×	×	1	1

× = 'Don't care' state.

(c)

Fig. 4.15. 8 channel multiplexer; (a) block diagram, (b) functional diagram, (c) truth table.

A variation on the code converter and multiplexer is a priority encoder. It consists of several input and output lines. Any number of input lines may be energized but only the one with the highest weight is operative. The output represents a code for this line. Suppose, for instance, that the circuit shown in Fig. 4.12 (c) is a ten input binary coded decimal output priority encoder in which lines 0 to 9 are inputs and lines A, B, C, D are outputs. If lines 1, 6 and 9 are energized the output will be 1001 since 9 is the input with the highest weight.

Priority encoders can be used for many applications, the commonest being as a means for keyboard data entry. For instance the 10 input lines in the encoder of Fig. 4.12 (c) can be connected to ten keys. The highest number key which is depressed generates the BCD code which is fed to the rest of the system. This is clearly a very simple keyboard encoder. More complex devices are commercially available which can accommodate almost 100 keys and produce the output code on about 10 lines. This code can be chosen to suit the user's requirements by programming an internal *read only memory*. These memories are described in chapter 5. In addition the encoder has such facilities as an on-chip clock for timing the internal operations, bounce protection and n-key rollover.

Another relatively complex device which is extensively used in data transmission applications is the universal asynchronous receiver transmitter or UART. It is shown in Fig. 4.16. The system consists of a separate transmitter and receiver section. The transmitter can accept data as parallel bits of 5, 6, 7 or 8 bits. These are stored in the transmitter buffer before being sent to the transmitter register. The data is then converted to serial format and the start, stop and parity bits added as required. The receiver section accepts the serial data and gates it through a system which accommodates the variable word length. The information is stored in the receiver register and checked for parity errors. It is then sent to the receiver buffer and put out in parallel form.

Both the receiver and transmitter have their own clocks and can operate asynchronously. Flags on the registers indicate when they are ready to send or receive the next word of data and parity errors are also indicated by flags. The control section is common to both the transmitter and receiver areas. It has the facility to vary the data word by coding the W_1 and W_2 lines, to clear or disable all flags, and to inhibit the parity generation or cause the system to operate on an even or odd parity bit. In addition the number of stop bits can be selected to be 1 or 2 according to whether a logic 0 or 1 is put on the stop bit select line.

4.7 Timing

Two forms of timing circuits are commonly used in electronic systems. The first is known as a single-shot or monostable multivibrator. It produces an output pulse, of controllable width, in response to a short duration input trigger pulse. The second type of timer is the astable multivibrator which produces a continuous series of output pulses and does not need to be triggered.

Timer circuits can be produced using counters. The output from a high accuracy clock feeds a binary counter and both the counter and clock are operated by control circuitry. An input trigger pulse frees the clock and counter. The output from the counter, which is gated, changes its state as the counter goes through its count and returns to its original position when this is completed. The duration of the output pulse is dependent on the number of counter bits and the clock frequency. The accuracy of the system depends on the stability of the clock

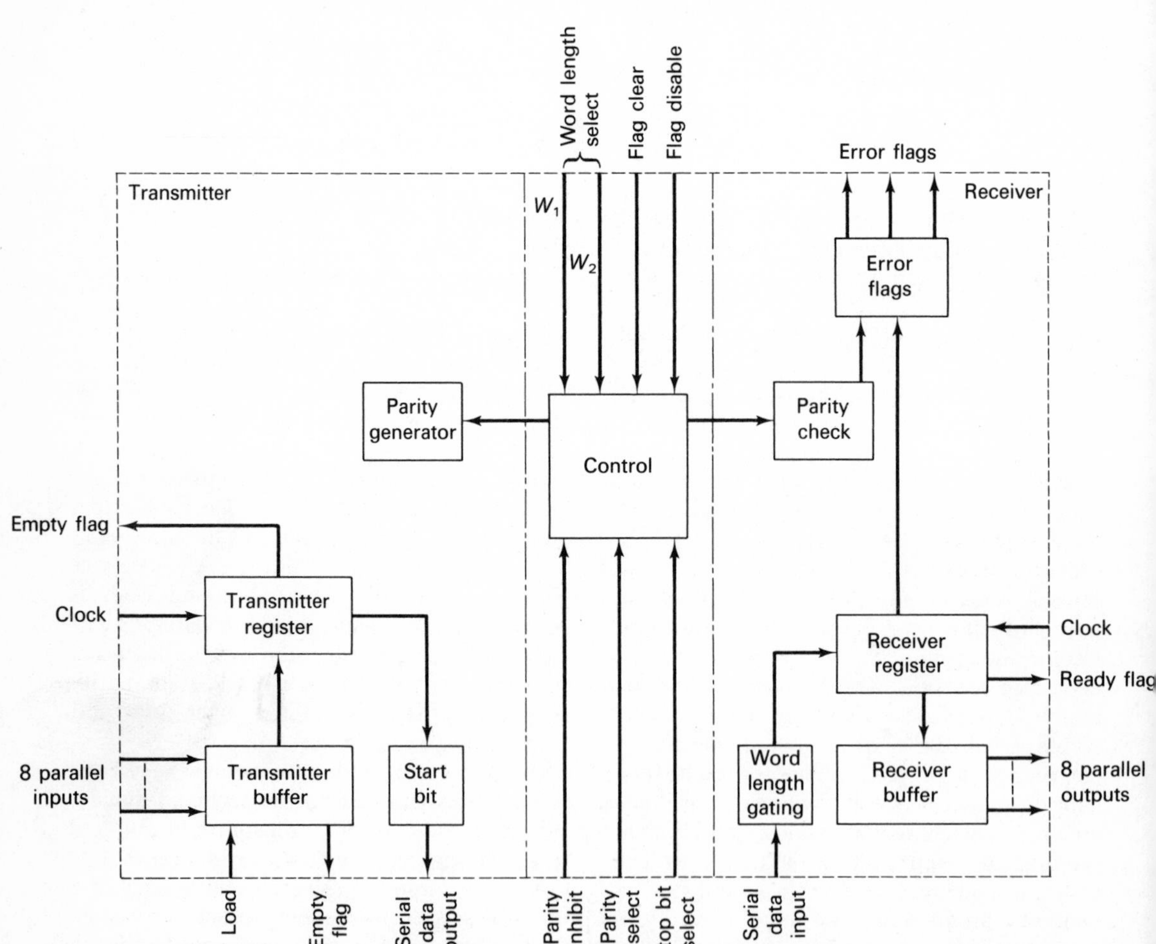

Fig. 4.16. Universal asynchronous receiver transmitter.

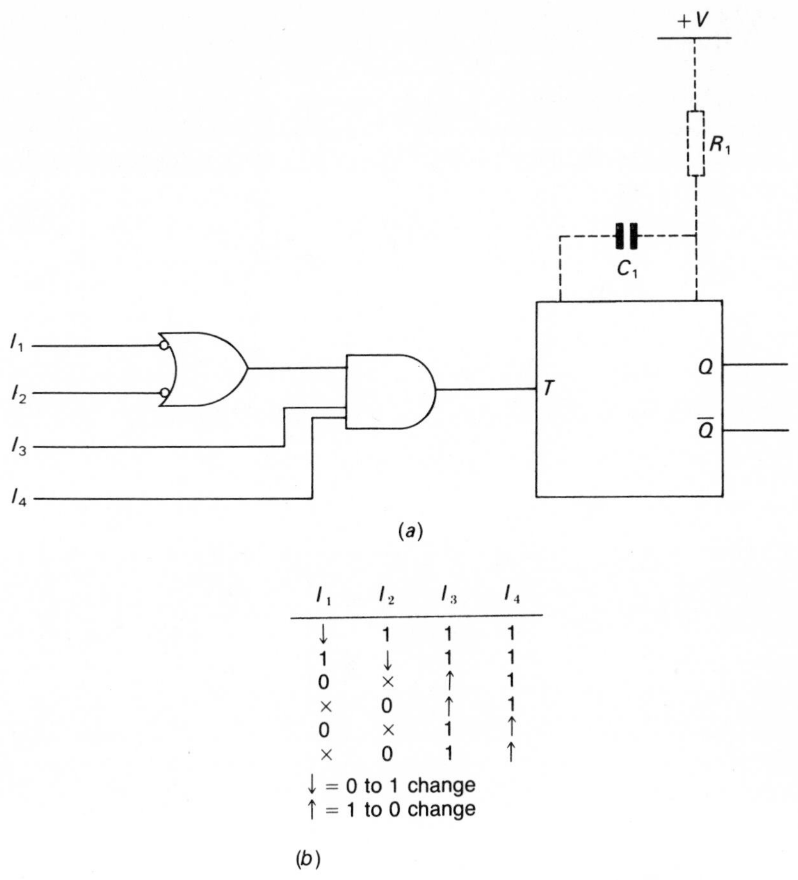

I_1	I_2	I_3	I_4
↓	1	1	1
1	↓	1	1
0	×	↑	1
×	0	↑	1
0	×	1	↑
×	0	1	↑

↓ = 0 to 1 change
↑ = 1 to 0 change

(b)

Fig. 4.17. Gated monostable with external timing components; (a) logic diagram, (b) trigger truth table.

frequency, which can be made very good, for example with a quartz crystal oscillator.

Besides counter timers it is also possible to produce circuits using R–C timing elements. Fig. 4.17 shows the logic diagram of a commercially available gated monostable circuit along with its triggering table. The circuit will trigger when the voltage at T goes to 1 and an examination of the truth table will confirm this to be the case. The timing resistor and capacitor are added externally to the package and can therefore be altered to give variable delay times. This monostable circuit can also be connected to behave as an astable multivibrator as in Fig. 4.18 (a). With both the inputs of the NOR gate at 0 its output is at 1. Therefore the output from the AND gate is equal to its input, i.e. to the value of $\overline{Q}$. When Q is at 1 the value of $\overline{Q}$ is 0. At the end of this period Q falls to 0 and $\overline{Q}$ rises to 1 which triggers the circuit back into the state Q = 1, $\overline{Q}$ = 0. Quite clearly the output consists of a series of pulses having very short pulse widths due to the rapid feedback trigger action of $\overline{Q}$. This feedback can be delayed by adding capacitor C_2 in the feedback path, as in Fig. 4.18 (b), to give a longer pulse duration.

There are many different commercially available integrated circuit configurations of R–C timers. A very versatile system is shown in Fig. 4.19. Initially the output is at logic 1 and TR1 is on so that C_1 is discharged. The trigger pulse will change the state of the flip-flop so that *TR*1 turns off and the output goes to logic 0. This is the active state and the output amplifier gives the circuit a large current sink capability. With *TR*1 off capacitor C_1 commences to charge through R_1. When its voltage exceeds V_{REF} the comparator sends a signal to the flip-flop, returning it to its original state, turning on *TR*1 to discharge C_1, and returning the output to the logic 1 state.

Although a monostable application has been shown, this timer circuit can be easily interconnected to produce an astable multivibrator and a variety of other systems.

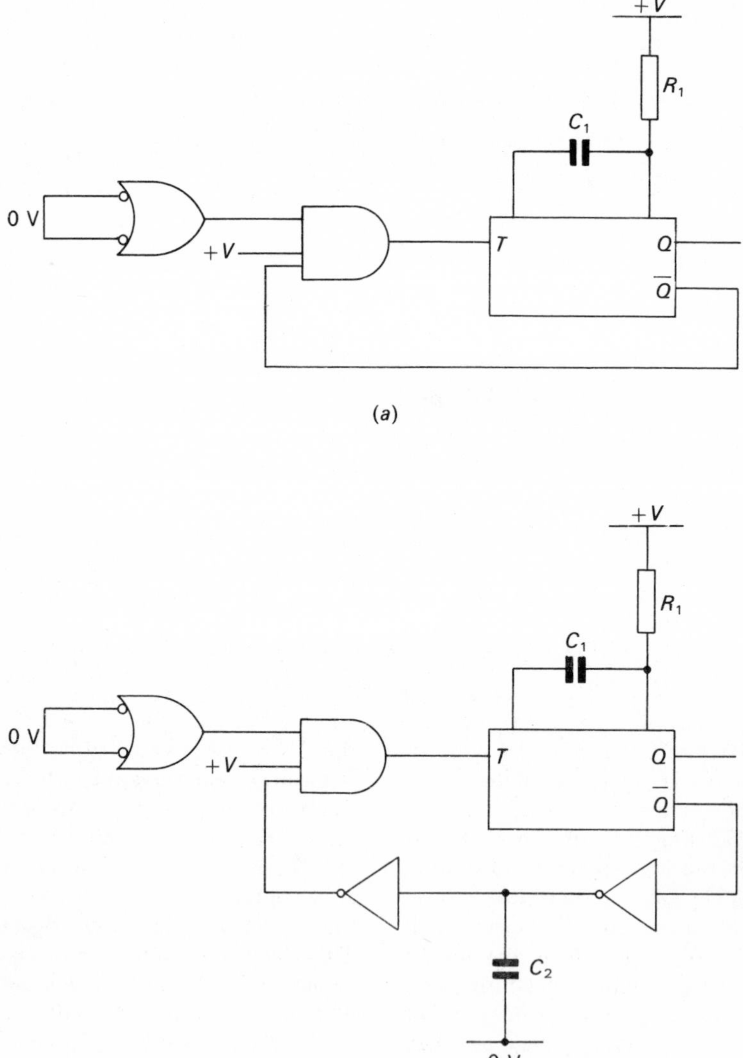

Fig. 4.18. Astable multivibrator using the monostable circuit of Fig. 4.17; (*a*) basic system, (*b*) system with longer output pulse.

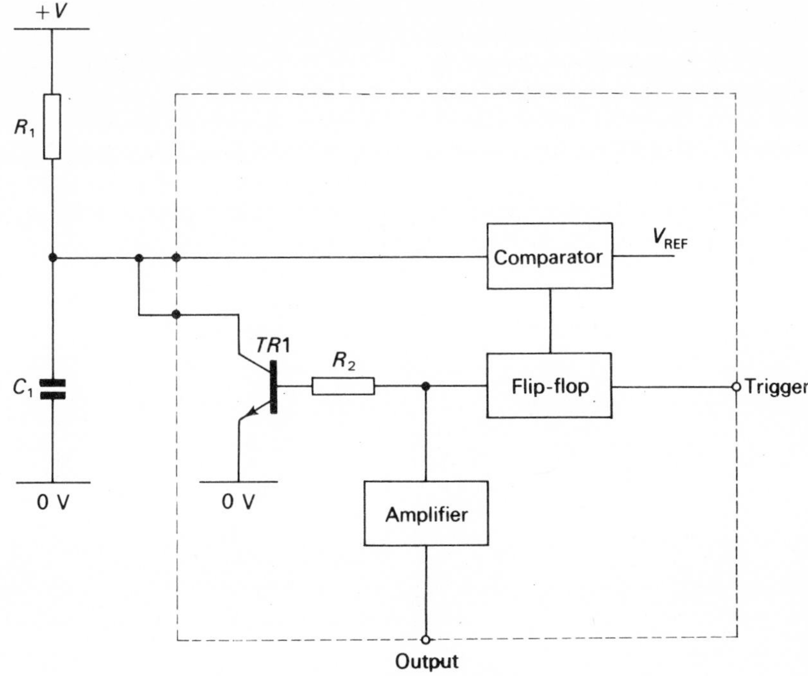

Fig. 4.19. A versatile timer circuit.

4.8 Arithmetic devices

There are many different integrated circuits which are available for use in arithmetic systems. In this section the basic operation of addition, subtraction, multiplication and division are described along with comparators and binary rate multipliers.

The most fundamental arithmetic device is the adder since most of the other operations are based on it. Fig. 4.20 shows the truth table for the sum S and carry C which are produced by adding two bits A and B. Inspection of this table will show that S is an EXCLUSIVE–OR function of A and B, and C is an AND function of these inputs. Therefore the logic diagram shown in

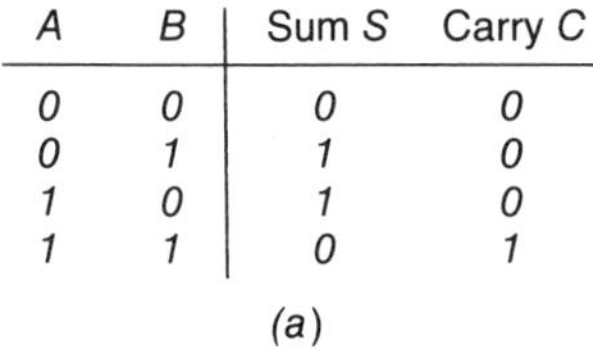

A	B	Sum S	Carry C
0	0	0	0
0	1	1	0
1	0	1	0
1	1	0	1

(a)

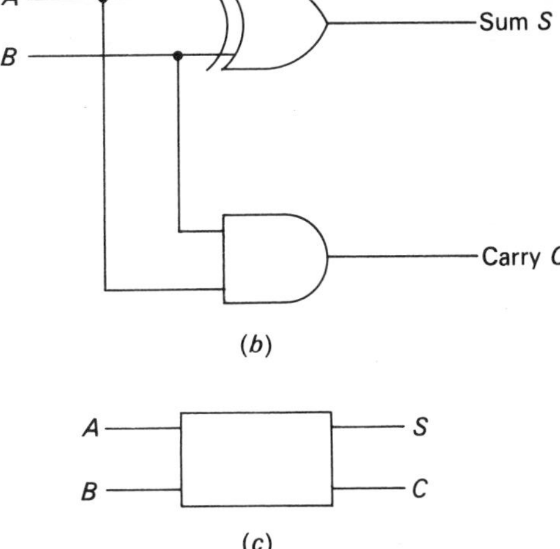

(c)

Fig. 4.20. A half adder; (a) truth table, (b) logic circuit, (c) symbol.

Fig. 4.20 (*b*) may be used to generate the sum and carry bits from inputs *A* and *B*. This circuit is called a half adder since it does not contain the facility to bring in a carry bit from a previous addition. Fig. 4.21 shows an adder circuit which has this carry in facility. It is now called a full adder.

The truth table for the full adder, shown in Fig. 4.21 (*a*), can be minimized to give the functions for S and C_0. These will be found to be

$$S = (A \oplus B) \oplus C_i \tag{4.1}$$

$$C_0 = A \,.\, B + (A \oplus B) \,.\, C_i \tag{4.2}$$

Fig. 4.21 (*b*) shows the logical implementation of these function. In the discussions which follow the logic symbol shown for the full adder in Fig. 4.21 (*c*) will be used.

Carry in C_i	A	B	Sum S	Carry out C_0
0	0	0	0	0
0	0	1	1	0
0	1	0	1	0
0	1	1	0	1
1	0	0	1	0
1	0	1	0	1
1	1	0	0	1
1	1	1	1	1

(*a*)

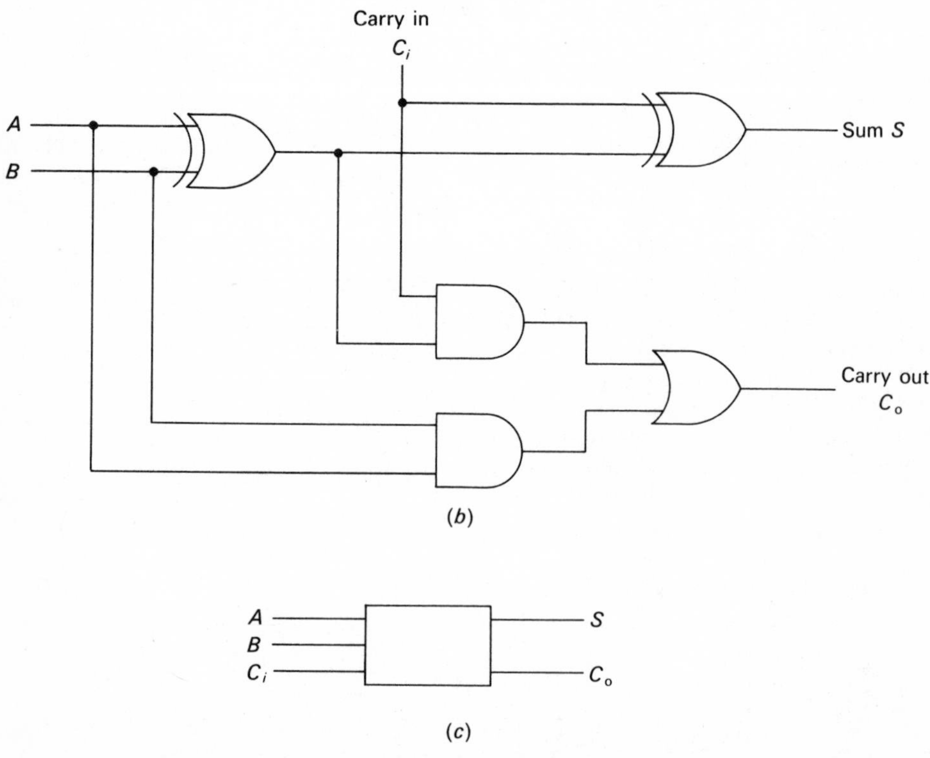

Fig. 4.21. A full adder; (*a*) truth table, (*b*) logic current, (*c*) symbol.

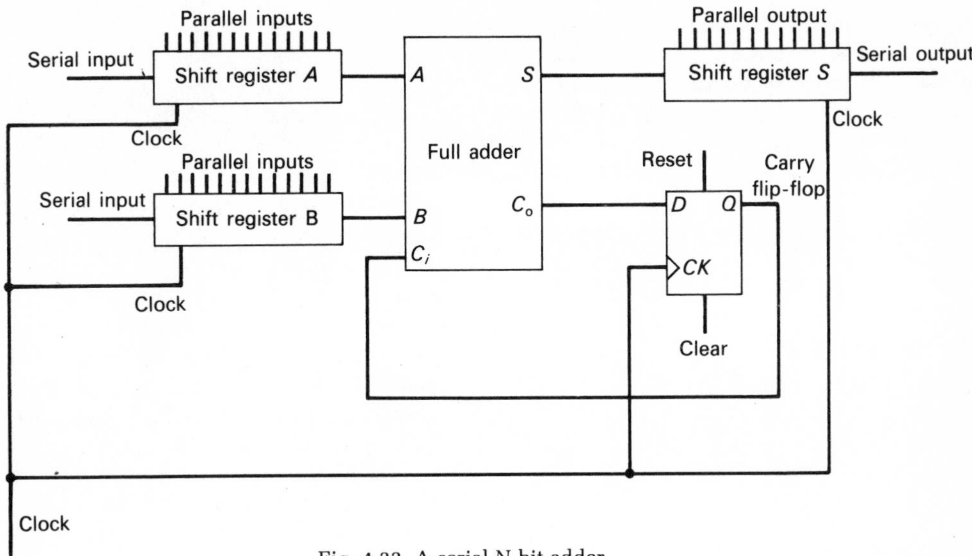

Fig. 4.22. A serial N bit adder.

Several techniques exist for adding two n bit numbers together. The simplest is the serial adder shown in Fig. 4.22. Shift registers A and B are loaded by their serial or parallel inputs with the two numbers to be added together. The carry flip-flop is initially cleared so that C_i is zero. The output from the addition is stored in shift register S. Initially the first bits in A and B are presented to the adder and their sum appears at S. Any carry generated is seen at C_0. The first clock pulse will feed the sum bit into shift register S while at the same time moving registers A and B along by one bit. The value of C_0 will also be fed into the carry flip-flop so that C_i now contains the carry output from the addition of the previous bits. In this manner all the bits in the two numbers are added together and stored in shift register S. Its output can be unloaded in parallel or serial form as required.

The serial adder is relatively slow since bits are added together one at a time. An alternative is to use a parallel adder, one type being shown in Fig. 4.23. In this arrangement the carry output from each bit is fed to the carry in of a more significant bit. The output from the most significant bit cannot therefore attain its true value until the carry generated in the least significant bit has had time to ripple through the system.

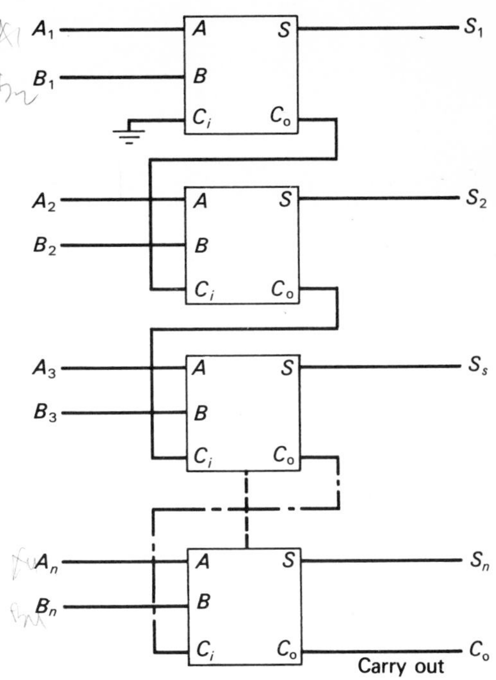

Fig. 4.23. A parallel adder with ripple carry.

These are therefore called ripple adders and although they are faster than serial adders they can still be too slow for some applications.

By reference to equation (4.2) the carry output from the nth bit is given by

$$C_{n+1} = A_n . B_n + (A_n \oplus B_n) . C_n \qquad (4.3)$$

Therefore the carry output from the second bit is

$$C_2 = A_1 . B_1 + (A_1 \oplus B_1) . C_1 \qquad (4.4)$$

For the third bit it is

$$\begin{aligned} C_3 &= A_2 . B_2 + (A_2 \oplus B_2) . C_2 \\ &= A_2 . B_2 \\ &\quad + (A_2 \oplus B_2) . [A_1 . B_1 + (A_1 \oplus B_1) . C_1] \end{aligned} \qquad (4.5)$$

Similarly

$$\begin{aligned} C_4 &= A_3 . B_3 + (A_3 \oplus B_3) . C_3 \\ &= A_3 . B_3 + (A_3 \oplus B_3) . \{A_2 . B_2 \\ &\quad + (A_2 \oplus B_2) . [A_1 . B_1 + (A_1 \oplus B_1) . C_1]\} \end{aligned} \qquad (4.6)$$

These equations show that the value of each carry can be anticipated by using AND, OR and EXCLUSIVE–OR circuitry, so long as the values of the various bits to be added and the first carry in C_1 (if any) is known. Circuits which generate these carries are known as look ahead carry generators and they can be used to give very fast addition in look ahead adders. Fig. 4.24 shows such an arrangement. It is seen that the value of the carry input for each bit is not dependent on the result of the addition of the previous bits but

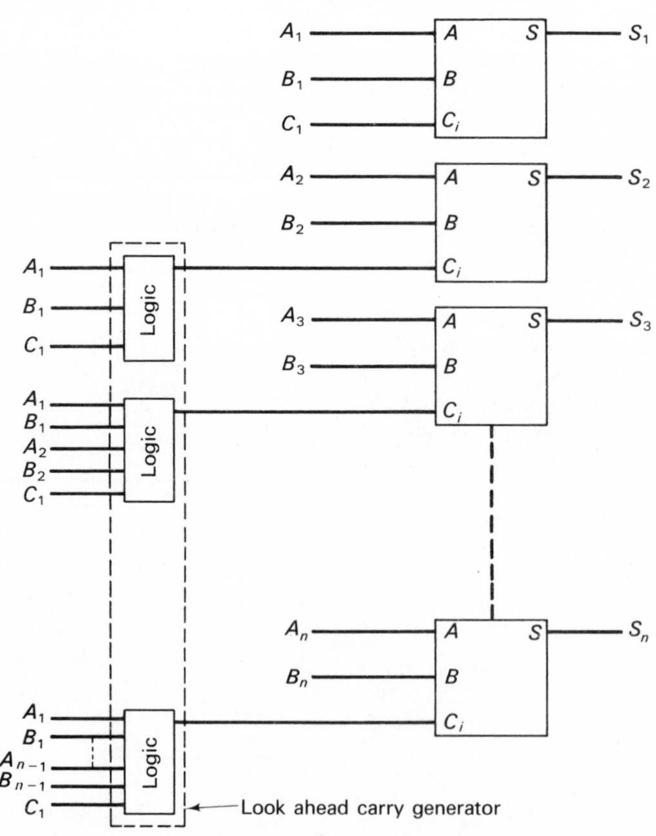

Fig. 4.24. A parallel adder with look ahead carry.

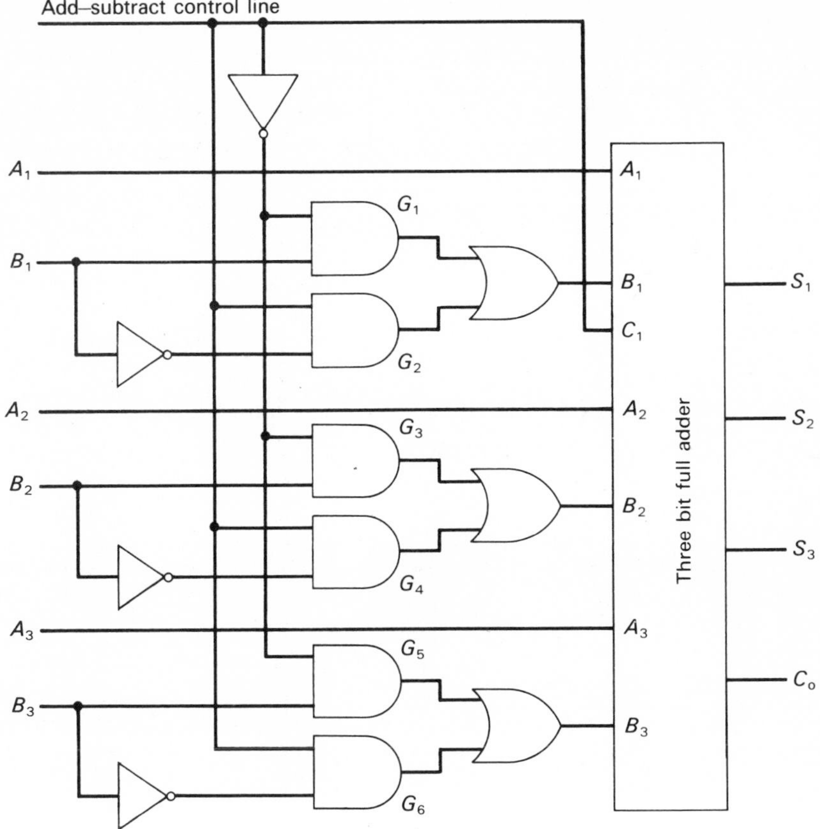

Fig. 4.25. A three bit parallel adder–subtractor.

can be generated from the inputs. The maximum carry delay is therefore independent of the number of bits being added together and is equal to the delay of three series gates which occur in the construction of the look ahead carry generator logic circuit.

Although look ahead parallel adders are very fast they also require a considerable amount of circuitry which can make them expensive when numbers with many bits are being added. A hybrid system may now be used. In this circuit look ahead techniques are used for a group of bits but these blocks are interconnected by ripple carry methods. This therefore represents a compromise between circuit complexity and system speed.

Binary subtraction was introduced in chapter 3 and it was shown there that this can be achieved by taking the 2's complement of the number to be subtracted and then adding. Resulting carries are ignored. The 2's complement is obtained by inverting each bit and then adding 1 to the least significant bit. Therefore a subtractor can be obtained from the basic adder. Fig. 4.25 shows a three bit adder–subtractor arrangement. A logic 0 on the control line would activate gates G_1, G_3 and G_5. This means that B_1, B_2 and B_3 are fed through to the adder terminals so that its output is equal to the sum of A and B. If the control line is at logic 1 gates G_2, G_4 and G_6 are operational so that $\overline{B}_1$, $\overline{B}_2$ and $\overline{B}_3$ are fed into the adder. At the same time C_i is equal to logic 1 so that this is equivalent to adding the 2's complement of number B to A.

The output is now the difference between A and B so the circuit is a subtractor. Negative numbers may be represented by a separate sign bit which is, say, logic 0 for positive numbers and logic 1 for negative numbers. This bit can be generated by comparing the sign bits of the two input numbers A and B and their relative magnitude.

Multiplication was shown in chapter 3 to be obtained by a process of shifting and addition. Fig. 4.26 (a) illustrates this more clearly and Fig. 4.26 (b) gives the logic diagram. The first product term is equal to the AND of the first bits A_1 and B_1. The second product term is the sum of $A_1 . B_2$ and $A_2 . B_1$ and this is generated by two AND gates and an adder. M_3 is the sum of $A_3 . B_1$, $A_2 . B_2$ and $A_1 . B_3$ so that two add

		A_3	A_2	A_1
		B_3	B_2	B_1
		$A_3 . B_1$	$A_2 . B_1$	$A_1 . B_1$
	$A_3 . B_2$	$A_2 . B_2$	$A_1 . B_2$	
$A_3 . B_3$	$A_2 . B_3$	$A_1 . B_3$		
M_5	M_4	M_3	M_2	M_1

(a)

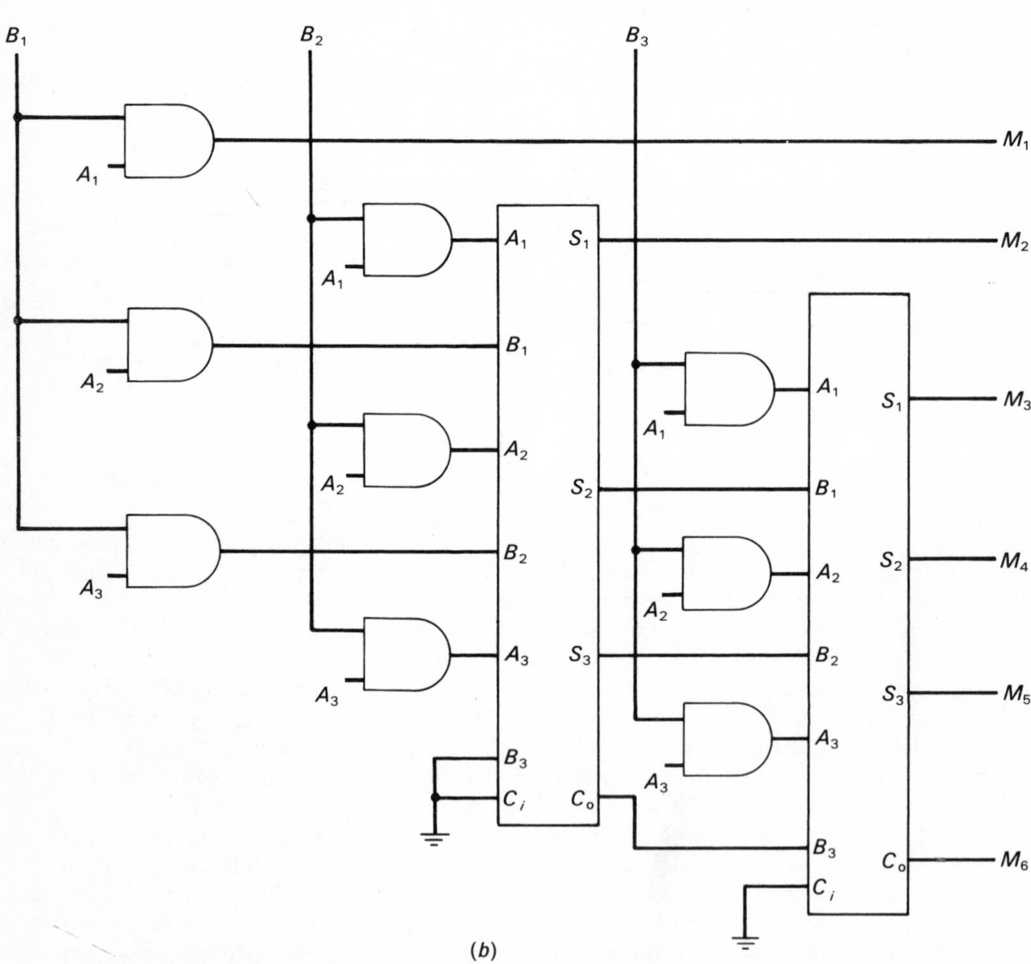

(b)

Fig. 4.26. A three bit parallel multiplier; (a) operation sequence, (b) logic diagram.

operations are required as shown. Similarly M_4 can be produced. M_5 is obtained by adding $A_3 \,.\, B_3$ to any carry out which results from the addition of the previous bit. Note that this may give rise to a carry out which appears as the sixth product term M_6. Therefore two three bit numbers when multiplied together will produce a six bit output.

Although parallel subtractor and multiplier circuits have been illustrated serial systems can be used as was done for the serial adder of Fig. 4.22. The bits are now handled one at a time so that although the circuit complexity is reduced the operating speed is also very low. Division can be obtained by a modification of the multiplication process in which subtraction and shifting is used. This is not illustrated here.

Another logic circuit commonly used in arithmetic systems is the magnitude comparator. A one bit comparator is shown in Fig. 4.27. From the truth table it is seen that the various magnitude conditions are satisfied by the values of A and B according to the function given in equations (4.7),

$$\left.\begin{aligned} A > B &\equiv A \,.\, \overline{B}, \\ A < B &\equiv \overline{A} \,.\, B, \\ A = B &\equiv A \,.\, B + \overline{A} \,.\, \overline{B} \end{aligned}\right\} \qquad (4.7)$$

These functions can be realized as in the logic diagram of Fig. 4.27 (b). The symbol for this one bit comparator is shown in Fig. 4.27 (c).

The single bit comparator can be expanded to compare n bit numbers by using serial comparator techniques as in Fig. 4.28. The two numbers which are to be compared are loaded

A	B	A > B	A < B	A = B
0	0	0	0	1
0	1	0	1	0
1	0	1	0	0
1	1	0	0	1

(a)

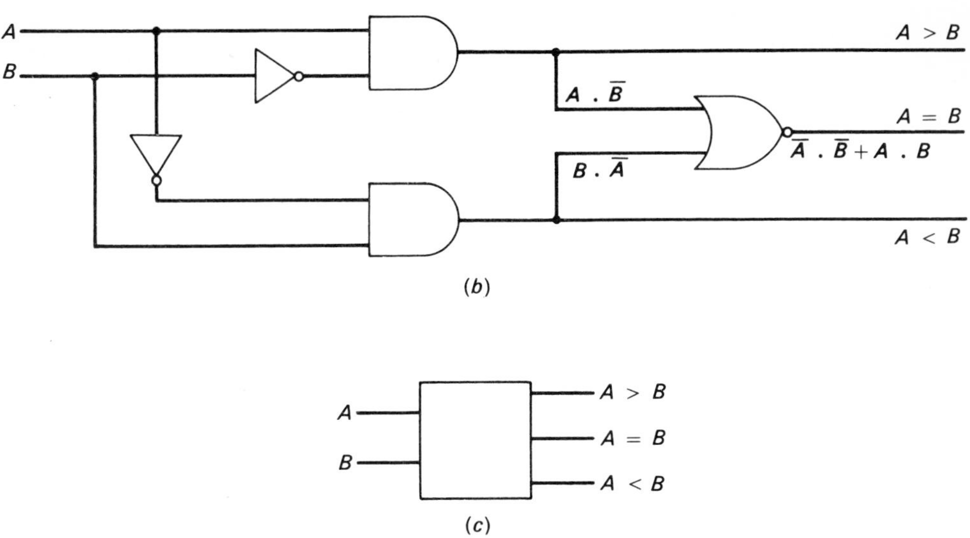

Fig. 4.27. One bit magnitude comparator; (a) truth table, (b) logic diagram, (c) symbol.

into shift registers A and B and the three flip-flops FA, FB and FC, which act as status flags, are cleared. The numbers are compared one bit at a time by comparator CA, starting from the most significant bit. An output on the $A > B$ or $A < B$ line will set its flip-flop and then disable the clock. For $A = B$ the clock is allowed to run until the last bit. If neither of the other two flip-flops are set at this stage then the numbers are equal. The serial comparator is relatively simple in construction but it is slow since the bits are compared one at a time at each clock pulse. Faster comparisons can be obtained in parallel comparators. For instance if A and B have three bits each $A_3A_2A_1$ and $B_3B_2B_1$ where bit 3 is the most significant then the condition that $A_3A_2A_1 > B_3B_2B_1$ is given by

$$(A_3 > B_3) + (A_3 = B_3) \,.\, (A_2 > B_2) + (A_3 = B_3) \,.\, (A_2 = B_2) \,.\, (A_1 > B_1) \tag{4.8}$$

where the cross and dot symbols represent OR and AND functions. It was seen from equation (4.7) that each of the expressions in equation (4.8) can be reduced further. This now gives

$$A_3A_2A_1 > B_3B_2B_1 \equiv$$

$$A_3 \,.\, \overline{B}_3 + (A_3 \,.\, B_3 + \overline{A}_3 \,.\, \overline{B}_3) \,.\, (A_2 \,.\, \overline{B}_2) + (A_3 . B_3 + \overline{A}_3 . \overline{B}_3) \,.\, (A_2 . B_2 + \overline{A}_2 . \overline{B}_2) \,.\, (A_1 \,.\, \overline{B}_1) \tag{4.9}$$

From expression (4.9) a logic system can be constructed which will compare the three bits of A and B to indicate the state when $A > B$. Similarly equations and logic circuits can be constructed for $A < B$ and $A = B$. This is not considered further. The logic complexity involved is much greater than that of the serial comparator but it is much faster and, for up to about four bits, can be implemented in present day MSI technology. Fig. 4.29 shows how the single bit comparator of Fig. 4.27 can be expanded to three bits. The interconnection for

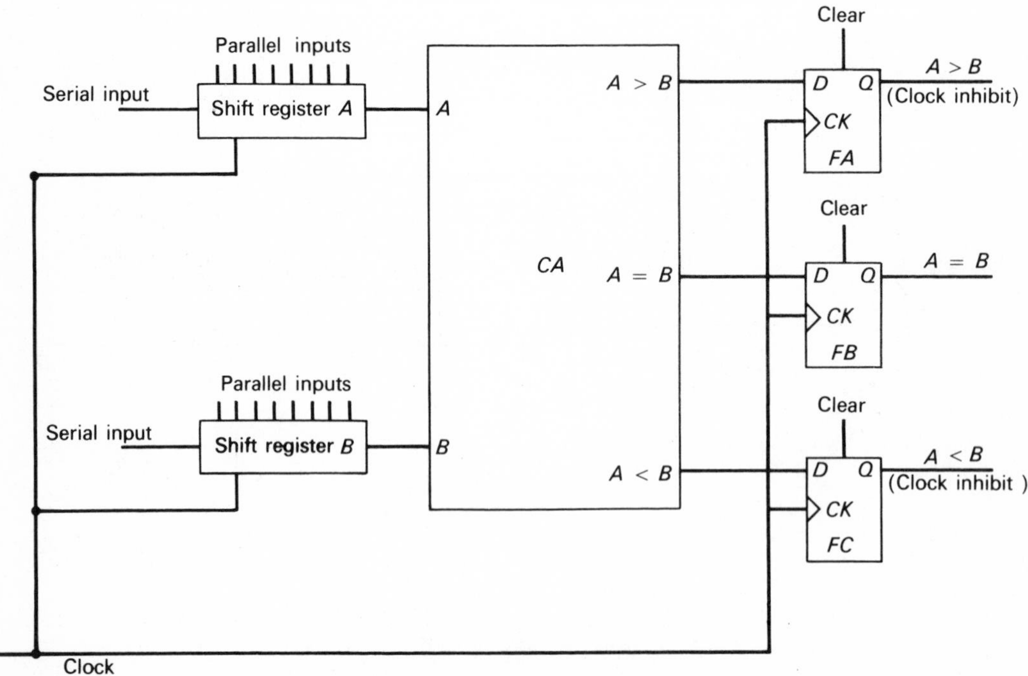

Fig. 4.28. Serial comparator.

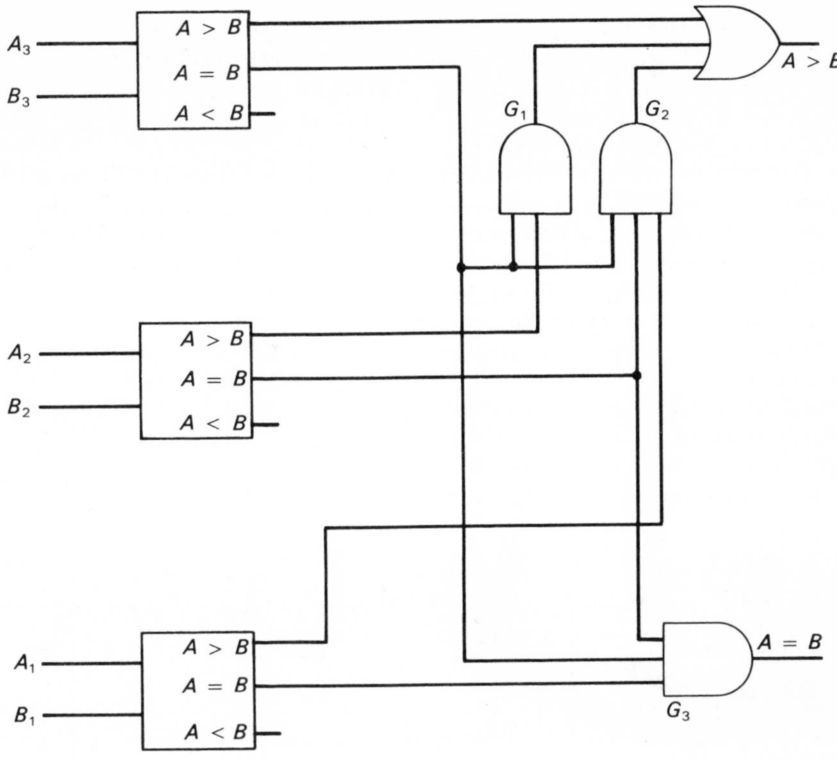

Fig. 4.29. Three bit magnitude comparator using the circuit of Fig. 4.27.

$A < B$ is not shown but it is similar to $A > B$. If $A_3 > B_3$ or $A_3 < B_3$ then that output is energized and all further outputs are disabled. If $A_3 = B_3$ then the next stage is enabled via gates G_1 and G_2. Now if $A_2 > B_2$ or $A_2 < B_2$ then an output is produced on these lines and further action is prevented. If $A_2 = B_2$ then the third stage is enabled via gate G_2. Now if $A_1 > B_1$ or $A_1 < B_1$ then outputs occur on these lines. If $A_1 = B_1$ then G_3 is enabled and an output occurs on the $A = B$ line.

The third class of device to be described in this section is the rate multiplier. Fig. 4.30 shows a simplified arrangement of a binary rate multiplier and Fig. 4.31 illustrates its waveforms. *FA*, *FB* and *FC* are *J–K* flip-flops and T_1, T_2 and T_3 are monostable circuits which are triggered on the rising edge of the input waveform. It is seen from the circuit waveforms that the input clock R is halved at S_1, divided by four at S_2 and by eight at S_3. Therefore

$$S_1 = \frac{R}{2}$$

$$S_2 = \frac{R}{2^2}$$

$$S_3 = \frac{R}{2^3}$$

Therefore for an n bit binary rate multiplier

$$S = \frac{R}{2^n}$$

Inputs C_1, C_2 and C_3 are control bits and represent the control word. When these are at logic 1 their corresponding gates are activated so that the pulses produced by R are summed by the OR gate to give output 0. Therefore

$$0 = \frac{R \cdot C.}{2^n} \tag{4.10}$$

The ripple counter arrangement of the rate multiplier shown in Fig. 4.30 is very slow and the monostables are difficult to construct in MSI circuitry. Therefore commercially available devices use alternative parallel counting techniques. These are not described here. The counter can also operate in binary or BCD format to give a binary rate multiplier or a decimal rate multiplier.

Rate multipliers are very versatile devices and can be used to perform a variety of arithmetic operations. Fig. 4.32 (*a*) shows how it can be used as an adder. The input clock R is split into three phases φ_1, φ_2 and φ_3. These separate phases are essential to ensure that pulses do not arrive at the OR gate G_1 or at the up-down counter at the same time. The pulses produced at R_3 and R_4 feed the up and down inputs of the counter and, ignoring the slight oscillations around the equilibrium point, which can be filtered out, the counter will stabilize when $R_3 = R_4$. Now applying equation (4.10) to the separate binary rate multipliers gives

$$R_1 = \frac{R \cdot X}{2^n}$$

$$R_2 = \frac{R \cdot Y}{2^n}$$

$$\text{and } R_4 = \frac{R \cdot Z}{2^n}$$

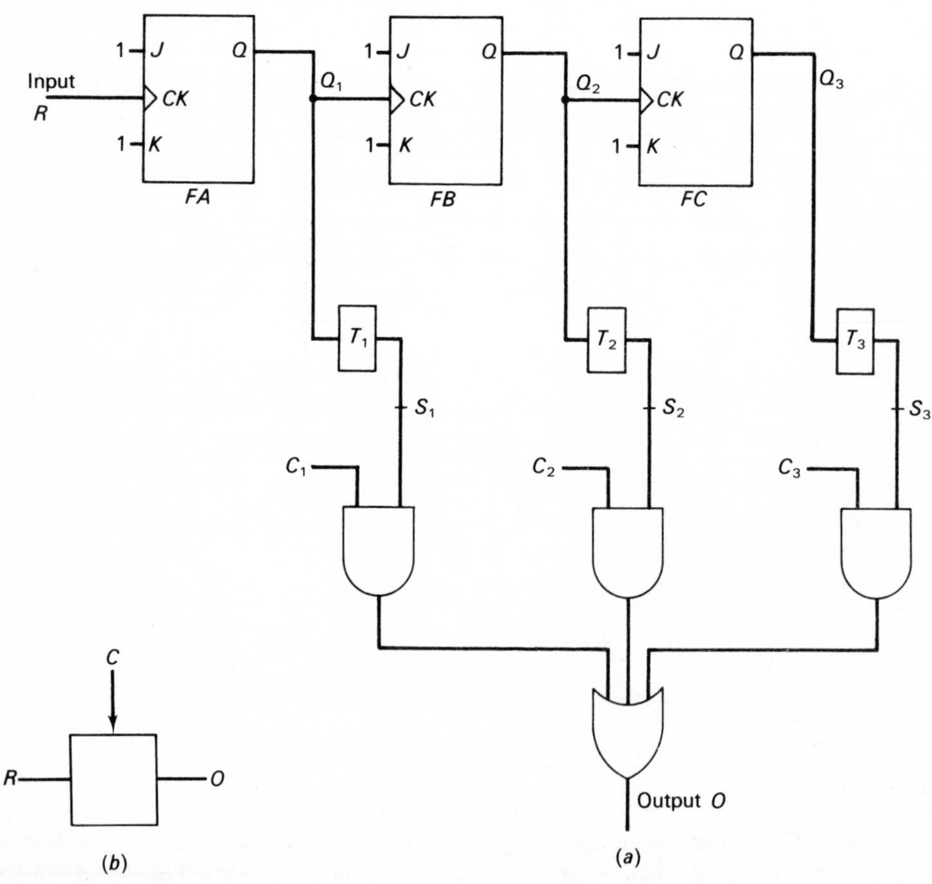

Fig. 4.30. Simplified representation of a binary rate multiplier; (*a*) logic diagram, (*b*) symbol.

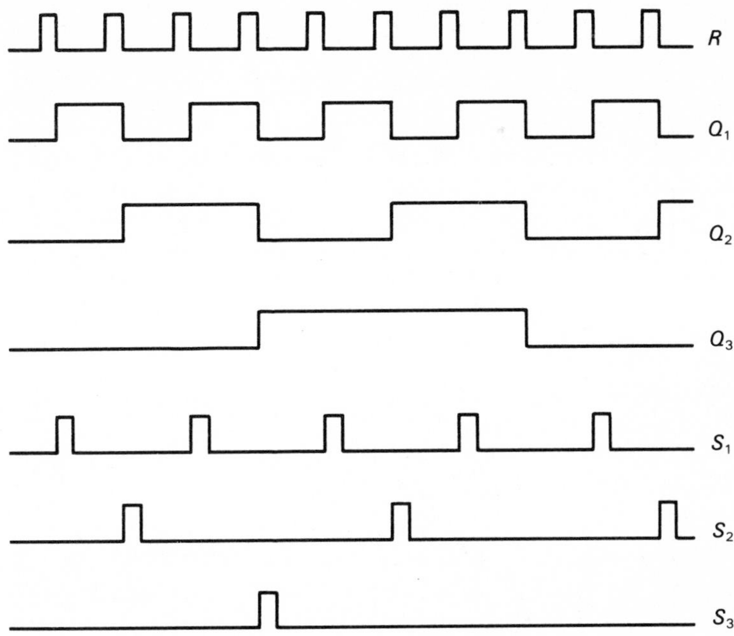

Fig. 4.31. Waveforms for the binary multiplier of Fig. 4.30.

Furthermore

$$R_3 = R_1 + R_2$$
$$= \frac{R}{2^n}(X + Y)$$

Therefore equating R_3 and R_4 gives

$$Z = X + Y$$

Similarly for the subtractor shown in Fig. 4.32 (*b*)

$$R_1 = R_4 \quad \text{and} \quad R_2 + R_3 = R_4.$$

$$\text{Now } R_2 = \frac{R\,.\,Y}{2^n}$$

$$R_3 = \frac{R\,.\,Z}{2^n}$$

$$R_1 = \frac{R\,.\,X}{2^n}$$

Therefore

$$Y + Z = X$$
$$\text{or } Z = X - Y$$

The addition and subtraction circuits can be combined as in Fig. 4.32 (*c*). Working through the system as in the previous two cases will give

$$Z = X + Y - P$$

A multiplier circuit using binary rate devices is shown in Fig. 4.33. Once again $R_2 = R_3$ at equilibrium;

$$R_1 = \frac{R\,.\,X}{2^n}$$

$$R_2 = \frac{R_1\,.\,Y}{2^n} = \frac{R\,.\,X\,.\,Y}{2^{2n}}$$

$$R_3 = \frac{R\,.\,Z}{2^n}$$

Therefore equating R_2 and R_3 gives

$$Z = X\,.\,Y$$

In the divider circuit of Fig. 4.33 (*b*) an ordinary divide by 2^n system, as obtained by, say, a counter, is introduced to balance the two arms.

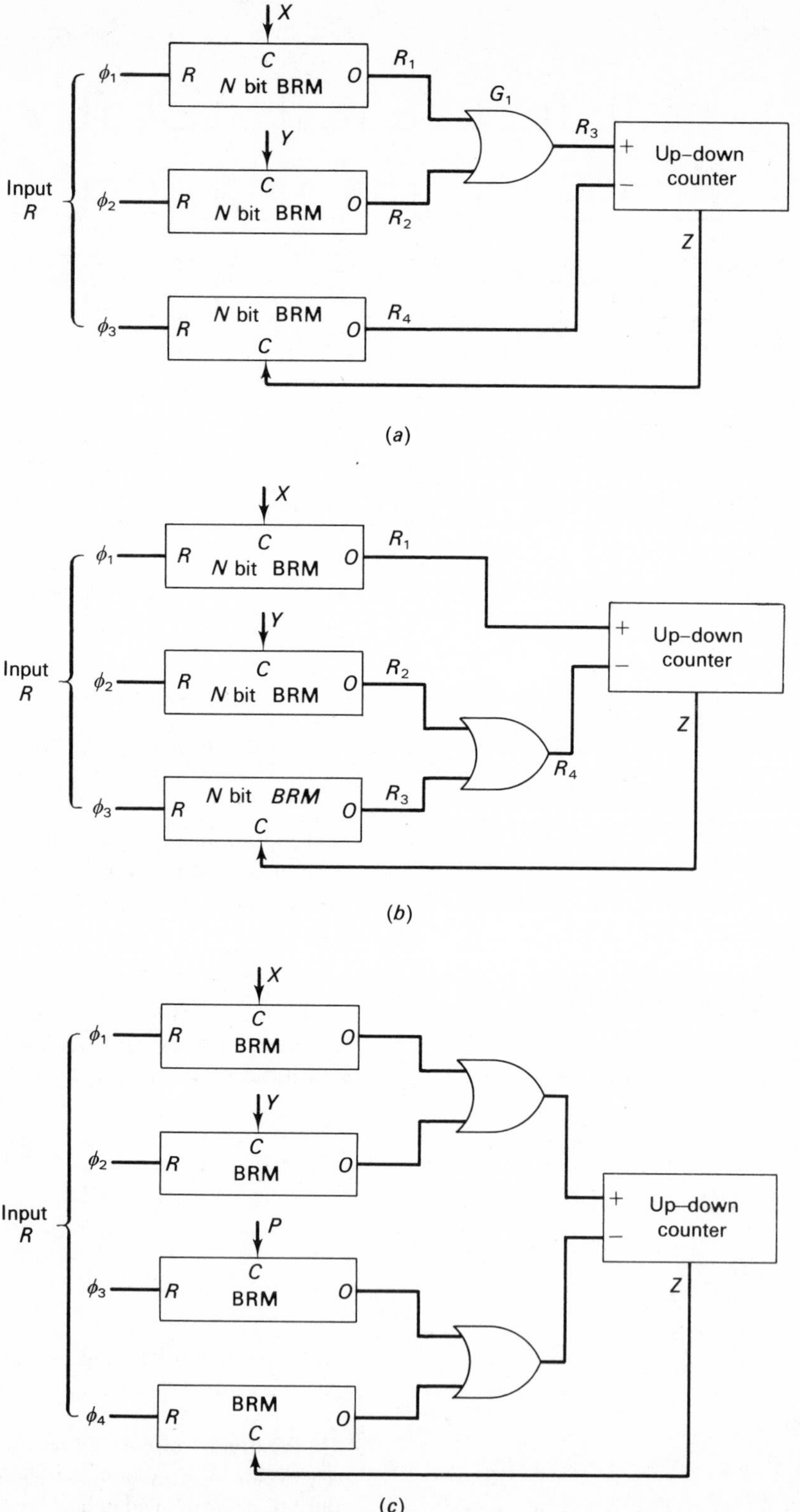

Fig. 4.32. Using the binary rate multiplier for addition and subtraction; (a) addition, (b) subtraction, (c) addition and subtraction.

As before

$$R_1 = \frac{R \cdot X}{2^n}$$

$$R_2 = \frac{R_1}{2^n} = \frac{R \cdot X}{2^{2n}}$$

$$R_3 = \frac{R \cdot Y}{2^n}$$

$$R_4 = \frac{R_3 \cdot Z}{2^n} = \frac{R \cdot Y \cdot Z}{2^{2n}}$$

Therefore, since $R_2 = R_4$

$$X = Y \cdot Z$$

or

$$Z = \frac{X}{Y}$$

The multiplier and divider circuits are combined in Fig. 4.33 (c). Working through this system as above it will be seen that

$$Z = \frac{X \cdot Y}{P \cdot Q}$$

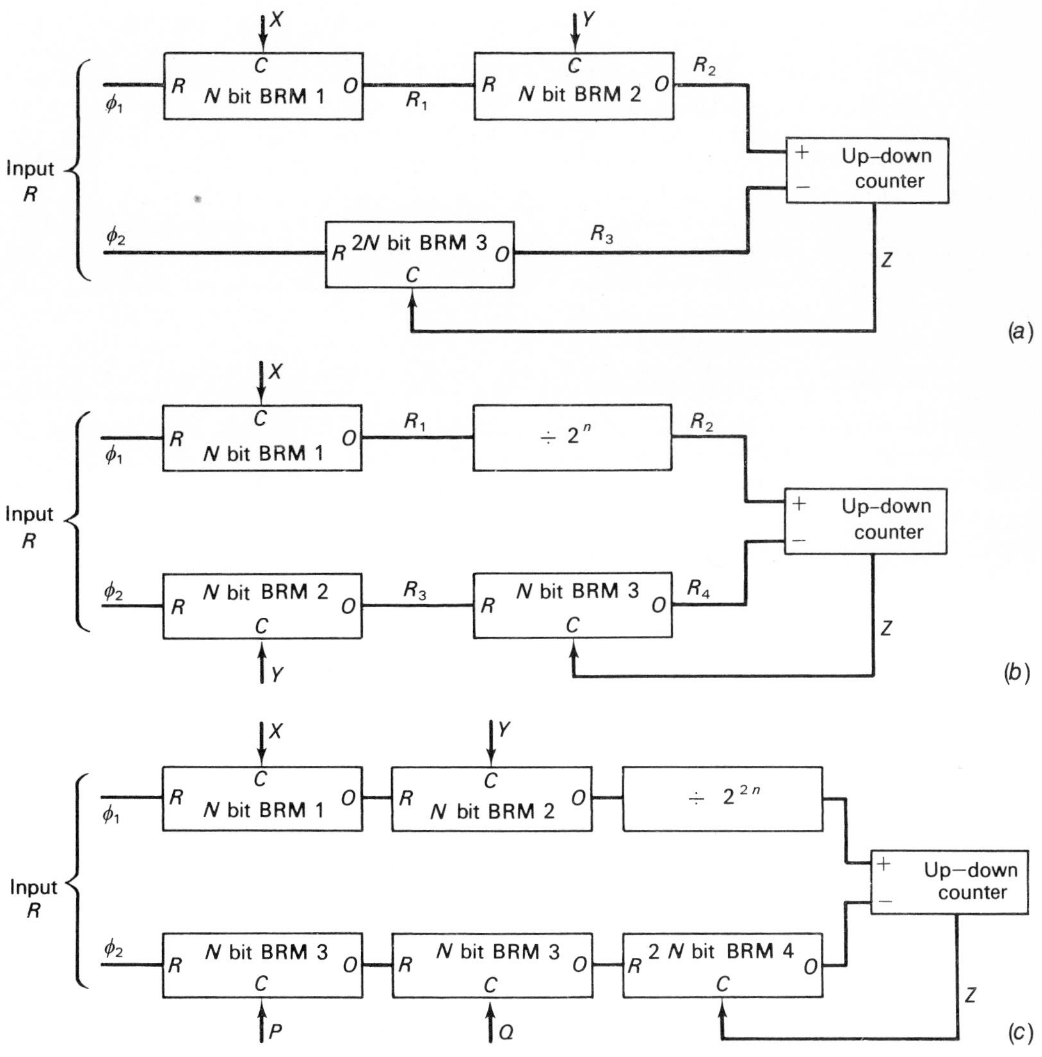

Fig. 4.33. Using the BRM for multiplication and division; (a) multiplication, (b) division, (c) combined multiplication and division.

4.9 Interface circuits

Most integrated circuits are only capable of providing relatively low values of sink or drive currents and of operating only up to modest voltage levels. Large currents and voltages are, however, often required in many applications. Interface circuits are designed to provide this extra power and are also often required to work in noisy environments or to sense low signal levels. As such they are generally composed of a mix of linear and digital circuitry.

Any circuit can be converted into an interface system by giving it additional drive capability. For instance BCD to seven segment or BCD to decimal decoders are often designed with high voltage or high current output stages so that they can drive electronic displays directly without requiring additional interface circuitry. A more general-purpose interface circuit could consist of a seven transistor array in which the emitters are all connected together. Allowing for an additional power supply line the circuit can be placed in a sixteen pin package. Generally each transistor is capable of operating at a high voltage and of sinking a large current. However to limit package dissipation there is usually a restriction on the number of transistors which can be conducting at any one time. Fig. 4.34 shows another interface circuit which is commonly used to drive inductive loads such as a relay. D_1 and D_2 are free wheeling or catching diodes and both the diodes and transistors are capable of carrying large currents. In addition to the power transistors this interface circuit is shown with an internal amplifier gate. It has four inputs for logic control and due to the internal amplification the input drive current requirements are very low.

Fig. 4.35 shows a class of circuits which is very popularly used in digital data transmission systems. The line *driver* consists of a digital input stage which, depending on the state of the logic inputs, switches on transistor *TR*1 or *TR*2. These two transistors form part of a constant current differential stage which is therefore a linear circuit. The line driver is capable of operating into low impedance lines and usually has an inhibit input which switches both the outputs into a high impedance mode. The line receiver has a differential input stage, for low level signal sensing and noise rejection, and a logic output. A strobe input is often provided to allow the output to be disabled.

Fig. 4.36 (*a*) shows a simple transmission system using the line driver and receiver. System noise is primarily common-mode. Therefore a balanced or two-wire system is generally used since the common-mode noise can be rejected by the line receiver provided it has adequate common-mode rejection properties. The

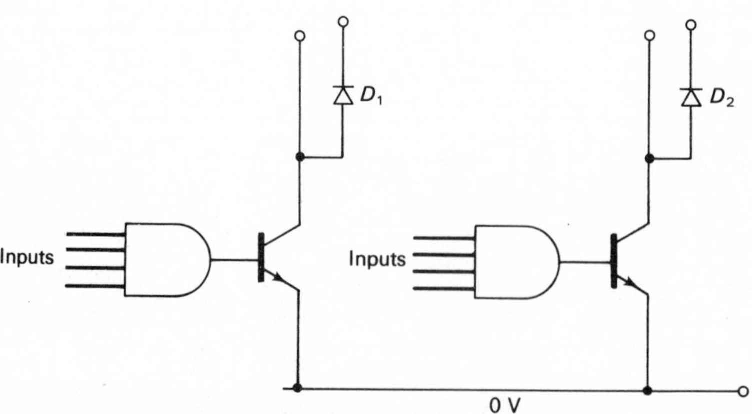

Fig. 4.34. Dual inductive load-driver.

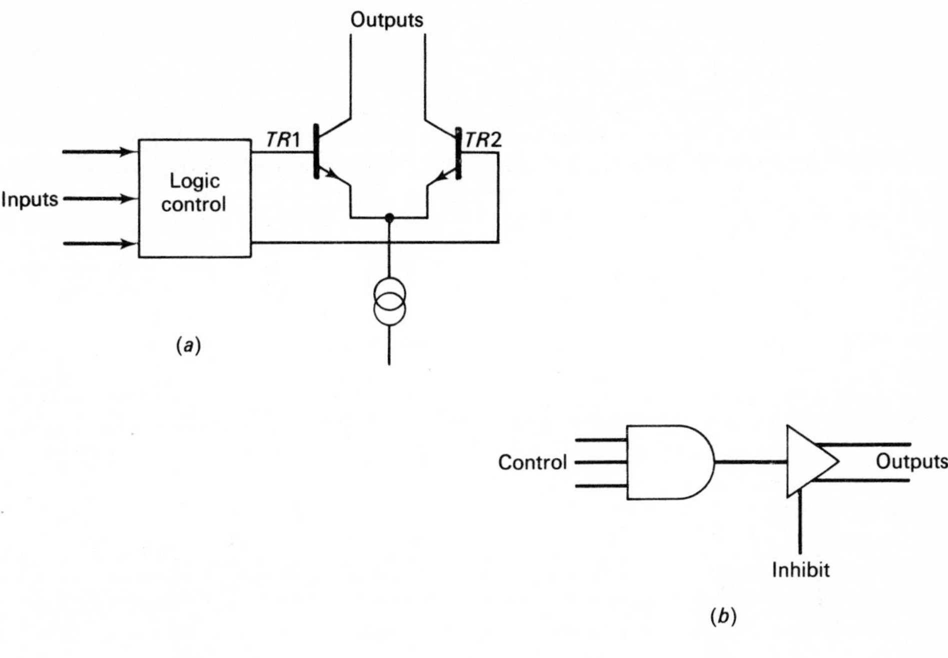

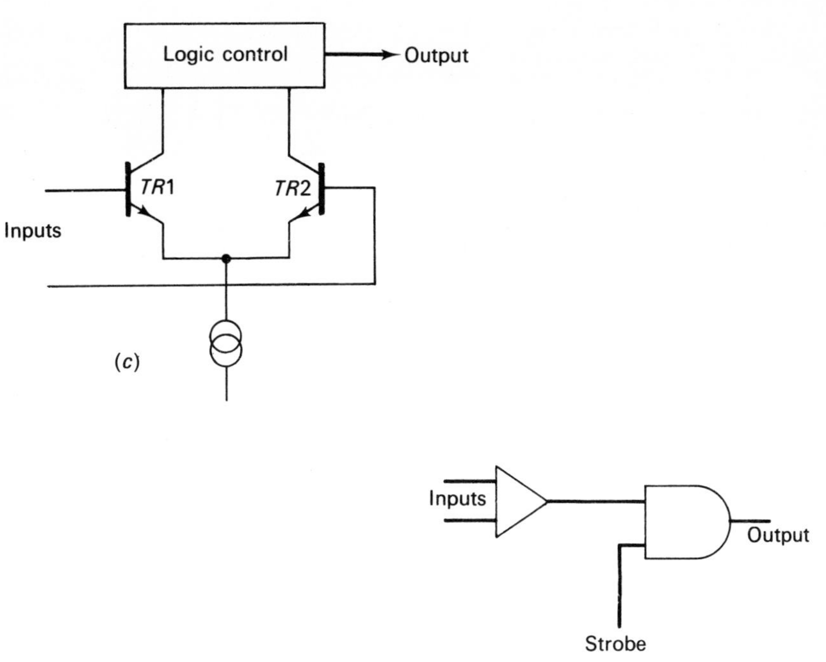

Fig. 4.35. Line drivers and receivers; (a) line driver, (b) line driver symbol, (c) line receiver, (d) line receiver symbol.

receiver is usually designed to have high signal sensitivity, a high input impedance and a large common-mode rejection at its input. The line driver in Fig. 4.36 (*a*) converts the input logic levels to a current switching action which unbalances the voltage on the transmission line and sends these to the receiver. The receiver senses the polarity of the differential input signal and converts this into a logic level output.

The system of Fig. 4.36 (*a*) uses a dedicated line for each transmitter–receiver pair. When a

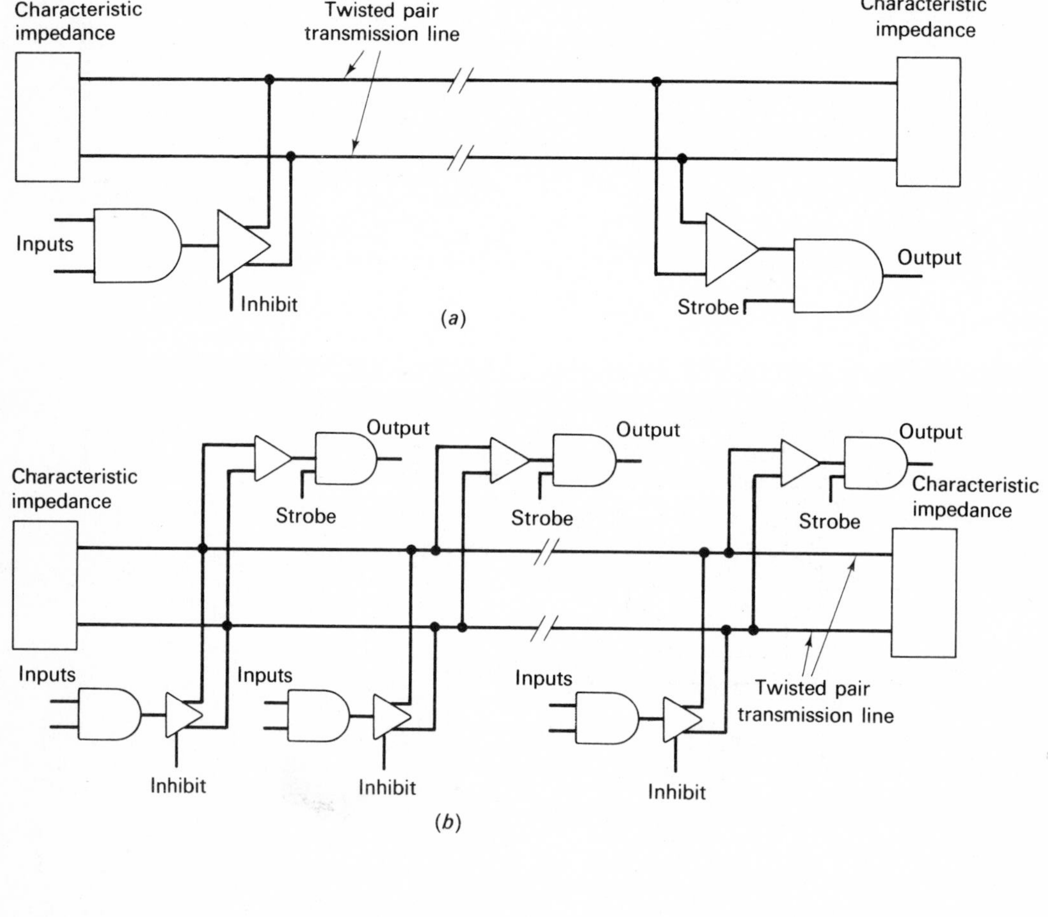

Fig. 4.36. Transmission systems; (*a*) dedicated pair line, (*b*) party line, (*c*) single line.

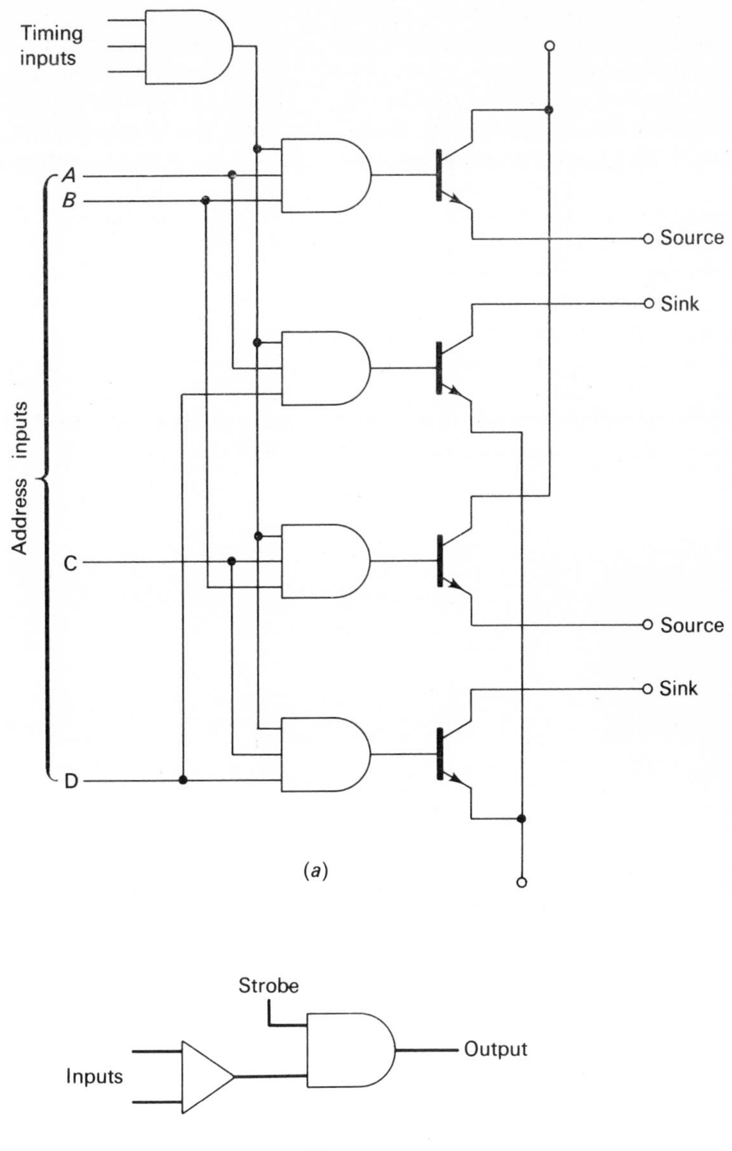

Fig. 4.37. Memory driver and sense amplifier; (a) driver, (b) sense amplifier.

large amount of interconnection is required between various transmitters and receivers such a system can be expensive due to the cabling involved. Fig. 4.36 (b) shows a party line connection in which one or more transmitters can transmit to one or more receivers, as required. The transmitters and receivers which are not operating at any time are disabled by their inhibit and strobe lines respectively. Clearly only one message can be transmitted on the line at any time so that the various signals need to be time multiplexed. In addition to the data lines all such systems also need control lines to operate the individual inhibit and strobe inputs at the correct times.

For very short lines, and in applications where environmental noise is low, the unbalanced or single-line transmission system can be used as shown in Fig. 4.36 (c). A reference d.c. level is applied to one of the differential inputs of the sense amplifier and the transmited signal is compared against this. For maximum immunity to pick up noise the magnitude of this reference voltage needs careful selection. Since only one line is used the cost of cabling is minimized with this system.

Another type of integrated circuit, which is used in computers with magnetic memories, is the memory driver and sensor. The driver needs to be capable of providing large pulses of currents in both the source and sink modes and one form is shown in Fig. 4.37 (a). Input gating allows the use of timing pulses and address lines. Inputs *B* and *D* select the source or sink modes and lines *A* and *C* choose the device pair which is to be operated. The sense amplifier of Fig. 4.37 (b) is very similar to that used in line receivers. The input to the amplifier can be as low as 10 millivolts so that high sensitivity is essential. The output can be strobed and have open collector stages for maximum flexibility. In addition storage registers are also sometimes incorporated to enable the data read out from the magnetic store to be retained until required.

5. Semiconductor memories

5.1 Introduction

The concept of digital storage was introduced in the last chapter with the bistable or flip-flop element. These basic elements may be combined in a serial format to give a shift register or they may be placed in a matrix to form a *memory array*. In this chapter the array memory concept is described in greater detail. These devices are generally referred to as LSI (large scale integration). The term is however very ambiguous since there are many small memory arrays which use fewer gate elements and occupy a smaller chip area than the MSI (medium scale integration) components of the previous chapter.

Memories are primarily of three types:

(1) Serial memory or shift register. These were described in the last chapter and will not be considered further here.

(2) Random access memory (RAM). These consist of a matrix array of *memory cells*. They are called random access since the time taken to address any individual cell in an array, in order to write information into it or to read it out, is the same regardless of its position in the array. This is clearly the fundamental difference between a RAM and a shift register since in the latter case, because information must pass serially from one cell to the next, the access time to a cell depends on its position in the chain.

(3) Read only memory (ROM). This is also a matrix array of memory cells, like the RAM. Once again the access time to any of the cells is independent of its position within the array. Therefore the ROM is random access in structure. However whereas it is possible to read or write into a RAM within a fraction of a second, and to do this for almost an unlimited number of times, the ROM is intended to be used in applications where the stored information is required to be read many times, but is written into or altered very infrequently. In these cases the write time is many orders of magnitude longer than the read time, and in many cases the information may only be altered by relatively complex means such as exposure of the silicon to ultraviolet light. In the conventional read only memory the stored information is written into the cells during manufacture and cannot be subsequently altered. They were therefore referred to as fixed programme memories (FPM). The advent of the reprogrammable memory cell has largely made this term obsolete and replaced it by yet another, the *read mostly memory* (RMM). In this chapter we will refer to both devices as read only memories and qualify them as mask programmable (FPM) or reprogrammable (RMM).

The three parameters of greatest importance in a memory system are: (i) minimization of power dissipation, (ii) maximization of operating speed, (iii) reduction in cost by a reduction of the silicon area used and the number of output pins required. To meet these requirements there have been many developments in memory structure and organization. In this chapter the various cell structures used in random access and read only memories are first described. This is then followed by a description of the organization of the cells within a memory chip. The content addressable memory (CAM) is a special case within this group. Although the CAM cell is similar to a RAM since it is used in a read write mode, its organization is such that it is capable of being *addressed* by the information it contains rather than the position that this cell occupies within the array. The CAM has not been as popularly used in the past as either the RAM or ROM. Its operation is described in detail in a later section.

Semiconductor memories are strong contenders with magnetic devices in low to medium size stores. They currently suffer from the disadvantage of being *volatile*, i.e. the memory stored in a RAM disappears once the supply voltage is switched off. Developments are taking place to make the RAM non-volatile and these are described in this chapter. Also included are sections on typical memory applications and a brief guide to the would be user regarding the variety of different devices which are currently on the market.

5.2 Read only memory

The operation of a read only memory can best be explained by means of the diode matrix shown in Fig. 5.1 (*a*). This is an array of W by B lines, referred to as words and bits respectively. It is possible by addressing X–Y co-ordinates to pin point any diode within this matrix in which the information stored is programmed, by omitting certain diodes. For instance, if we assume that +5 volts represents a logic 1, then if the word line 1 goes to a logic 1 level the diodes connected to bit lines 1, 3 and 4 will conduct and these will also be raised to a voltage close to +5 volts, i.e. a logic 1. This therefore represents the output from the ROM for any input. Similarly when word line 2 is addressed the output is represented by a logic 1 on bit lines 2 and 3. The input lines may be accessed in sequence or at random as many times as required. The ROM keeps a permanent record of the required information which is available for reading within a fraction of a second, the actual time being called the memory access time. The information store can be changed only by the laborious process of physically removing a diode, or connecting a new one between the two lines. Once this change is made however the device has been materially altered and can be considered to be a new memory. The disadvantage of this ROM arrangement is that it is necessary to bring out a total of $W + B$ lines to the outside world via pins on its package. The greater the number of external pins the higher the component cost and the lower its system reliability. Therefore it is usual to include auxiliary circuitry on the memory array chip to reduce the numbers of outputs. Fig. 5.1 (*b*) shows an 8 by 8 array in which the number of output lines has been reduced to six since three bits may address eight lines in binary code. In general terms an input of I lines can be decoded into 2^I lines for addressing the memory array.

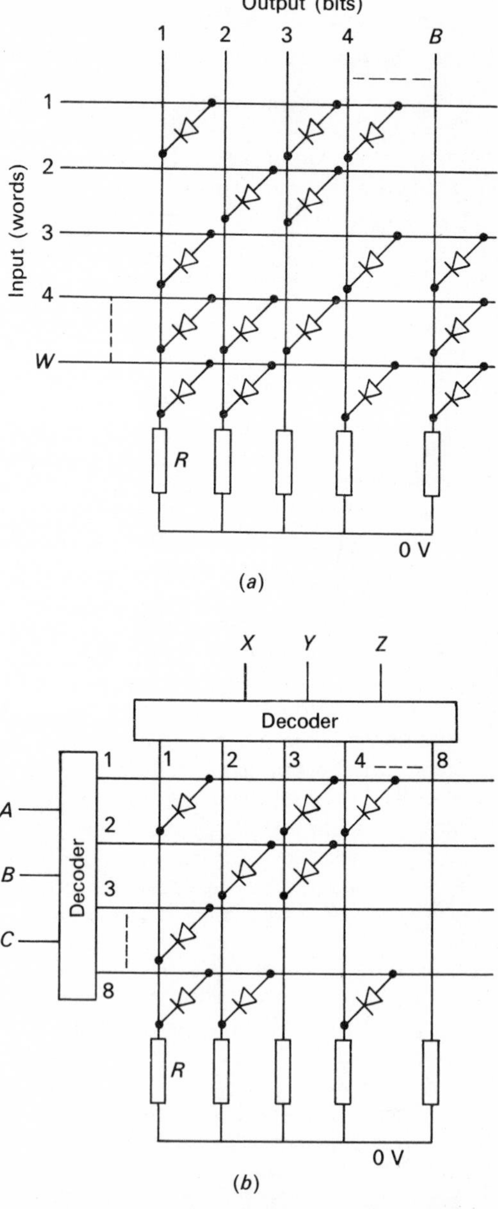

Fig. 5.1. Read only memories; (*a*) diode array, (*b*) array with decoding.

Read only memories may be of three types:

(1) Mask programmed. In this the stored information is programmed during the manufacture stage and cannot be subsequently altered by the user.

(2) Field programmed. The vendor produces memories which are identical in having a logic 0 (or logic 1) written into all locations. However the memory construction is such that the user can, after delivery, change the content of any of these store elements so as to produce his required bit pattern. Therefore the memory has been programmed in the 'field'. The programming process is irreversible. Once the user has changed any of the cell locations he cannot alter them again. However, locations which still contain the manufacturer's original information can still be changed at a later date.

(3) Reprogrammable. In this memory the content of each cell, when the user receives the device from the manufacturer, is not important. The user can erase the whole store and program it with his required information. Programming is now no longer a once for all operation. The memory may be erased and reprogrammed many times during its life. The erase and write processes are relatively complicated so that this is not a read–write or random access memory in the usual usage of that term. Furthermore the programmed information is non-volatile and may be kept throughout the useful life of the equipment.

5.2.1 *Mask programmed ROM*

The diode matrix shown in Fig. 5.1 can be produced in monolithic integrated circuit form and programmed by omitting connections to the anodes of diode cells which are to be left open circuited. However transistor cells are more common for ROMs and both the MOS and bipolar structures are in use. Fig. 5.2 (*a*) shows a MOS ROM which has been programmed by omitting transistors in certain memory locations (shown dotted). Transistors T_1 and T_2 have a fixed gate bias and are essentially load devices. When word line 1 is addressed bit line 1 is taken to ground level whereas bit 2 is close to V_{DD}. For word line 2 both bits are at ground. The actual presence or absence of a transistor may be accomplished by one of two methods. In

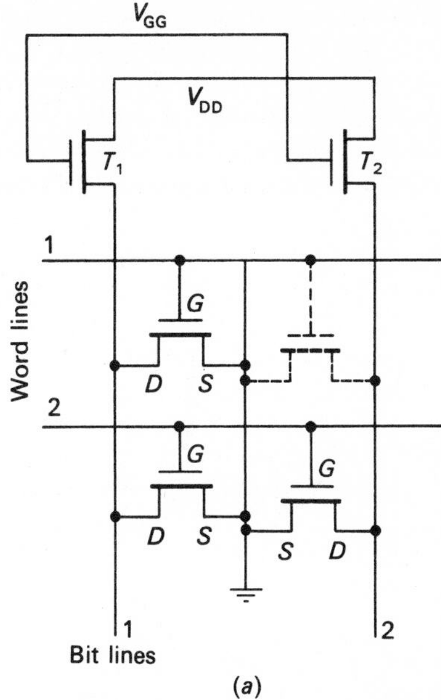

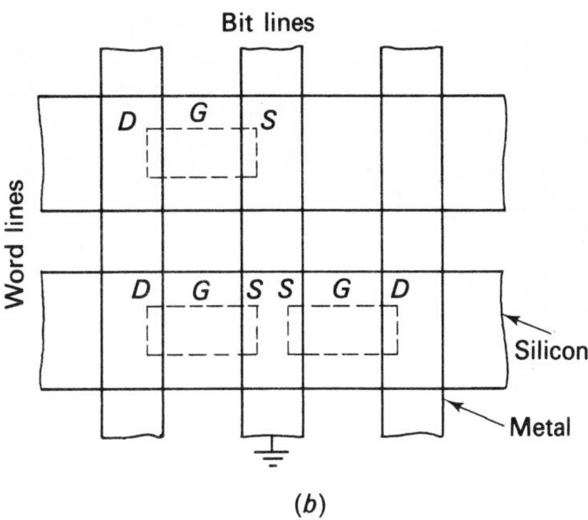

Fig. 5.2. MOS mask programmed ROM; (*a*) schematic, (*b*) layout.

the first the silicon wafer is processed right up to the metallization stage on the assumption that there is to be a transistor at every cell location. Programming is done by selecting the metallization mask such that only those gates are connected to the metal word lines where a transistor is required to exist. This method has the advantage that the turn round time between the customer specifying his requirements and the manufacturer producing the devices is reduced since memories may be processed up to the metallization stage and held in stock until required. The disadvantage, however, is that since the metallization is omitted from several thin oxide regions, at locations where there is to be no transistor, it represents a potential source of unreliability. Surface charge accumulation can occur on the oxide and it could be sufficient to turn the transistors on.

An alternative mask ROM technique uses a custom designed oxide mask to produce the devices. In this a thick oxide is maintained over

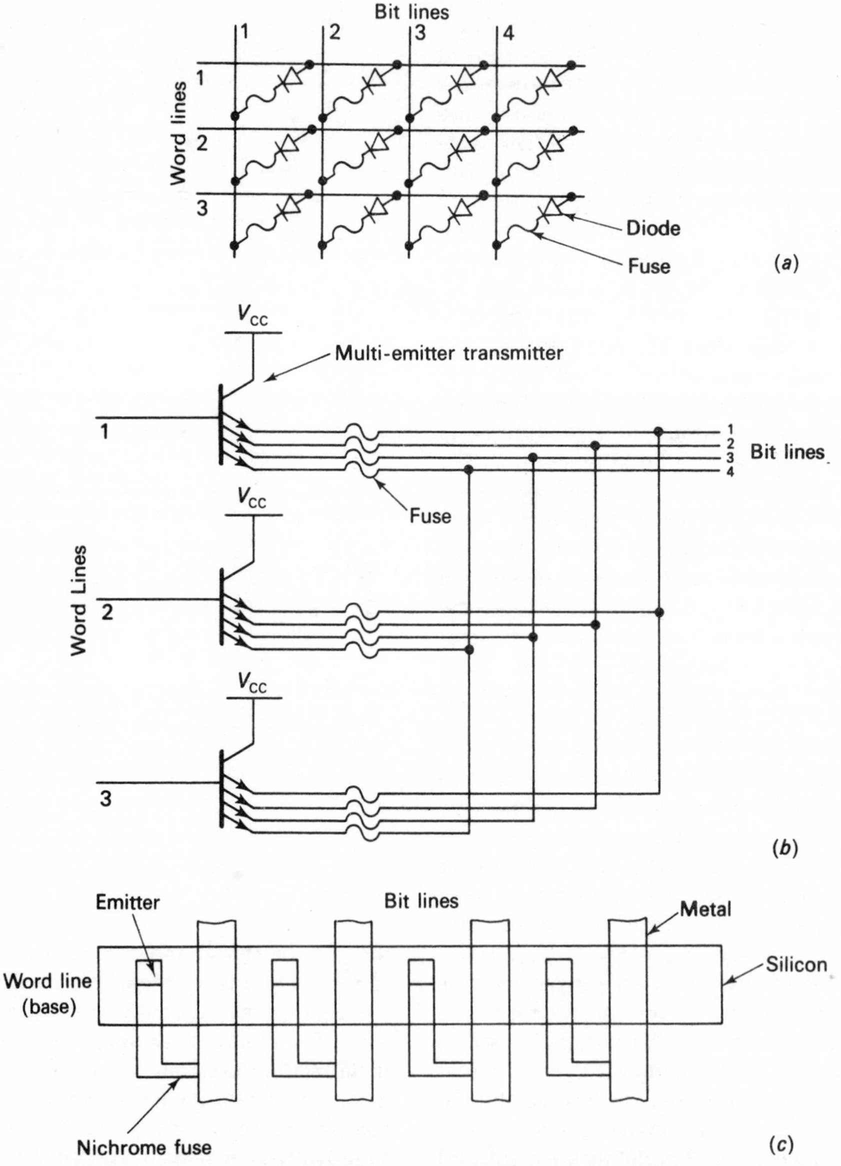

Fig. 5.3. Electrically programmable ROM with fusable link; (a) diode matrix, (b) transistor arrangement, (c) transistor cell construction.

the total area except where transistors are to occur and in these regions a thin gate oxide is produced. Apart from giving a much higher reliability this masking technique also produces cells which are some 25 per cent to 50 per cent smaller in area than the metal mask programmed devices. Therefore this is the system most frequently used. Fig. 5.2 (*b*) shows the layout of the memory cells. If two layer metallization is used then both the word and bit lines may be made from aluminium. It is more usual however to use metal for the bit lines and to use heavily doped diffused word lines within the silicon, so that the memory structure is considerably simplified.

5.5.2 *Field programmed ROM*

The mask programmed ROM requires a relatively long programming time and incurs a high mask design cost. Field programmed ROMs are designed to overcome both these disadvantages. They are supplied as *off-the-shelf* items to the customer who can then program them with his required information in a matter of minutes using specially constructed equipment. No development costs are involved since the manufacturer produces many batches of identical ROMs for his various customers. The memory is called a *field programmable read only memory* (FROM) or *electrically programmable* ROM (EROM) to differentiate it from the mask programmed ROM. There are two versions of this device, one using a fusable link and the other a shorted diode.

Fig. 5.3 (*a*) illustrates a simple fusable link memory using a diode matrix. When the device is shipped to the user all the fuses are intact so that each output produces a logic 1, or logic 0 depending on the convention employed. The customer can now program the memory by passing high current pulses between selected bit and word lines in order to blow the corresponding fuse and to open circuit its diode. This programming is usually done automatically by means of a programming box which either operates from a punched tape containing the required bit pattern or copies the information from a master memory which has been programmed previously.

Fig. 5.3 (*b*) shows a transistor version of the fusable link memory. Each transistor base represents a word line and the multi-emitters are fused and brought out as bit lines. The collector of the transistor is taken to the supply voltage. When a word line is addressed its transistor turns on and supplies current to the emitters which have their fuses intact. To program the memory the required fuses may be blown by turning on the transistor and taking the corresponding bit emitter to a low impedance so as to pass the fusing current through it. The layout of this memory is shown in Fig. 5.3 (*c*). The metal base line runs across each emitter section which is connected via a fuse, in the present case made from Nichrome, to the bit line. Two layer metallization may be used or a diffused word line employed.

The current required to blow a Nichrome fuse is of the order of 20 milliamps to 50 milliamps and the fusing time varies from 10 milliseconds to 200 milliseconds. To achieve this large current pulse the transistor must present a low impedance. This cannot be achieved with MOS devices unless they are uneconomically large. Therefore field programmable ROMs are currently only made in bipolar technology.

Apart from Nichrome other materials may be used for the fuse. Some manufacturers have successfully employed polysilicon. This is deposited onto the oxide prior to the metallization stage, and then etched to form an hour glass shaped fuse. The metallization is then deposited to connect onto the ends of this fuse. The fuse may be blown at its centre with current pulses which are of the same order as those required for Nichrome.

A second version of the field programmed ROM is shown in Fig. 5.4 (*a*). Essentially it consists of two diode junctions connected back to back. In this case each cell is a high impedance until programmed, unlike the fusable link device which is initially a low impedance. To program the device a high voltage high current pulse is passed through the required word and bit lines. This causes avalanche breakdown in the corresponding program junction, and results in a permanent short circuit across it. Fig. 5.4 (*b*) shows the structure of this memory where in practice a npn transistor is used to

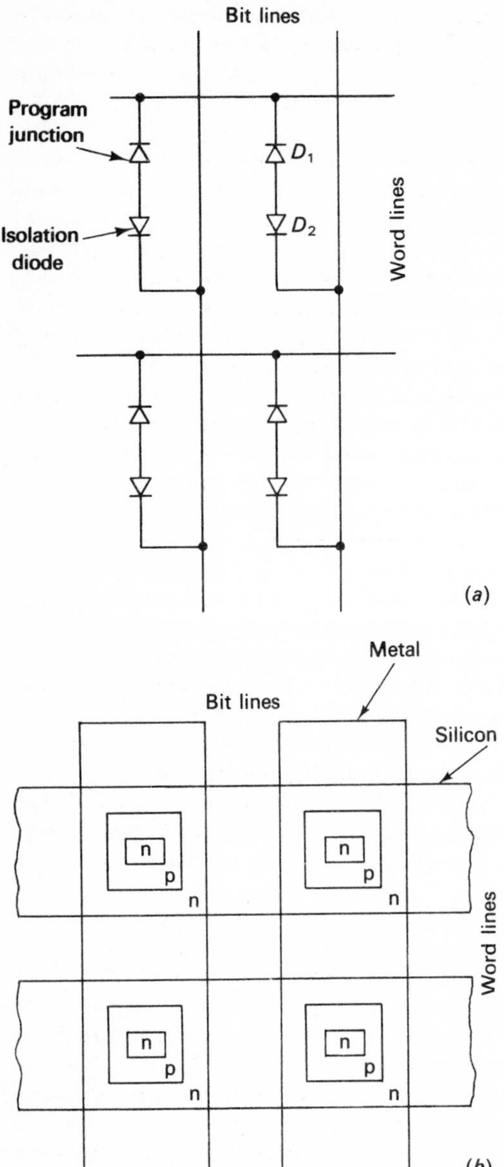

Fig. 5.4. Electrically programmable ROM with metal migration; (a) memory arrangement, (b) memory construction.

produce the two diodes. Diode D_1 in Fig. 5.4 (a) corresponds to the emitter–base junction of this transistor and D_2 to the base–collector junction. When the program pulse is applied across the collector–emitter, junction D_1 breaks down and the emitter metallization migrates to the base causing a permanent and solid short circuit between the emitter and base regions. Clearly the current pulse must be controlled to prevent the metal migration spreading to the collector region. As seen from Fig. 5.4 (b) several cells have a common silicon collector, which forms the word line, and a common emitter metallization, which forms the bit line.

The shorted emitter memory requires a higher programming voltage and current, in the region of 200 milliamperes to 300 milliamperes, than a fusable link device. However the shorting time is several orders faster, i.e. 0.02 milliseconds to 0.05 milliseconds so that programming a large memory is considerably quicker. Also, since no sputtering of metal (fuse) within the memory is involved, this device is claimed to have a higher long term reliability.

The field programmed ROM is more convenient to use than a mask programmed device and is more economical when low volumes are needed. However if the production requirements approach the medium to large volume range then the masked device is of the order of 2 to 10 times cheaper. This is primarily due to the fact that it has a smaller cell size and a simpler construction. A further disadvantage of the field programmed ROM is that the manufacturer cannot test it completely prior to shipment. He cannot, for instance, check that the bit output is a logic 1 because the corresponding fuse has not been blown or that there is a circuit malfunction. Indeed he cannot even be certain that all the fuses can be blown if required! This disadvantage is overcome in the reprogrammable memory described in the next section.

5.2.3 *Reprogrammable ROM*

The reprogrammable memory structure which is currently in most common use is the floating gate avalanche injection MOS (FAMOS) device illustrated in Fig. 5.5 (a). It has a structure very similar to a conventional silicon gate MOS transistor except that the gate is completely surrounded by oxide and has no electrical connections to it. The cell shown in Fig. 5.5 (a) is off since no current can flow between source and drain. To turn it on a high voltage, usually in excess of 50 volts, is applied between the source and drain. This causes avalanching of the pn junction at the source or drain, depending on the applied polarity, and an injection of electrons into the polysilicon gate. This is shown in Fig. 5.5 (*b*). Since the gate is surrounded by low conductivity oxide the charge remains trapped in it for a very long time even after the supply is removed. This results in an inversion of the silicon surface under the gate, as in normal gate action, so that the transistor cell is now on. To erase the stored information, by turning the transistor off again, it is exposed to ultraviolet light. This causes neutralization of the gate charge and a return to the state shown in Fig. 5.5 (*a*).

The FAMOS cell has an oxide thickness of about 0.1 micron below the gate and 1.0 micron above. Its substrate resistivity is between 5 ohm centimeters to 8 ohm centimeters. The gate current is about 10^{-7} amperes per square centimetre and this results in an accumulation of charge within the floating gate. This charge is dependent on the amplitude and duration of the programming pulse and clearly the larger these values the longer the memory is capable of maintaining its stored information. For instance a program pulse of 50 volts for 5 milliseconds would give a gate charge of about 3×10^{-7} coulombs per square centimeter and a decay to 70 per cent of this value after 10 to 100 years. In order to enable the memory to be erased it is normally enclosed in a package with a quartz lid through which the ultraviolet light may pass. If the lid was metal then erasure is still possible by using X-rays which however, cause radiation damage to the silicon and reduce the number of re-use operations. The required level of X-ray is of the order of 5×10^4 rad. This is many orders of magnitude larger than the value in the atmosphere so that the memory is generally not significantly affected during normal operation.

In the use of FAMOS memories there are two considerations regarding reliability:

(1) The loss of charge stored within the floating gate due to leakage must not be significant during the life of the memory, unless it is intended to be deliberately erased by ultraviolet light. There is now a considerable amount of accumulated data on this to suggest that storage times of the order of many hundreds of years are possible.

(2) The memory is read by applying a low voltage, of about 15 volts, across it to test if it is on or off. The application of this voltage could result in a gradual build-up of charge within the gate of a normally off transistor, turning it on. Once again tests have shown that many many read cycles, resulting in hundreds of years of normal operation, are possible without any significant charge accumulation.

Fig. 5.5 (*c*) shows the normal arrangements of FAMOS cells within a memory array. Since these transistors have no gates they are connected in series with a conventional MOS transistor which can then be addressed by the word lines. Therefore if transistor T_1 is on and T_2 and T_3 are off then when word line 1 is addressed it will turn on transistors T_4, T_5 and T_6 but will only produce an output on bit line 1.

Reprogrammable ROMs have the obvious advantage in prototype work that the information can be changed many times to correct system errors. It has the added attraction of being fully testable after assembly and prior to shipment to the user. There is however evidence to suggest that the life of these memories is affected by overdoses of ultraviolet light so that the reprogramming cycle is usually limited to a few tens of operations.

5.3 Random access memory cells

In this section the basic cell structures which have been used in random access (or read–write) memories are described. There are two ways in which a bit of information may be stored. The first usually consists in using a cross coupled form of device such as a flip-flop. The second method uses charge storage on a capacitor. These are referred to as static and dynamic memories respectively, the latter type clearly requires refreshing of the capacitor voltage since this will leak away in time.

Both static and dynamic memories can be built using bipolar or unipolar technologies. The differences between these various approaches are considered in the last section of this chapter.

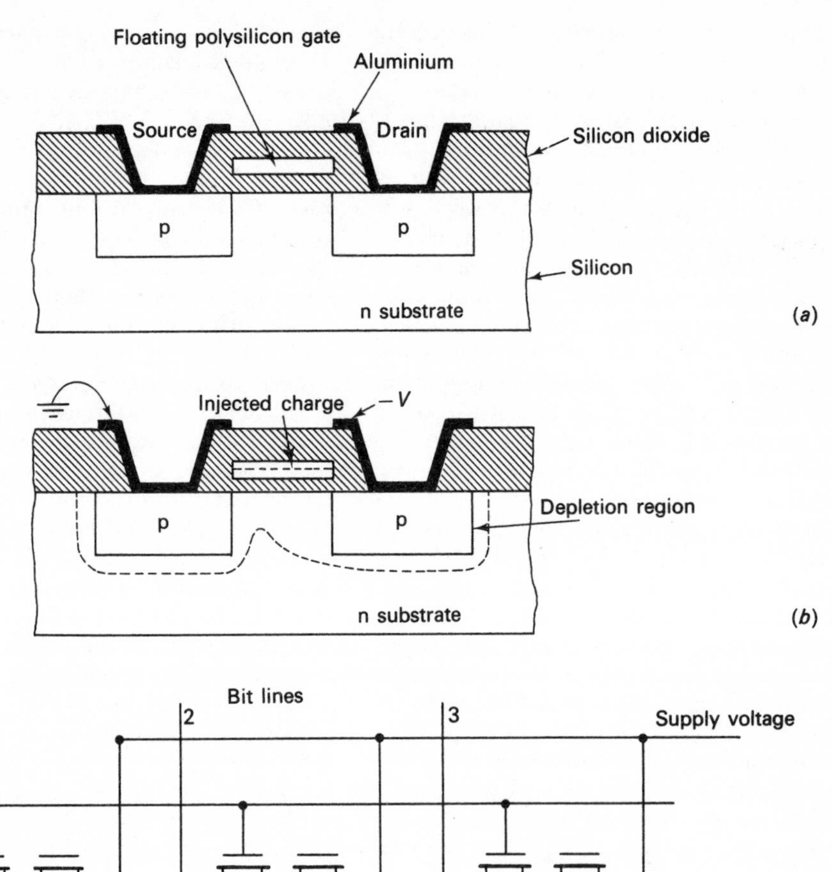

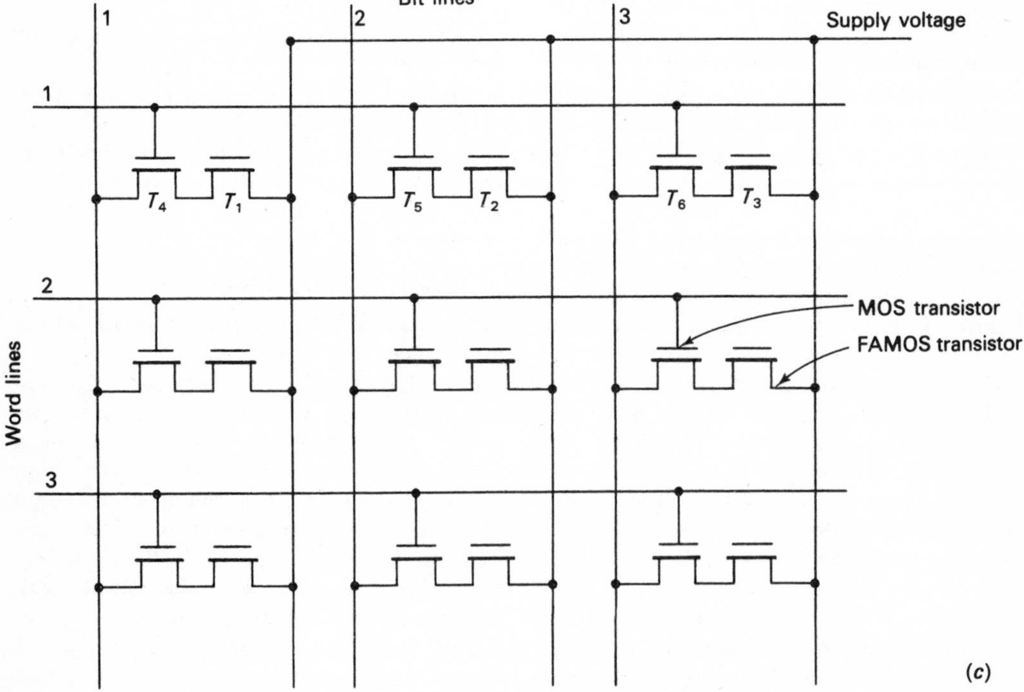

Fig. 5.5. Reprogrammable FAMOS ROM: (*a*) memory transistor, off, (*b*) memory transistor, on, (*c*) memory array.

5.3.1 *Bipolar static memories*

Fig. 5.6 (*a*) shows a basic flip-flop memory cell in which the transistors have multiple emitters for addressing. Ignoring these for the present, and supposing that when the supply is switched on transistor T_1 comes on, then T_2 will be off and this state will be maintained so long as the supply voltage is connected. Therefore the cell remembers the information or has memory. Capacitors C_1 and C_2 represent parasitic components and may be ignored at this stage.

The information stored in the cell is sensed by the 1 and 0 bit lines. If the word line is at zero volts then cell current will flow to ground through the lower emitter of the transistor and not through the bit lines. The cell has not been selected (or addressed or accessed) at this stage. To address the cell its word line is taken positive. If T_1 was originally on and T_2 off then this state will remain unchanged but there will be a larger current flowing in bit line 1 than in line 0. This differential current can be sensed and used to indicate the state of the stored information.

With the word line at zero volts the state of the bit lines have no effect on the storage cell. To change its information, that is to turn T_1 off and T_2 on in the present instance, it is necessary to raise bit line 1 and lower bit line 0 at the same time that the word line voltage is raised to select the cell. This will force the flip-flop into the desired state, which will then be maintained even after the word line is returned to ground potential.

There are many ways in which a cell may be fitted into a memory organization. These are described in greater detail in section 5.6. Organizations however often influence the cell structure. For instance if *X–Y* addressing is desired, in which a cell is not selected unless two lines in the matrix are activated, then the structure shown in Fig. 5.6 (*b*) is used. It is very similar to the word line addressed cell except that now both the *X* select and *Y* select lines must be taken to a positive voltage before the cell is accessed. When either line is at zero volts it provides a path for the cell current so that the bit lines have no effect on it. In both the multiple emitter cells described it is important that the cell resistors R_1 and R_2 be of low value so as to give a large cell current for high speed sensing. This often means that fast memories also dissipate relatively large power.

Returning to Fig. 5.6 (*a*) and considering the effect of capacitors C_1 and C_2 it is evident that by careful choice of the resistors and capacitors one could design the cell with unequal time constants such that on switching on the power, one transistor always comes on instead of the other. This is referred to as a latent image cell and is useful in many memory applications in which the system needs periodic access to a fixed information format. Where this format must occur when the supply is first switched on then nothing else is required. If the pattern must occur during the course of an operation it is required to pulse the memory supply temporarily off and on to obtain it. In these instances an alternative latent image memory, shown in Fig. 5.6 (*c*), may be used. It consists of a diode connected to the base of the transistor which is to be turned on in the memory cell. During normal operation the anode of this diode is at zero volts so that it is reverse biased and plays no part in the operation of the cell. To switch the cell into its latent image the anode is pulsed with a positive voltage so that T_1 is on irrespective of its previous state. Diode *D* may be fabricated in the Schottky process within the collector region of the main memory transistor so that it occupies no additional space on the silicon chip.

A cell arrangement which is capable of high speed and low power dissipation is shown in Fig. 5.6 (*d*). Voltage V_1 is greater than V_2 and the word line is normally at a value between these two. Therefore diodes D_1 and D_2 are reverse biased. The information stored in the cell, say T_1 on and T_2 off, is maintained so long as the supplies are present. To read the cell the word line is taken to ground potential. A heavy differential current now flows from V_2 through the bit lines and these can be sensed to indicate the polarity of the stored information. To write into the cell the word line potential is again lowered and one of the bit lines raised in potential to turn on the corresponding transistor. Since a large sense current flows via D_1 and D_2 the values of the cell resistors may be made large so reducing power dissipation but not

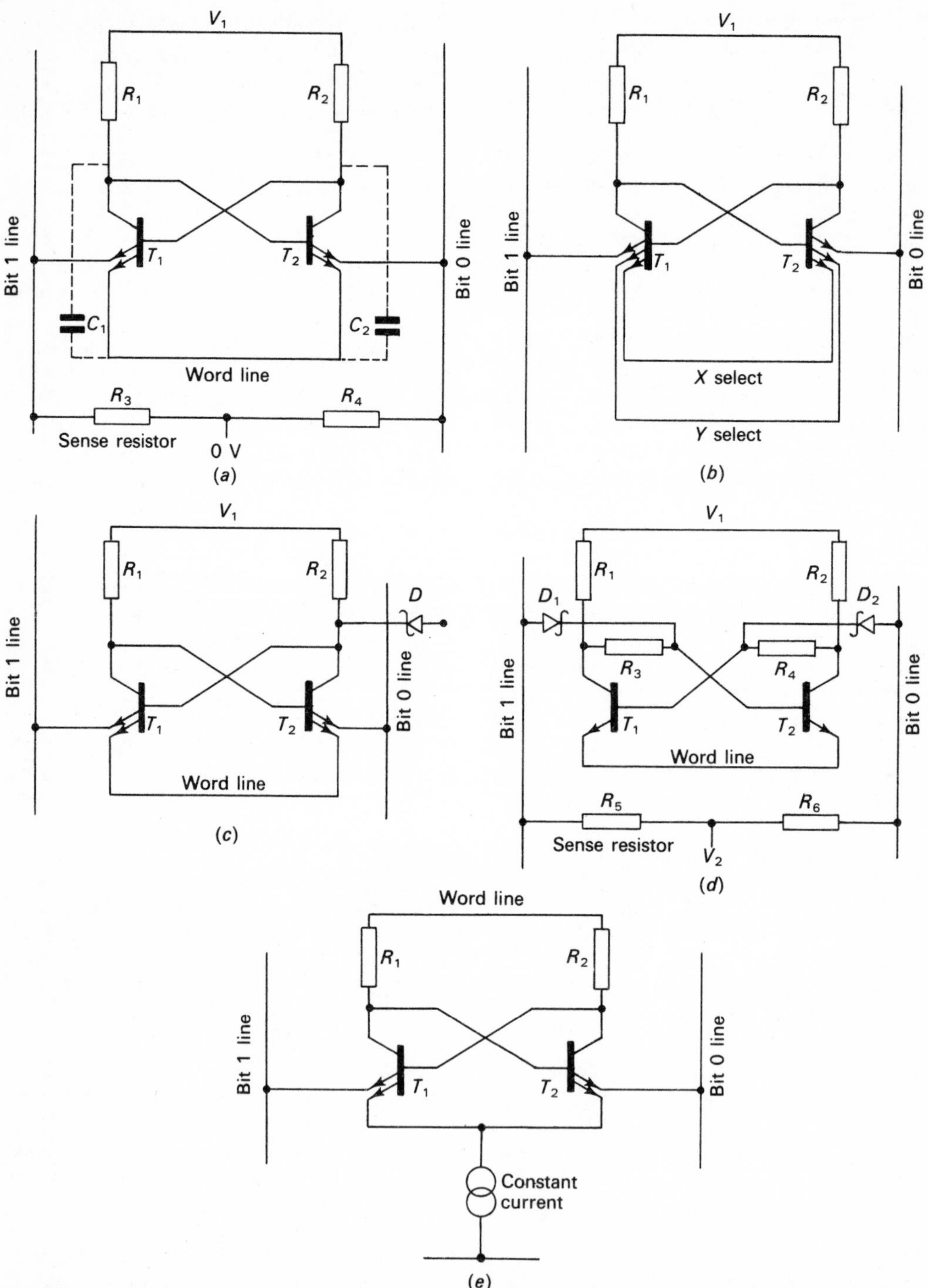

Fig. 5.6. Bipolar static memory cells; (*a*) multiple emitter cell, (*b*) multiple emitter cell with *X–Y* address, (*c*) cell with latent image, (*d*) diode coupled cell, (*e*) non-saturating (ECL) cell.

significantly affecting speed. Although word line addressing is shown for this cell it is of course also possible to include X–Y selection by means of multiple emitter transistors.

All the cells described so far use gold doped transistors and operate in the saturated mode. By employing Schottky devices speed improvements may be obtained. However, for very high speed operations non-saturating ECL cells are required. This is shown in Fig. 5.6 (*e*). Resistors R_1 and R_2 are selected such that, for the designed constant current, neither of the transistors operate in a saturated mode. The word line is at a relatively low voltage when the cell is not addressed so that the power dissipation is low. To access the cell the word line is taken to a high positive voltage. This causes a large differential current to flow in the bit lines, the polarity of which depends on the state of the transistors, without changing the original memory state. This current can be used to read the cell as before. To write into the cell one bit line is raised in potential and the other lowered at the same time as the cell is accessed so as to force the transistors into the desired memory state.

5.3.2 *Bipolar dynamic memories*

Dynamic memories operate on the same principle as the dynamic logic systems described in earlier chapters, that is the storage of charge on a capacitor. They have the advantage over static memories of lower standby power and smaller cell size. However since additional circuitry is required for *refreshing*, the devices are more difficult to use.

Generally unipolar technology is more suitable than bipolar for the design of dynamic systems. This is primarily due to the greater control which it is possible to exercise over the fabrication of gate capacitors and the high input impedance of the unipolar transistors resulting in a slower leakage of capacitor charge. Several bipolar dynamic cells have however been fabricated.

The cell shown in Fig. 5.7 (*a*) uses no resistors and has only three interconnection lines so that its size is relatively small. Assume that initially T_1 is on and T_2 off so that the parasitic capacitor C_1 has a greater charge than C_2. This charge will gradually leak away and must be refreshed by taking the bit lines to voltage V via the resistors R_1 and R_2. This will charge C_1 and C_2 to their initial value but will not affect their charge ratio so that T_1 will still be on and T_2 off. During the store state V is at ground potential so that the diodes are reversed biased. To read the cell V is again taken positive and the differential currents in the bit lines are sensed. To write into the cell one of the bit lines is forced to ground and V taken positive so that only one of the cell transistor's capacitor is charged. In all the above instances the cell is addressed, during reading, writing and refreshing, by taking the word line to ground potential. In any other position the cell is disabled but the state of its memory is stored on the capacitors. Fig. 5.7 (*d*) shows the cell layout, the diode being fabricated within the transistor collector.

The simplest cell arrangement is achieved by the single transistor cell of Fig. 5.7 (*c*). The transistor has no base connection so that cell complexity and size are further reduced. Storage occurs on the base–collector and base–emitter capacitors C_1 and C_2. To store a logic 0 a low voltage pulse is applied across the cell so that none of its junctions are avalanched and a charge is stored on both C_1 and C_2 strongly reverse biasing both junctions of the transistor. To store logic 1 a high voltage pulse is applied across the cell. The base–collector junction avalanches and gives rise to a large minority carrier flow which leaves both the transistor junctions only slightly reverse biased. To read the information a low voltage pulse is used and the current flow through the cell noted. For a stored 0 it will be low but for a 1 it will be large since the transistor will be turned on. Since the output of this cell is equal to the transistor gain times the stored charge, it can be made large by careful design.

Fig. 5.7 (*b*) shows a novel dynamic cell which uses only two diodes. D_1 is a conventional device whereas D_2 is a Schottky diode which has a very low charge storage time, assumed here to be zero, and appreciable capacitance. The cell stores information by virtue of the difference in charge recovery times between the two diodes. To write into the cell a current is passed in the forward direction storing charge in D_1. When

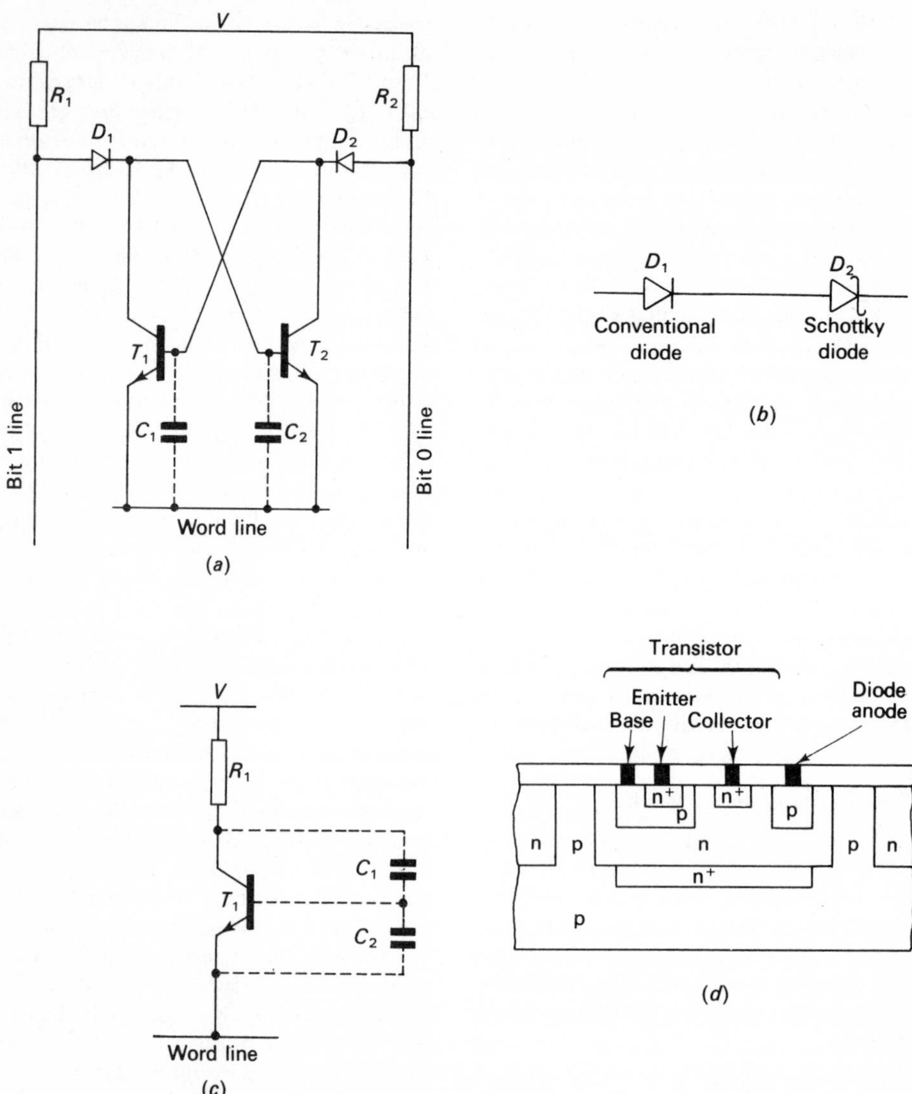

Fig. 5.7. Bipolar dynamic memory cells; (a) two transistor cell, (b) two diode cell, (c) single transistor cell, (d) part of two transistor cell construction.

the applied voltage is reversed a reverse minority current flows in D_1 but not in D_2 so that voltage builds up on the capacitance of the Schottky diode. This voltage represents a logic 1. To read the cell a signal with a constant rate of rise of voltage is applied in the forward direction across the cell. After a delay, which is dependent on the voltage stored on the Schottky diode, the cell will turn on. The time difference between the conduction of a charged and uncharged cell gives a measure of the polarity of the stored information. Like the previous two devices this cell is clearly dynamic in operation since the voltage stored on the Schottky diode will leak away and will need periodic refreshing.

5.3.3 *Unipolar static memories*

Like the bipolar devices the basic storage element of a unipolar static memory is the flip-flop. This is shown in Fig. 5.8 (*a*) in which T_3 and T_4 form the memory transistors whereas T_1 and T_2 act as load devices. The bit lines are connected to the cell by the select transistors T_5 and T_6. To read the memory the word select line is taken negative, for PMOS devices, so as to turn on T_5 and T_6, and the differential current is sensed in the bit lines. To write into the cell the word line is again taken negative and the bit lines taken to a 0 or 1 state so as to force the memory into the required mode. The gate of the load transistors may be taken to V_{DD} or to a separate supply V_{GG}. The advantage of this latter approach is that when the cell is not addressed V_{GG} can be pulsed so that T_1 and T_2 are on only for a fraction of the normal duty cycle, the memory state being stored in between these times on the parasitic capacitors associated with the gates of T_3 and T_4. The cell is now, operating in a partially dynamic mode and has a considerably reduced power dissipation. The basic memory cell of Fig. 5.8 (*a*) is reproduced in Fig. 5.8 (*b*) but *X–Y* addressing is now shown so that eight transistors, instead of six, are required, increasing the memory cell size. However the addressing circuitry is now reduced in complexity.

Although the memory cell shown in Fig. 5.8 (*c*) looks similar to the six transistor cell of Fig. 5.8 (*a*) it differs considerably due to the operation of its load transistors T_1 and T_2. In all these memory cells the information is stored on the capacitance associated with the gates of transistors T_3 and T_4 and the function of the load transistors is to compensate for charge leakage from these devices. This leakage is very small so that the load transistors may be made to have high impedances. However this usually means that they take up considerable space on the chip so that a compromise between cell size and power dissipation is required. The load transistors shown in Fig. 5.8 (*c*) operate on the charge pump principle which enables them to have large impedances while still occupying a small volume. Note that these transistors have no drain connection and that their gate is fed by an a.c. voltage, from an oscillator, of about 15 volts peak at between 500 kilohertz and 800 kilohertz. When the gate voltage is positive a channel forms under the gate, assuming that the device is NMOS. When the voltage goes negative recombination occurs and this channel disappears. During recombination some of the minority carriers are lost and result in a small current flow from substrate to source. This current is sufficient to compensate for capacitor leakage within the memory cell, while being low enough to give very moderate power dissipation without the need for conventional dynamic operation.

Fig. 5.8 (*d*) shows another low dissipation memory cell using complementary MOS technology. Since only one p- or n-channel device is on in any inverter pair the dissipation is reduced, while the low impedance of the transistors when conducting ensures that the operating speed is high. As before, word addressing or *X–Y* addressing techniques may be used. This also applies to the latent image cell of Fig. 5.8 (*e*) in which the impedance of T_1 and T_2 and the value of C_1 and C_2 may be selected to ensure that either T_3 or T_4 always comes on first when the power is switched on. Therefore the device is similar in operation to the bipolar version illustrated in Fig. 5.6.

5.3.4 *Unipolar dynamic memories*

Unipolar dynamic cells are very commonly used due to their reduced cell size and power dissipation. Fig. 5.9 (*a*) shows a four transistor cell which has two fewer transistors than the best static memory. T_1 and T_2 perform the dual role of load and select transistors. Assume that initially the state of the charge on C_1 and C_2 is such that T_3 is on and T_4 is off. This charge will leak away so that periodically it will need refreshing by applying a voltage on the two bit lines and turning on T_1 and T_2 so that C_1 is recharged while C_2 still maintains its low charge. To read the cell the bit line capacitors, which are normally parasitic components, are charged to a voltage and T_1 and T_2 then turned on. Since T_3 is on, bit line 0 will be discharged more than bit 1 and the polarity of this differential voltage can be sensed to indicate the state of the cell. To write into the memory the bit lines are forced to the required high or low

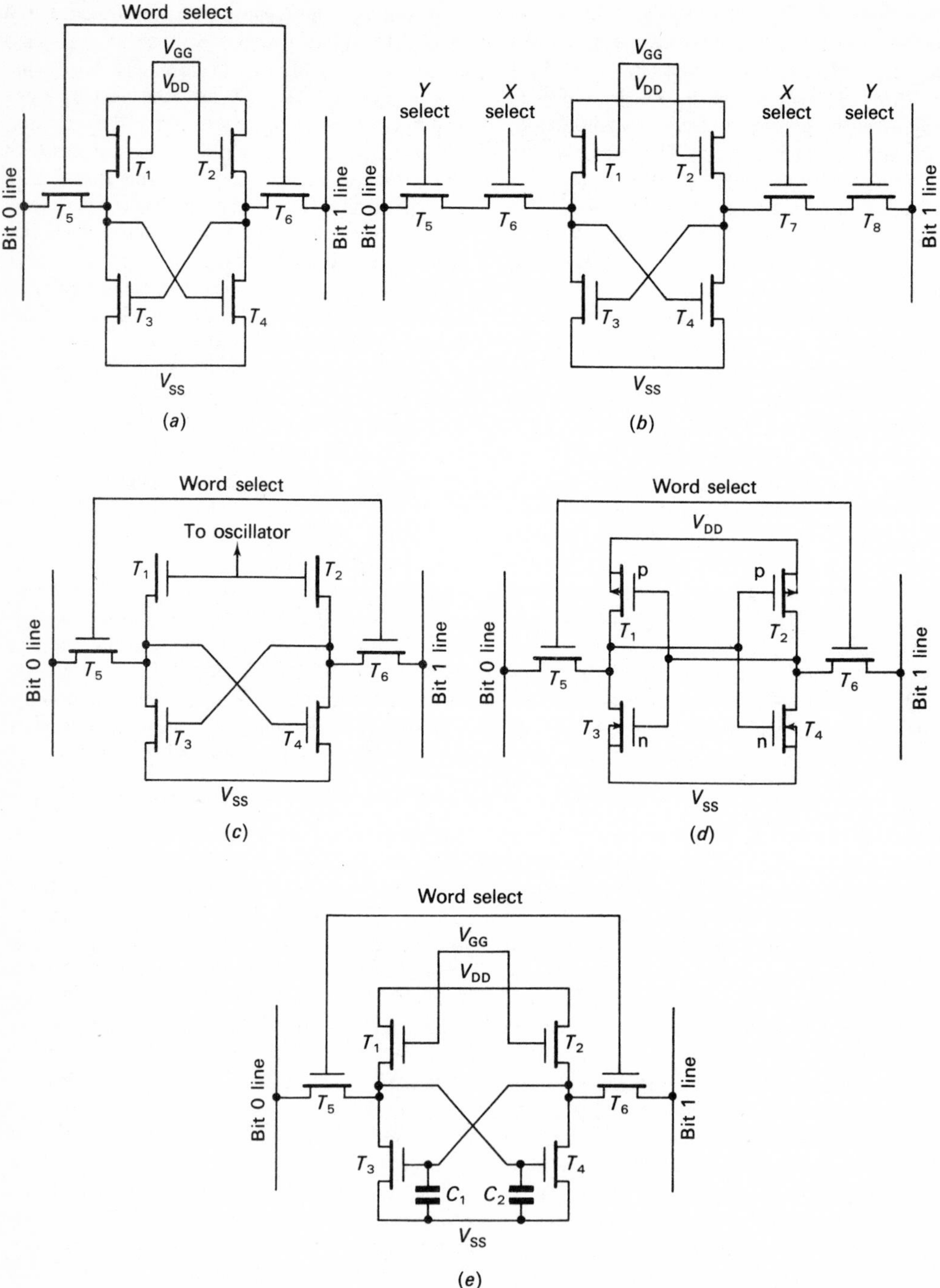

Fig. 5.8. Unipolar static RAM cells; (*a*) word addressed cell, (*b*) *X–Y* addressed cell, (*c*) charge pump cell, (*d*) CMOS cell, (*e*) latent image cell.

voltage level and T_1 and T_2 then turned on. This memory needs careful design trade-offs between the impedance ratios of transistors T_1/T_3 and T_2/T_4. It has the advantage of a simple refresh mechanism and good speed but is still relatively complex.

A three transistor cell is shown in Fig. 5.9 (*b*). The memory storage capacitor C_1 is associated with the gate of T_2. To write into the memory transistor T_1 is turned on so that the input data line can charge or discharge C_1. To read the cell the data out line is first precharged and T_3 is

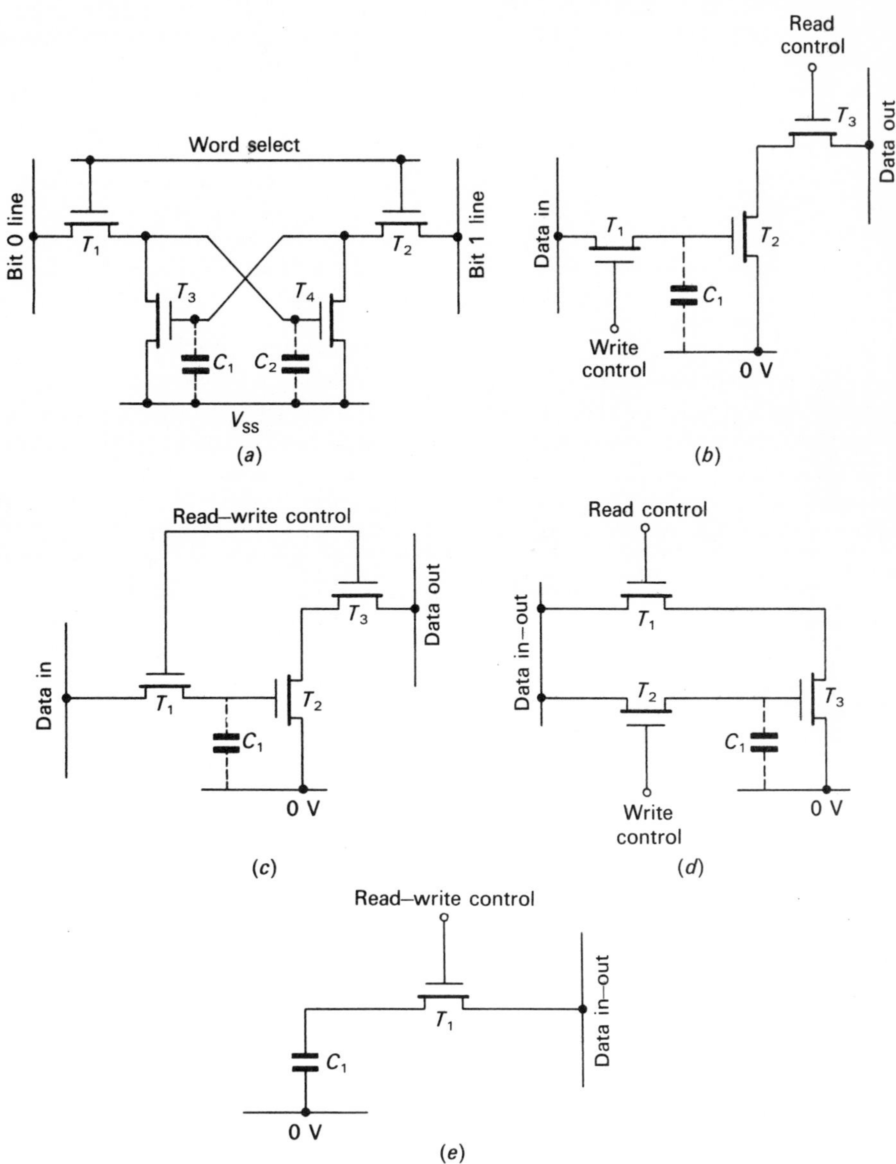

Fig. 5.9. Unipolar dynamic RAM cells; (*a*) four transistor cell, (*b*) three transistor cell, (*c*) three transistor cell with common address line, (*d*) three transistor cell with common bit line, (*e*) single transistor cell.

then turned on. If a logic 1 was stored on C_1 then T_2 will be on and the data out line will be discharged. Therefore sensing this line indicates the state of the memory. The cell is refreshed by first reading out to indicate its state and then reading in via the data in line after the output has been amplified. This memory cell has a small size and a relatively fast operation since separate read and write lines are used. It is very popular for several memory products and has been used in several forms. Fig. 5.9 (*c*) for instance shows a common read–write control line while Fig. 5.9 (*d*) uses a common data in–out line. In both instances this reduces cell size, the latter arrangement also making refresh easier since the input and output are already looped together.

The ultimate in cell simplicity is the single transistor cell of Fig. 5.9 (*e*) in which both the read–write control and data in–out lines are commoned. There is now no amplification between the charge stored on C_1 and the voltage readout on the data lines, and this can present sensing problems. For instance the data lines have considerable capacitance associated with them which is comparable in size to that of the memory capacitor C_1. Therefore during readout voltage sharing occurs and the output is considerably reduced, which can result in poor speed and noise immunity. Design techniques are used to increase cell capacitance, reduce data line capacitance and increase the sensitivity of the sense amplifiers. This however usually means a larger cell. A further disadvantage of the single transistor cell is that all readout is destructive so that the information must be immediately written back. Therefore although the single transistor design presents many advantages it has associated with it a considerable amount of peripheral circuitry which makes it uneconomical in small memory arrays.

5.4 The non-volatile RAM

Semiconductor random access memories are strongly challenging magnetic devices in all applications except those which require non-volatility of store. Considerable work has been done to overcome this last hurdle. Primarily there have been two lines of approach. The first accepts the volatility of the semiconductor RAM but designs it to have a very low stand-by power consumption so that it can run from battery supplies during power-down phases. The second approach is to design a memory cell which is truly non-volatile. There has been some success in this area. However most of the memories developed have tended to have relatively long write times so that they may be considered more as reprogrammable read only memories rather than RAMs. One such device, the FAMOS memory was described in section 5.2.3. In the present section other such devices will be covered. All these however are electrically reprogrammable so that they show potential for use in the non-volatile RAMs of the future.

5.4.1 *Battery powered RAMs*

In section 5.3 many cell arrangements were described which used dynamic techniques to reduce power dissipation. Memory arrays made from these cells can be operated in a stand-by mode where only the refresh cycle is in use. By employing a low duty period for this cycle the cell dissipation can be made very small, the mean current drawn from the supply being a few tens of milliamperes. Fig. 5.10 shows such an arrangement. During normal operation the battery is charged from the a.c. supply and the stand-by refresh logic is disconnected. The memory array, decoders and drivers operate in their usual mode. If the a.c. supply fails then the circuit used to detect power failure activates the refresh logic system. This disables the complete memory so that only the refresh logic is on continuous supply from the battery. The logic activates the memory decoder and drivers only during the short refresh periods in order to recharge the relevant memory storage capacitors within the array. This ensures that power dissipation is minimal. Since a refresh operation is required to replace charge lost by internal leakage within the memory array, and this leakage is a function of ambient temperature, for optimum operation the refresh times could also be varied with temperature. This is however not frequently done.

The batteries used in this application must clearly be rechargeable. If the required mains-

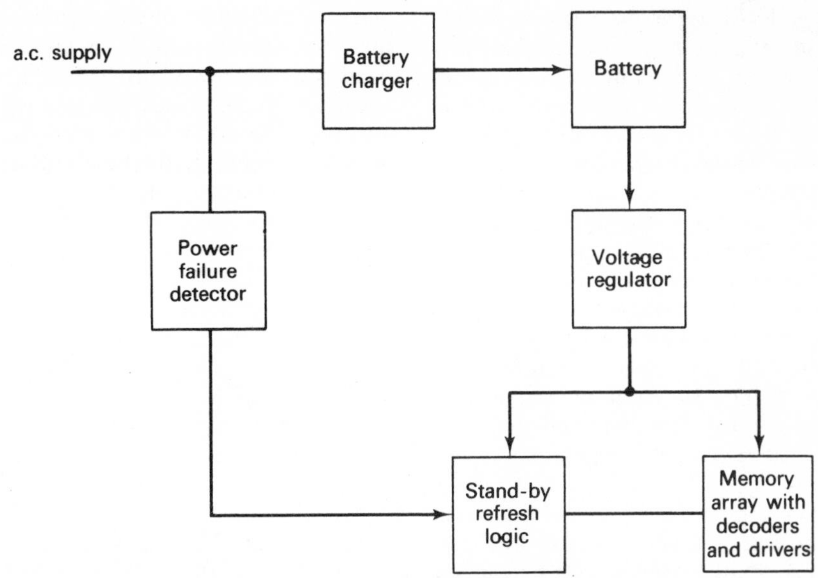

Fig. 5.10. Memory with mains-failure facilities.

failure operating time is short then they can be nickel–cadmium types, but for periods longer than a month sealed lead–acid batteries are preferred. The voltage regulator shown in Fig. 5.10 also needs careful design. It must be capable of operating over the full range of battery voltage variation and of handling the peak operating and stand-by power of the memory. Since the regulator is on even during the stand-by mode it is essential that it is efficient even though the current is low, typically about 30 milliamperes.

5.4.2 *The MNOS memory*

Of the three reprogrammable memory cells described in the section the metal–nitride–oxide-semiconductor (MNOS) device is perhaps the most promising. The structure of this cell is shown in Fig. 5.11 and it is seen to be very similar to the more common low threshold nitride transistor.

However the dimensions of the various layers are different. The nitride is formed by chemical vapour deposition in which silane (SiN_4) is decomposed in oxygen. This layer is about 700 ångströms thick whereas the oxide layer is in the region of 40 ångströms. The thickness of the gate oxide is perhaps the single most important parameter of this cell.

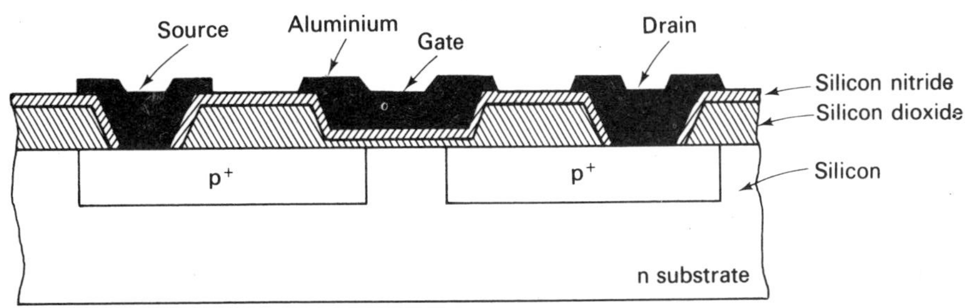

Fig. 5.11. MNOS cell construction.

The MNOS transistor is normally in the off mode and presents a high impedance between source and drain. To write into the cell, that is to switch it into a low impedance mode, a positive voltage exceeding a critical value (typically 30 volts) is applied to the gate electrode. This causes electrons to tunnel through the thin oxide layer and charge to build up at the oxide–nitride interface. This charge is prevented from escaping by the insulating oxide, even after the gate voltage pulse is removed. Providing the charge is relatively close to the silicon surface it will cause an inversion layer from source to drain so that a low impedance exists between these terminals and the transistor is on. To erase the stored charge the gate is pulsed with a negative voltage of the same magnitude as before. This reverses the write process. The cell is read by applying a gate voltage of magnitude in between the thresholds of the low and high impedance states. A current flow indicates that a charge was previously stored in the cell. During reading the transistor behaves as a conventional MOS device so that its speed, which is limited by gate and drain capacitances, is comparable to that of a mask programmed ROM.

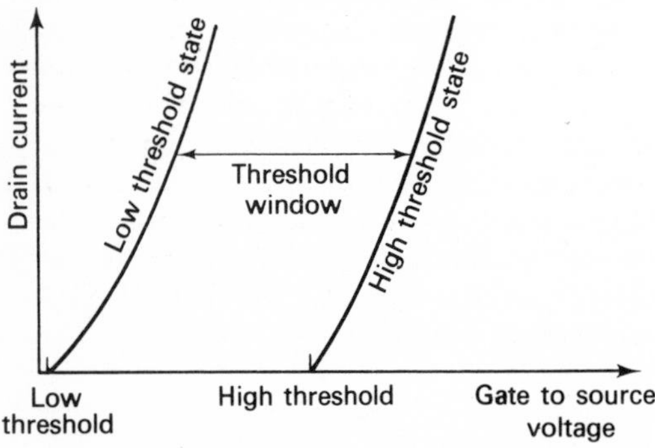

Fig. 5.12. Two states in a MNOS memory cell.

Fig. 5.12 shows the transfer characteristics of the cell during its two operating states. The threshold voltage determines the ability of the cell to distinguish between a stored logical 1 and logical 0. The magnitude of this window is determined by the write pulse energy. This is illustrated in Fig. 5.13. The greater the write voltage the shorter the write pulse width can be made for a given threshold window. In practice the peak gate voltage is determined by cell breakdown characteristics so that a write voltage of 35 volts is typical giving a write time of about 5 milliseconds and an erase time of about 50 milliseconds. Both these are too long for RAM operation.

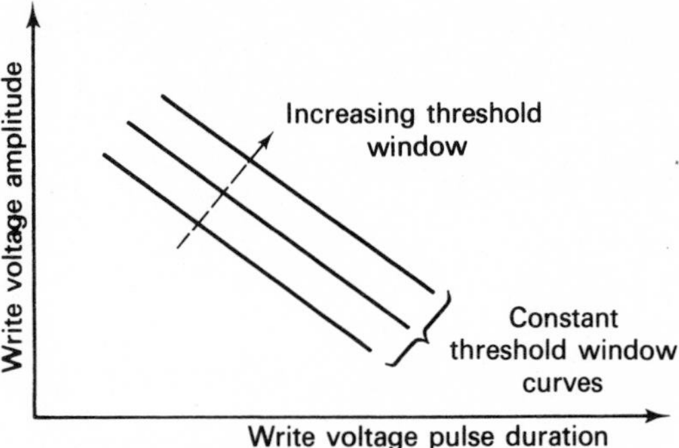

Fig. 5.13. Effect of write pulse energy on the threshold window of a MNOS memory.

The physical properties and thickness of the oxide layer determine the voltage, speed and retention time of the MNOS memory. The stored charge will leak away so that the threshold window will gradually decrease. This is shown in Fig. 5.14. The acceptable or retention life of the memory is determined by the ability of the cell to distinguish between its two logic states. Generally this period is many years. It should however be remembered that the initial window size is determined by the pulse energy. Therefore for a retention of a few weeks a write time of a few microseconds may be acceptable whereas for a retention of several years the pulse must be many milliseconds long. A serious limitation of the MNOS memory is that the charge in an off transistor cell will gradually build up with each read operation. This again causes a narrowing of the threshold window. At present about 10^{11} to 10^{12} read accesses are possible before there is any significant reduction in this window. It should be remembered that the FAMOS device could tolerate many read cycles but was limited to a relative few erase operations.

The MNOS device uses only a single transistor per memory cell so that it gives a high

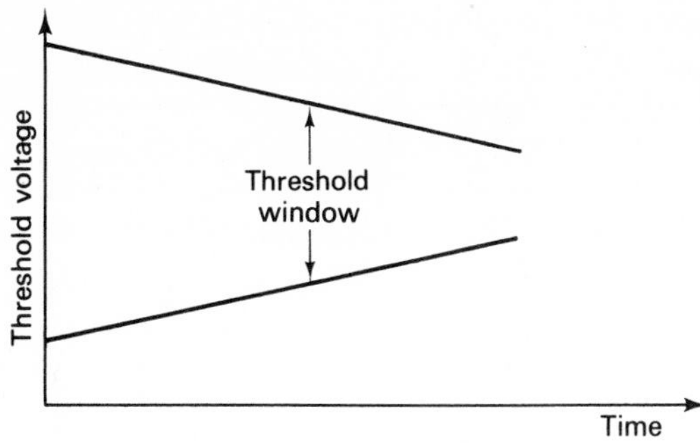

Fig. 5.14. Decay of stored charge with time in an MNOS memory.

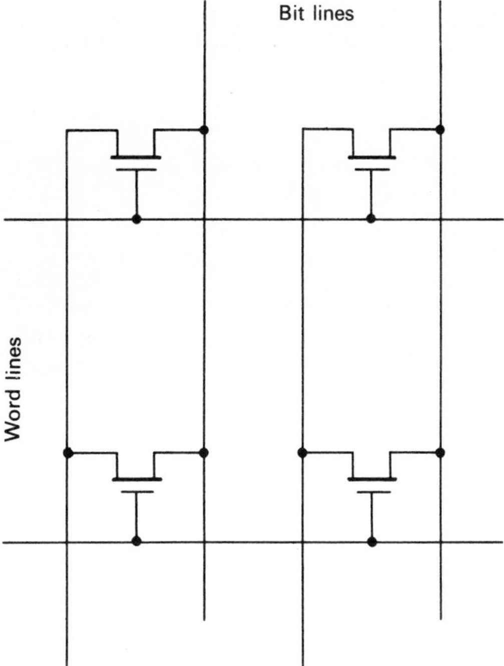

Fig. 5.15. MNOS transistor array.

density array. Since the device has a gate control lead an auxiliary transistor is not required for addressing as in the FAMOS memory. Fig. 5.15 shows a MNOS array. It is seen to be comparable in simplicity to a mask programmed ROM.

5.4.3 *The MAOS memory*

The basic difference between the MAOS (metal–alumina–oxide-semiconductor) and the MNOS cell is the replacement of the gate nitride by an alumina layer. The memory is shown in Fig. 5.16. It consists of a gate oxide layer of about 60 ångströms thickness followed by an alumina layer of 1000 ångströms. The thick oxide on top of this exceeds 4000 ångströms. n-channel devices are commonly used, as shown, and high impurity channel stops are diffused around memory cells to increase surface concentration of p ions and so suppress the cell leakage currents during the relative high voltage write periods. In principle the cell operates very much like an MNOS device but since it is n-channel its states are reversed. Normally the transistor is on (low threshold). To turn it off a positive gate voltage, above a certain critical value, is used. This causes electron tunnelling into the oxide–alumina interface and a raising of the threshold voltage, since the electrons oppose a channel formation. To erase the stored information a high negative gate pulse is used to reverse the process. Other characteristics of the transistor, such as window threshold, are similar to the MNOS device.

5.4.4 *Chalcogenide memories*

Chalcogenide memories are not based on the properties of silicon. Instead they depend on the switching action exhibited by thin films of the group IV, V and VI semiconductors called chalcogenide glasses. These films are capable of stability in any one of two states when operated below normal glass transition temperatures. In the first state the device has an amorphous structure and a relatively high resistivity of the order of 5×10^4 ohm centimeters. The other stable state is polycrystalline and in this the cell has low resistivity of about 0.3 ohm centimeters. The amorphous state is of a higher energy than the crystalline state and the transition between the two is explained by considering the effect of heat on an amorphous material. If the temperature is gradually raised a point is reached at about 130°C when an exothermic reaction occurs causing the material to shift from its amorphous to polycrystalline form. However when the temperature approaches about 300° C an endothermic reaction occurs, the material

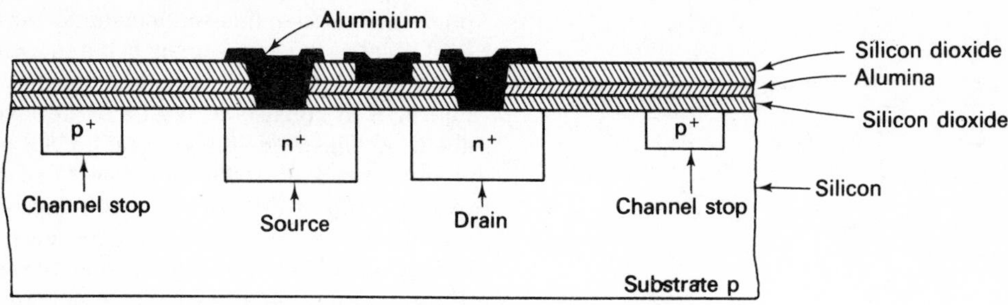

Fig. 5.16. MAOS cell construction.

begins to melt, and the above process is reversed. On cooling two processes can occur. If cooling is rapid the material maintains its disordered amorphous state, but if cooled slowly it will recrystallize and maintain this state even at room temperature. Therefore switching between the two states can occur by controlling the temperature–time profile on the chalcogenide element. In a memory this is done by current pulses.

A second important property of chalcogenide films is that they exhibit a threshold voltage in the region of about 15 volts. If a voltage in excess of this value is applied across an amorphous device it will switch to a low impedance state. This is not a thermal effect and occurs very quickly, within a few microseconds. The switching speed is dependent on the applied voltage so that a value of about 25 volts is normally used. However if the switching pulse is now removed the material will return to its original high impedance state. To change it from the amorphous to the polycrystalline state not only must it first be switched by a high voltage pulse but this must then be followed by a low current pulse of about 7 milliamperes for about 10 milliseconds. The cell will now remain in the low impedance state for an almost indefinite time. To return to the high impedance amorphous state high temperature and rapid cooling is required. This is achieved by pulsing the device for about 5 microseconds with about 150 milliamperes. Fig. 5.17 shows the memory characteristics. It is capable of remaining in either memory state for very long times unless pulsed into the other mode as described above. Reading is accomplished by a low current of about 1 milliampere which, in a low impedance cell, will cause a typical volts drop of about 0.6 volts. The resistance of an 'off' cell is normally about 300 kilo-ohms.

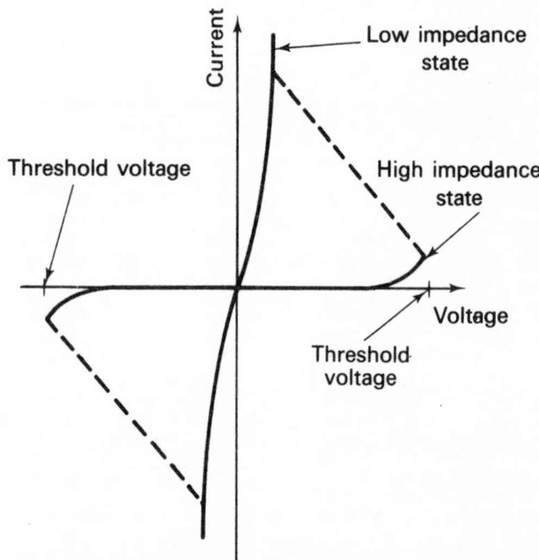

Fig. 5.17. Amorphous memory switching characteristic.

The chalcogenide switches are used as cross points in a memory array. However to eliminate 'sneak' paths it is usual to include a diode in series with each switch as shown in Fig. 5.18 (*a*). Conventional photolithographic techniques may be used to produce the diode and memory switch as shown in Fig. 5.18 (*b*). The memory switch can be made by sandwiching the thin chalcogenide film between the two electrodes or by placing these side by side. In either case the electrode in contact with the chalcogenide must be made from refractory material for best results. Molybdenum is the material most commonly used. However since this metal has a slow deposition rate it is still preferable to use aluminium for the main interconnection lines, specially since this provides a better bonding surface.

5.5 Content addressable memories

In the random access memory or read only memory described earlier a storage cell within a matrix may be addressed only if its exact location is known. In a content addressable memory or CAM, also known as associative memory, the location of the cells within a matrix is not of much interest. In these memories a cell is addressed by calling on the information stored within the cell. If the cell has the required pattern it will identify itself, irrespective of its location in the memory matrix. Quite clearly the CAM cell must have the ability to store information, as in a RAM, as well as to be able to compare the required information with the stored data. Each cell therefore has associated with it some peripheral circuitry.

Perhaps the commonest CAM cell is the unipolar device shown in Fig. 5.19. Three operating modes are possible – 'read' and 'write' as for a normal RAM, plus 'search', which is a special characteristic of the CAM

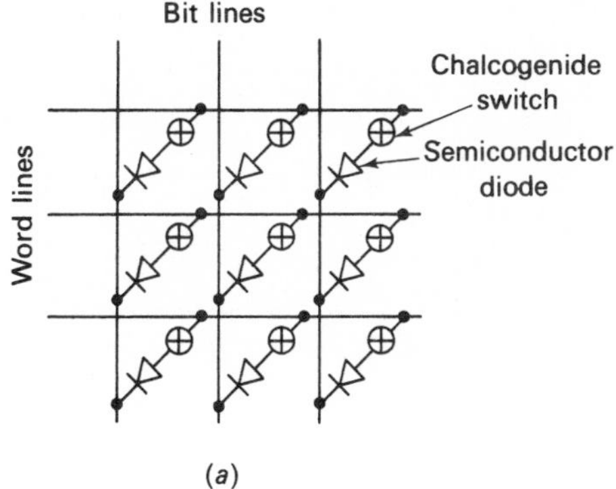

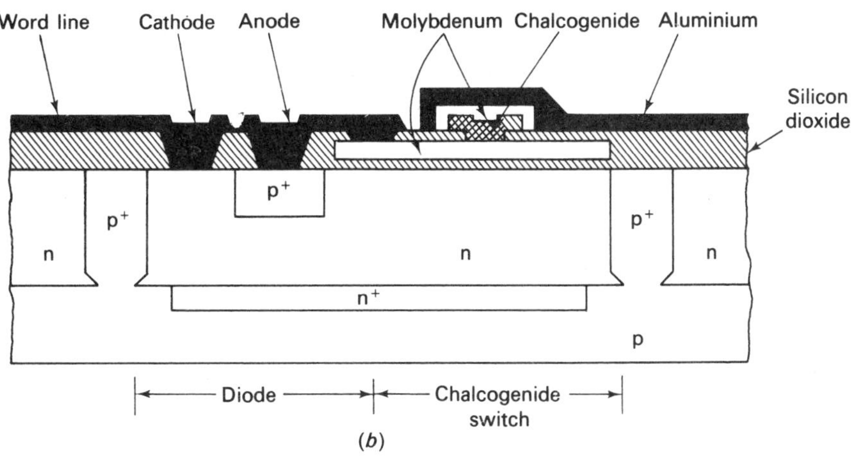

Fig. 5.18. Chalcongenide memory; (a) memory arrangement, (b) single cell.

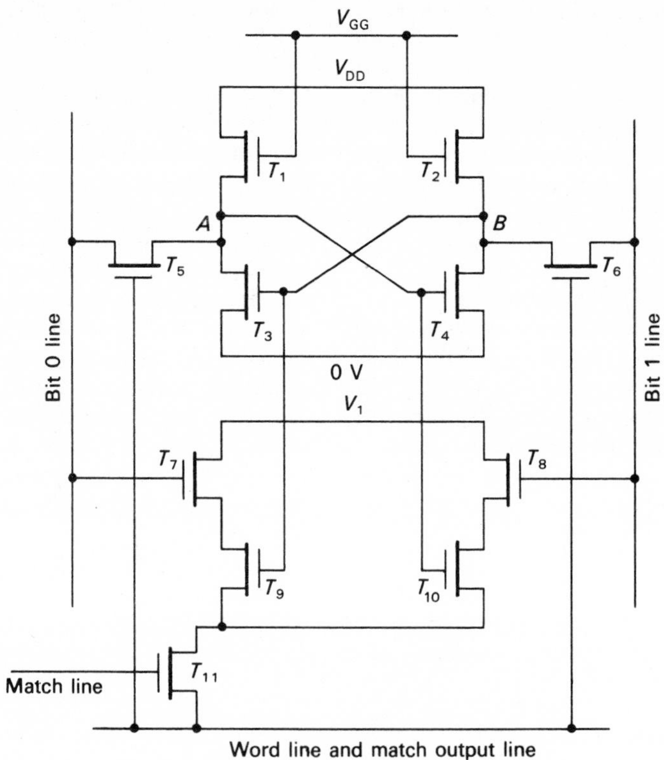

Fig. 5.19. Unipolar content addressable memory.

memory. T_1 and T_2 are load devices and along with T_3 to T_6 they form the usual RAM configuration described in section 5.3. During read and write the word line is addressed to select the cell within the matrix. Therefore a CAM can always be used as a conventional RAM if so desired. During the search mode all the cells within the memory are simultaneously accessed so that the word lines are not utilized to turn on T_5 and T_6. Transistors T_7 to T_{11} are now involved. Supposing the cell is storing logic 1 data so that, working in negative logic, point B is at a 1 (negative) and A is at 0. If we are now searching the cell for logic 1, bit 1 line is set to a negative voltage and bit 0 line to zero volts. Therefore T_8 and T_9 are biased on whereas T_1 and T_{10} are off. Therefore when the match line is pulsed to turn on T_{11} no current will flow into the match output line. If on the other hand one were searching for a logical zero then bit 0 line would be made negative and bit 1 line is put to zero volts. This means that whereas T_9 is still on and T_{10} off, now T_7 is also on and T_8 off, so that the current can flow in the match output line from supply V_1. Therefore current in this line indicates a mismatch of cell information. If one were comparing a word of information with several bits then a perfect match on all bits would result in negligible match output current whereas the magnitude of this current would give the degree of mismatch, that is the number of bits which do not match. Details of the organization of CAM memories are described in section 5.6. It is important to appreciate that many words are simultaneously searched (accessed) within the memory so that a CAM can operate at several times the speed of a RAM in this mode. During the read and write operations a CAM is behaving as a RAM and it obviously operates at the same level of speed.

Fig. 5.20 illustrates a bipolar CAM cell which, when implemented using Schottky transistors and diodes, is capable of very high speeds. The memory cell, made from T_1 and T_2, operates as a conventional RAM. To search the cell its word select line is not activated, that is it is kept at zero volts as in a normal store phase. This means that the bit lines cannot alter the stored information. If logic 1 is stored in the cell then point B is at positive volts, using positive logic conventions, and A is at zero. To search for logic 1 the bit 1 line is made positive and bit 0 line zero. Therefore, since D_3, D_4 and D_2, D_1 each form an OR input, T_3 and T_4 are both held off and when T_5 is driven on there is negligible current in the match output line. If a search is now made for logic 0 then the bit 0 line is made positive and since D_1 and D_2 are now both reverse biased transistor T_3 is turned on by R_5 and current flows in the match output line from supply V_{CC}, when T_5 is on.

5.6 Memory organization

There are many ways in which the memory cells, described in earlier sections, may be incorporated into a complete semiconductor memory. The number of possible variations is so large that this section can only introduce a few of these. In all instances, however, a static memory will consist of an array of the individual cells along with some method of addressing each individual section. Random access memories will also have provisions for reading and writing whereas a read only memory, unless of the programmable type, need only be capable of being read. In addition to the

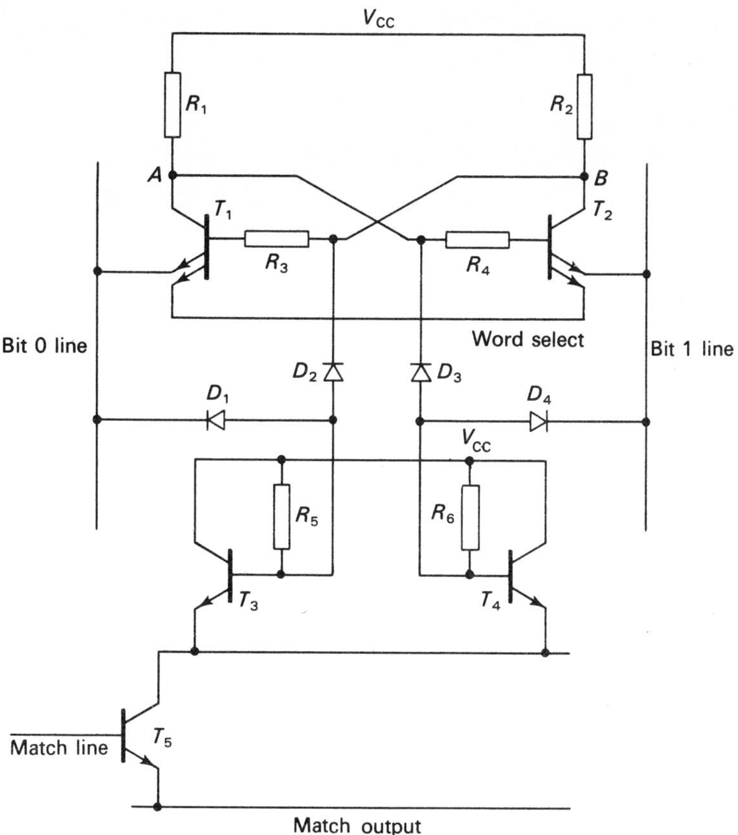

Fig. 5.20. Bipolar content addressable memory.

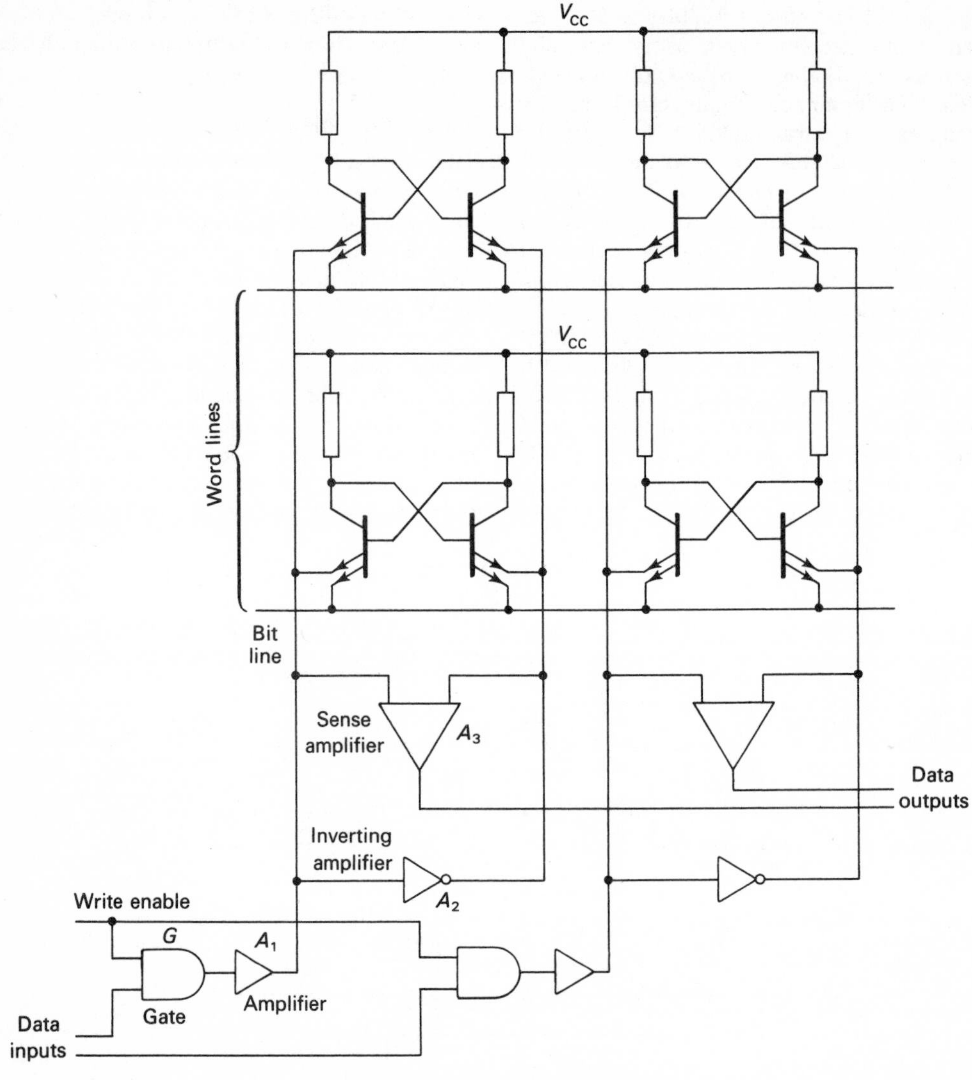

Fig. 5.21. Bipolar memory array.

above a dynamic RAM needs a refresh cycle to recharge its storage locations. As explained in section 5.2 for a dynamic ROM only the peripheral circuitry is operated in an intermittent or pulsed mode. Refreshing, of the form used with RAMs, is therefore not present. This section illustrates a few typical examples of RAM, ROM and CAM memories.

5.6.1 *RAM organization*

Fig. 5.21 illustrates a 2 by 2 array of bipolar cells using the structure described earlier with reference to Fig. 5.6 (*a*). With the word lines at ground potential the memory cells are disabled and retain their original state. To address a row of cells their word line potential is raised. To write, gate G is now closed via the write enable line and the bit lines are forced to the required state by means of the data input line. For reading, the write enable line is used to disable the data input gate. The sense amplifiers provide the data output. Quite clearly one word is addressed at a time and the number of input and output lines equals the number of bits per word.

X–Y addressing can also be used in an array as shown in Fig. 5.22 (*a*). This uses a memory cell of the type illustrated in Fig. 5.8 (*b*) although only the gating transistors are now shown in Fig. 5.22 (*b*). The memory elements are drawn as a box and numbered to indicate their position within the array. Therefore 1–1 means first column first row, 1–2 means first column second row, and so on. Note that in this arrangement only one *Y*-address transistor is

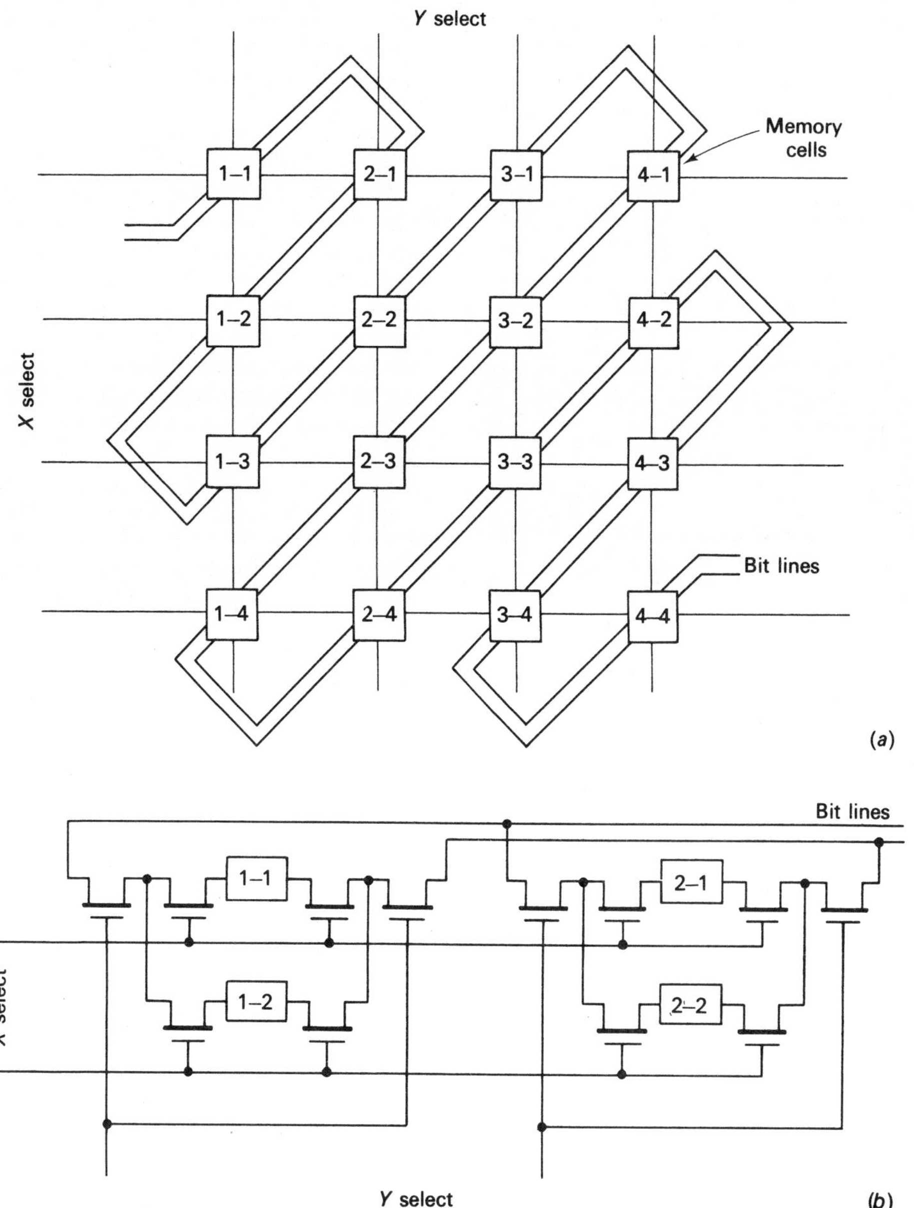

Fig. 5.22. Unipolar memory array; (*a*) 16 by 1 cell arrangement, (*b*) 4 by 1 symbolic representation.

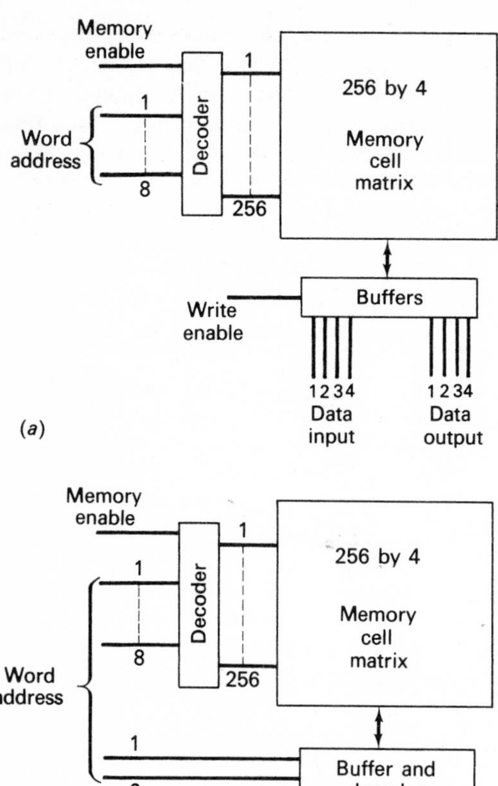

Fig. 5.23. Memory organizations; (a) 256 by 4, (b) 1024 by 1.

used per column. Any cell within the memory may be addressed by energizing its X and Y line. The output bit lines of all cells have also been commoned so that the net result in Fig. 5.22 is a four word by one bit memory. Fig. 5.22 (a) shows a common way of drawing such a memory array, which is of course also applicable to an X–Y addressed bipolar memory.

One of the problems with big memories is the necessity of getting a large number of leads into and out of the package. For instance to build a memory of 256 words by 4 bits will need 256 word lines, 4 data input lines, 4 data output lines and a write enable line. To get this memory into a conventional integrated circuit package it is now necessary to use binary decoders. Eight inputs can uniquely address 256 (2^8) lines. The arrangement for such a memory is shown in Fig. 5.23 (a). A memory enable line has been added which disables the whole memory if energized. Further reduction of the number of lines is possible by coding the data lines as well, as in Fig. 5.23 (b). Although the memory array in Fig. 5.23 (b) is still in a 256 by 4 format there is now only one bit line so that effectively this memory is organized as 1024 words by one bit. A more convenient organization of array cells would now be 32 by 32 using X–Y addressing.

Memories can be increased in size either by expanding the number of bits or words. Generally this requires connecting smaller memories to common lines so that devices with tri-state outputs are required. Fig. 5.24 (a) shows how two 64 by 1 memories may be made to share common address lines but have separate bit lines so as to produce an overall 64 by 2 memory. Note that a common line is used in this illustration for data input and output although separate lines may also be employed if required. In Fig. 5.24 (b) the bit lines from the two memories are also commoned. However the chip enable lines of each memory block are now used to give effectively two more address lines so that the overall system behaves like a 128 by 1 bit organization. Increasing the memory size generally reduces speed since the capacitance of interconnections has been introduced. Decoding also introduces propagation delays so that large memories generally tend to be slower, for the same technology, than small ones.

The organization of the popular three transistor dynamic memory cell, illustrated in Fig. 5.9 (b), is shown in Fig. 5.25. The data lines are precharged prior to a read cycle. For a read operation the read line turns on T_3 and, depending on the state of the stored charge on C_1, the node at B will or will not be discharged. Writing occurs by turning on T_1 so that C_1 is taken to the state of node A. If there is a logic 1 (negative voltage) at A then C_1 is charged to a logic 1 and during a read operation T_2 is turned on so that node B is discharged to zero volts. Therefore there is inversion between A and B. For refresh a read cycle is followed by a write cycle during

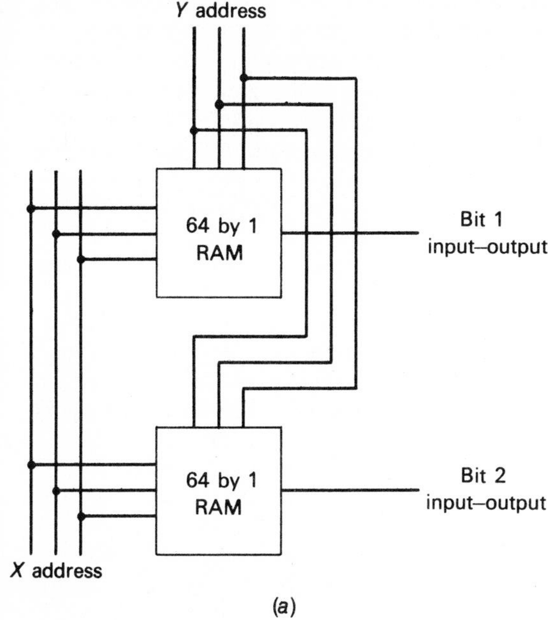

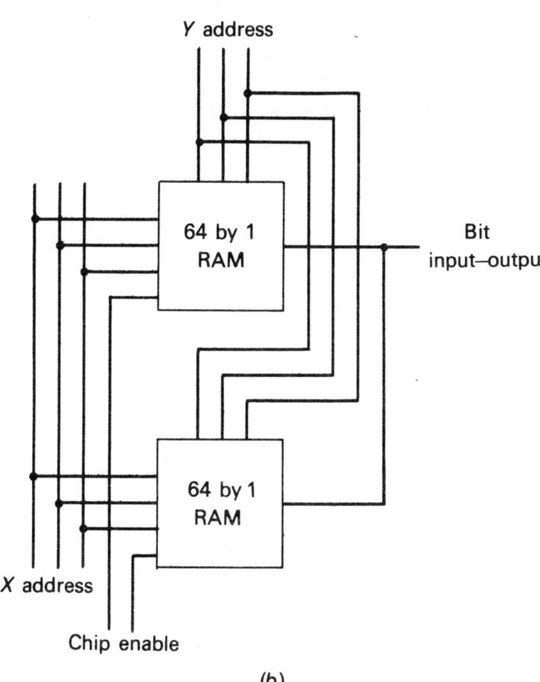

Fig. 5.24. Memory expansion; (a) increasing bits, (b) increasing words.

which the write enable line is energized turning on T_4. A logic 1 at B will now turn on T_5 and give a logic 0 at A so that true refresh of the capacitor C_1 occurs. As seen from Fig. 5.25 a complete row of cells is refreshed at a given time so that the operation can be accomplished relatively quickly even for large memories.

The organization of a three transistor common bit line cell (of Fig. 5.9 (d)) is shown in Fig. 5.26. Since a single bit line is used the overall arrangement is considerably simplified. The memory information is stored on C_1. During writing, T_2 and T_4 are turned on and the data is impressed on the bit line and on C_1. For reading, the bit line is precharged and its state sensed via T_5 after T_1 is turned on. If C_1 contained a logic 1 then T_3 is turned on and the bit line discharged (logic 0). Otherwise it will maintain its charge. To refresh the cell a read cycle is followed immediately by a write cycle during which the data input line is disabled. After reading, the output from C_1 is temporarily stored on C_2 and this is fed back to C_1 (refresh) on the next write operation. A similar principle exists in the single transistor memory of Fig. 5.27. Although the array is in the form of 2 by 2 cells by X and Y addressing the memory is made to function as a 4 by 1 device. During read the X and Y select lines are energized so that the output from one of the storage capacitors C_1 appears on the data line. For writing the required cell is again addressed and the write line is energized so as to feed the information on the data line into the memory capacitor. For refreshing, a read cycle is followed immediately by a write. The Y select is now inoperative so that refreshing occurs a completed row at a time. Following the read command the charge on C_1 is temporarily stored in C_2 and this is then fed back onto C_1 on the refresh (write) instruction.

5.6.2 *ROM organization*

The read only memory is organized in a similar format to a RAM with an array of memory cells, decoded inputs for X and Y addressing and an output data amplifier. Such an arrangement is shown in Fig. 5.28. However ROMs are also used for special applications, such as character generation, when a less conventional organization is preferred. A 'tree' arrangement is

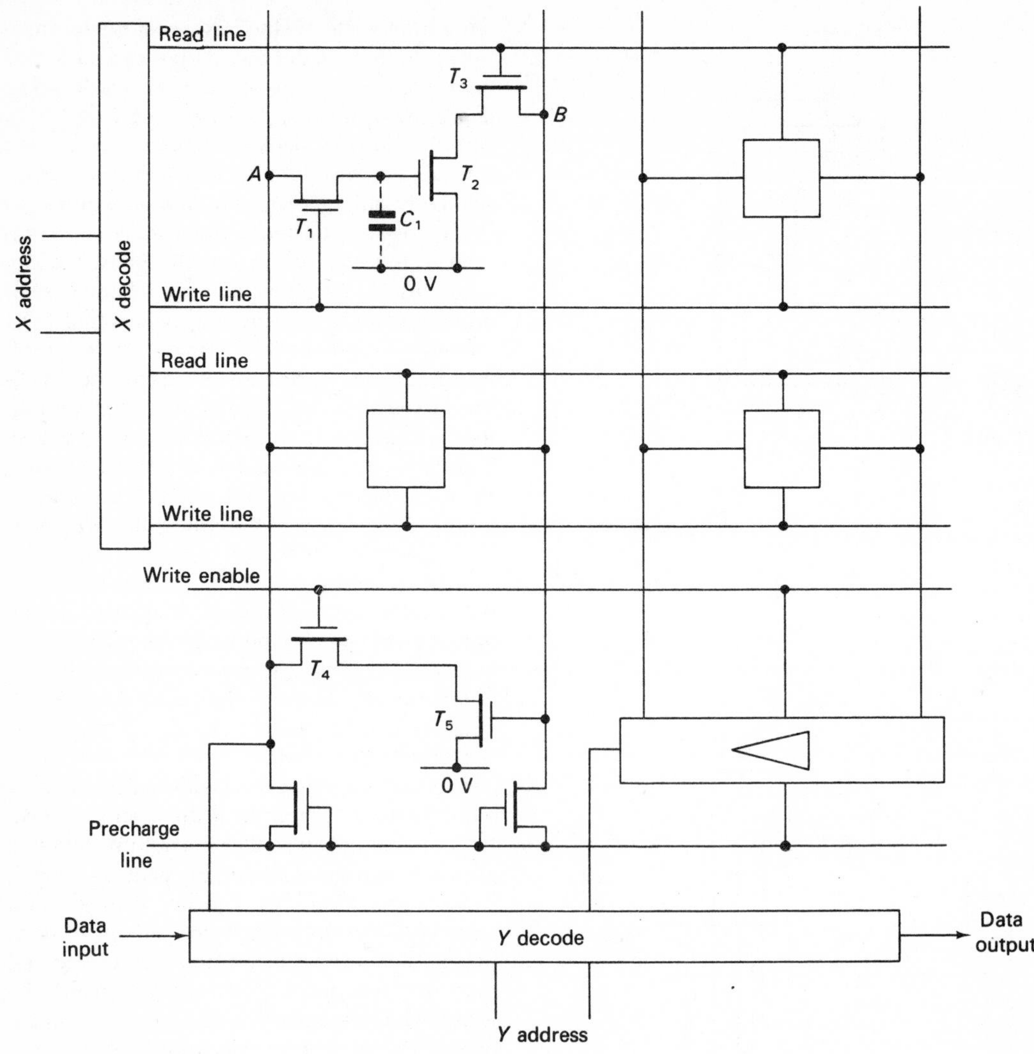

Fig. 5.25. Three transistor dynamic 2 by 2 RAM.

shown in Fig. 5.29. T_1 acts as a low valued resistor so that output nodes may be rapidly charged, giving this arrangement a high operating speed. Clearly the output is a logic 1 unless all transistors in any vertical column of the tree are simultaneously addressed.

Although a ROM generally need only have a read mechanism a PROM or reprogrammable ROM must be capable of write operations as well. Fig. 5.30 shows the organization of a 265 by 8 FAMOS memory. Word addressing occurs via a decoder with eight input lines and the eight bit lines are all brought out. For programming (write) the memory is operated essentially in a read mode. However the data which is to be written into each word is applied at the data bit output lines, and the words then selected by high voltage pulses.

5.6.3 *CAM organization*

As explained earlier a CAM may be operated as a RAM but it has the additional capability of searching through its memory, with a word held in the search register, and indicating which locations match the word. Search occurs a word at a time, with all bits in parallel, and the degree of match in any location may be obtained if required. It is often also possible to mask out certain bits of a word and only search for unmasked bits. For example, if a memory contains codes representing the name, age and height of a group of people then we may mask out the name and height and search through the memory and output the name and height of everyone who is, say, thirty years old.

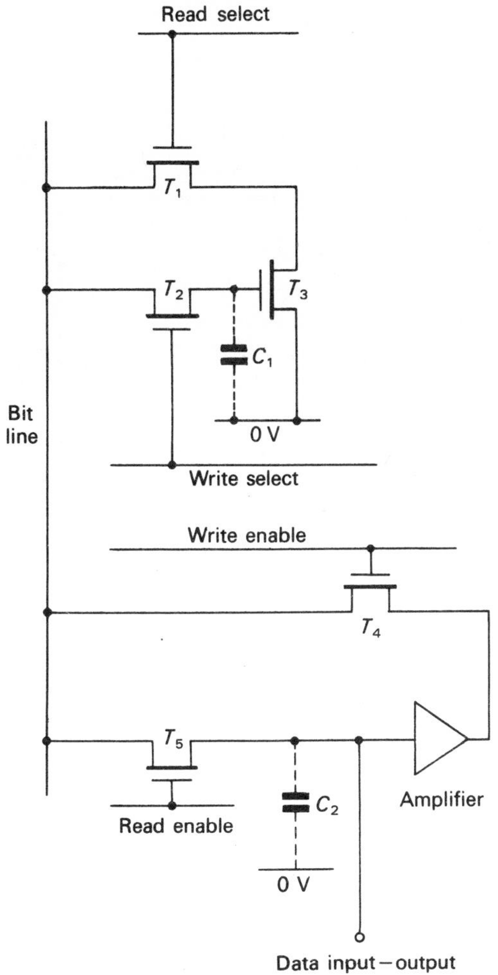

Fig. 5.26. Arrangement for a three transistor common bit line cell.

Fig. 5.31 (*a*) shows the arrangement of a CAM memory and its operation is illustrated by Fig. 5.31 (*b*). Supposing that the search word is 1100 0101 0011 (twelve bits long). If we are only interested in the last four bits then the first eight may be masked out by an AND operation with the word 0000 0000 1111. Now only the last four bits of the search register and each word in memory are operative. The whole of the search operation is dictated by a control system which has in it a word select register. This enables only certain words to be searched. For instance in the example words 1 and 8 are not searched so that even though their last four bits match that of the search register, a negative search result is output. In the example the search result register contains a match for words 3 and 7 only. It is now possible to output the whole of the content of these words from the memory.

It will be seen from Fig. 5.31 (*a*) that a large part of the CAM contains items such as the memory matrix, address decoder, amplifiers and drivers and data input–output lines. These are identical to those used in a RAM memory.

5.7 Memory applications

Three types of memories have been described in this chapter, i.e. RAM, ROM and CAM. The CAM is not at present widely used. One of the reasons for this is that it was relatively expensive to fabricate prior to the introduction of dense silicon technologies and even now it represents the highest priced memory. It has however, a unique role to play in data storage and retrieval systems and there is little doubt that its popularity will grow steadily in this field.

The present section describes a few typical examples of ROMs and RAMs. The applications are very general and no attempt has been made to differentiate between memory types such as PROMs or reprogrammable ROMs.

5.7.1 *RAM applications*

The random access memory is used as a storage system in a myriad of different applications. Even within a single system, such as a computer, many different size and types of memories may be found. Fig. 5.32 indicates this division for a large computer system. The scratch pad memory is a super fast, relatively

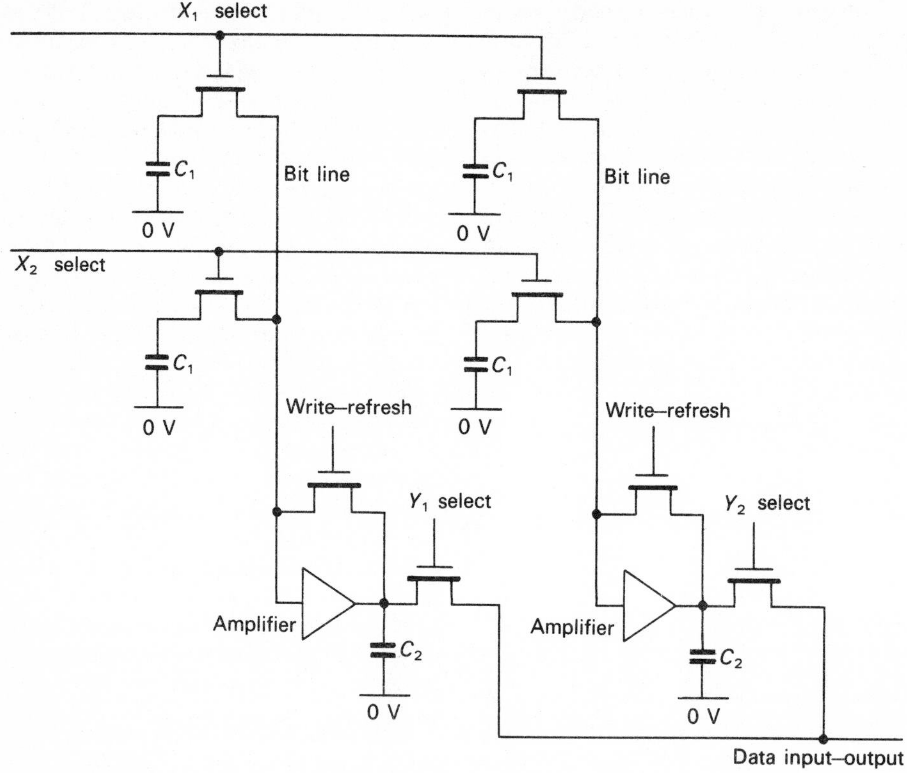

Fig. 5.27. Single transistor 4 by 1 RAM.

small system which acts primarily as a temporary store for use with fast arithmetic and control logic in computers. It is concerned only with the data of immediate interest to the processor. The caché memory is a larger memory which may also be slower in operation than the scratch pad memory. It acts as a buffer between the main store and the central processor unit (CPU) of the computer. The most frequently addressed information is transferred from the main store to the caché at regular intervals so that the CPU may access the caché to obtain the required data at a faster rate. Clearly as the program proceeds different blocks of data are accessed at varying frequencies by the processor so that the information held in caché is constantly changing.

The next memory, in order of decreasing speed, is the buffer used to interface the relatively slow peripheral devices to the computer. The main memory store forms the heart of a modern computer. It is a compromise in size and speed between the other memories and at the present time it forms the battleground between semiconductor and magnetic (core) devices. Depending on the main memory size it may interface with the bulk system either directly or via a backing store. The *bulk memory* is normally sequential in nature and currently almost exclusively magnetic (disc and tape) in construction. Although semiconductors, in the guise of technologies such as charge coupled devices, are attempting to enter this field their success is at present uncertain.

An interesting application of RAMs, and one which is likely to be more widely used in future, is their replacement of shift registers. Fig. 5.33 (*a*) shows a typical recirculating shift register in which the output sequence is 1 – 0 – 0 – 1. This simple four bit register may be replaced by the RAM–counter arrangement of Fig. 5.33 (*b*). *A* and *B* represent the RAM address lines so that information may be written into any bit as required. The counter cycles through the entire memory in sequence and produces the one bit output at C. With the RAM programmed as shown in the table of Fig. 5.33 (c) this system is equivalent to the shift register illustrated earlier. Even for a large system the counter is relatively small since n bits can address 2^n bits of RAM. It is possible to increase the speed of a RAM system by multiplexing, as for a shift register. This is shown in Fig. 5.33 (d) for two RAMs. The operation of gates 1 and 2 is controlled by the flip-flop such that during the first clock pulse, counter 1 is operated and the output from RAM 1 is sampled, whilst during the next clock pulse, counter 2 and RAM 2 are involved. The memories are now operating at half the clock speed.

5.7.2 *ROM applications*

A read only memory may be organized in any way and programmed with any desired bit pattern. Consequently it is basically an input–output converter since the input acts as an address

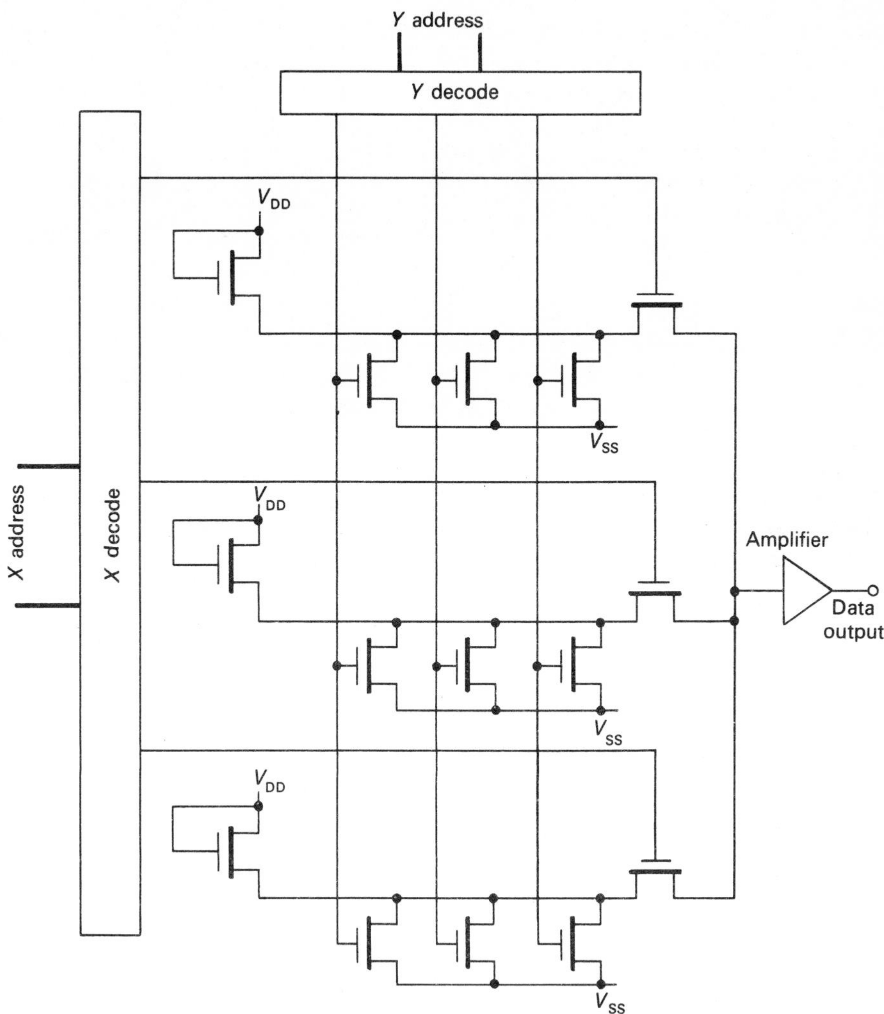

Fig. 5.28. A 9 by 1 ROM.

and the output is the data stored in that word. This basic principle has, however, many different uses. Fig. 5.34 (a) shows a simple 4 word by 2 bit ROM and 5.34 (b) gives the truth table of how it has been programmed. For the present output line *D* has been ignored. Examination of this table will show that *C* is the product of *A* and *B* so that the ROM is acting as a multiplier. Furthermore since no shift and add operations are involved, as for a conventional multiplier, the overall speed is very fast being limited only by the memory access time. Clearly the inputs *A* and *B* may have any number of bits. If these are equal to *X* and *Y* respectively then the ROM size required is $2^{(X + Y)}$ words by $(X + Y)$ bits. For large numbers this can prove to be uneconomical. For example $X + Y = 12$ would need a memory with 4096 words each of 12 bits. It now becomes more attractive to split the system into a number of smaller ROMs. For instance if:

$$X = a + b$$

$$Y = c + d$$

Then

$$X \,.\, Y = (a + b) \,.\, (c + d)$$
$$= a \,.\, c + b \,.\, c + a \,.\, d + b \,.\, d$$

This last relationship may be obtained by using 4 ROMs and several adders.

The ROM of Fig. 5.34 (a) may have any desired code programmed into it. Therefore if one wished to convert the code given by *A B* in the table of Fig. 5.34 (c) to that shown by *C D*, it is a simple matter of programming the four memory words (of 2 bits each) with the desired outputs. Code conversion is also frequently associated with character generation where a coded input will produce, from the ROM, a bit pattern corresponding to the actual shape of the character. Such an arrangement is shown in Fig. 5.35 for a 7 by 5 dot matrix alphanumeric character generator. The input address lines select the location of a word in the ROM. This produces a signal on the seven output lines corresponding to the first column of the desired character. By sequentially running through the column select lines the data for the next six columns is also obtainable as output from the ROM.

It is possible to have several entirely different

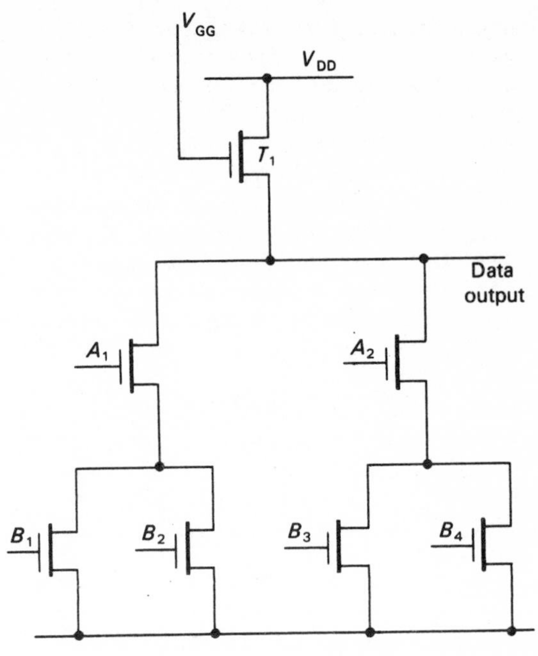

Fig. 5.29. Tree organization of a ROM.

types of codes simultaneously in a ROM. For instance both the codes illustrated in Fig. 5.36 (b) may be stored in an 8 word by 2 bit ROM. There are now three address lines and the third input is used to select either the first four, or last four words of the ROM in which the two different sets of output data are stored. This is illustrated in Fig. 5.36 (a). Since the input–output codes are entirely programmable, code conversion is often used for emulation in which the *software* of one range of computers is made to look like that of another.

ROMs are also commonly used to generate special trigonometrical functions using 'look up' (stored data) techniques. However, ROM sizes may now be large unless splitting techniques are used as before. For example suppose that it is desired to generate the sine function within the range of zero to 90 degrees. If one requires a resolution of 1 per cent per degree and 16 bit accuracy (i.e. 1 part in 65536) then it is necessary to have a ROM which is 9000 words by 16 bits. However, let $X = x_1 + x_2$ where x_1 is in integer degrees and x_2 is a fraction so that x_1 varies from 1 to 90 and x_2 in 1 per cent steps. Then:

$$\begin{aligned}\sin X &= \sin(x_1 + x_2)\\ &= \sin x_1 \cos x_2\\ &\quad + \cos x_1 \sin x_2\end{aligned}$$

Therefore 4 ROMs are required, each of 128 words by 16 bits, along with several multipliers, which may themselves be ROMs.

A further ROM application is that of microprogramming, illustrated in Fig. 5.37. This was originally developed for use in computers but is now widely used in control and instrumentation. It consists of storing a sequence of *instructions* (the microprogram) on a ROM which may then be called up when required. In Fig. 5.37 the arrangement is such that the initial input to the instruction register starts the ROM off at the selected location. The output of the ROM is in two parts. The first represents the desired control data to the rest of the system. The second output modifies the ROM address so that the output from the ROM is now taken from a new location representing the next instruction. This sequence is repeated until the end of the microinstruction is reached when the signal resets the instruction register and waits for the next input.

5.8. Memory selection

With such a large number of technologies to choose from the selection of the optimum memory for an application is often a difficult process. Generally there are many different grades of memories in any technology so that by trading off one parameter against another it is possible to obtain the type of device required. Due to the many different considerations which need to be taken into account in selecting a memory, this section can only introduce this complex subject.

Of all the parameters which have to be considered the organization of the memory is often fixed by system requirements. However even here there is often a question of choice. For instance a 512 word by 4 bit system may be built from a single 2K memory chip or from two 512 word by 2 bit chips, or from two 256 word by 4 bit devices, and so on. Other factors, such as speed and cost need to be considered before a final choice can be made. Perhaps the next most important technical parameters are memory speed followed by power dissipation. As memory sizes get larger it is essential that their dissipation be kept low, not only to reduce power supply requirements but also to keep package dissipations down and enable *chip densities* to

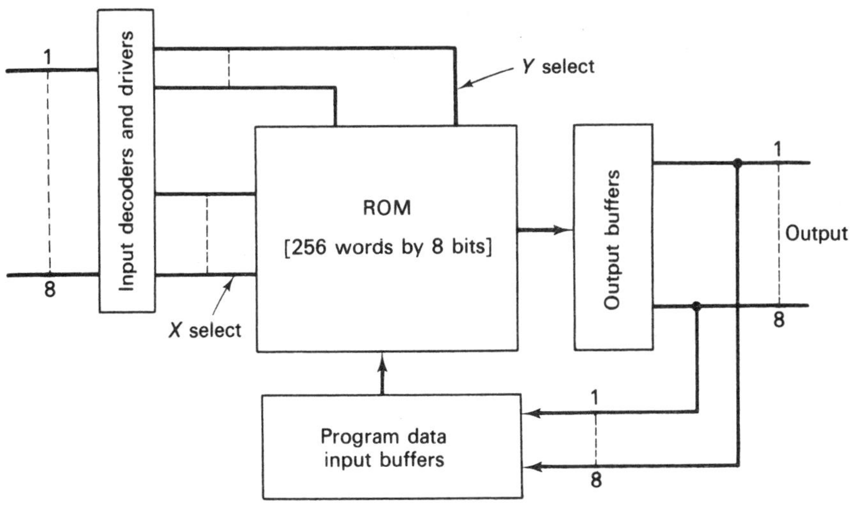

Fig. 5.30. FAMOS memory organization.

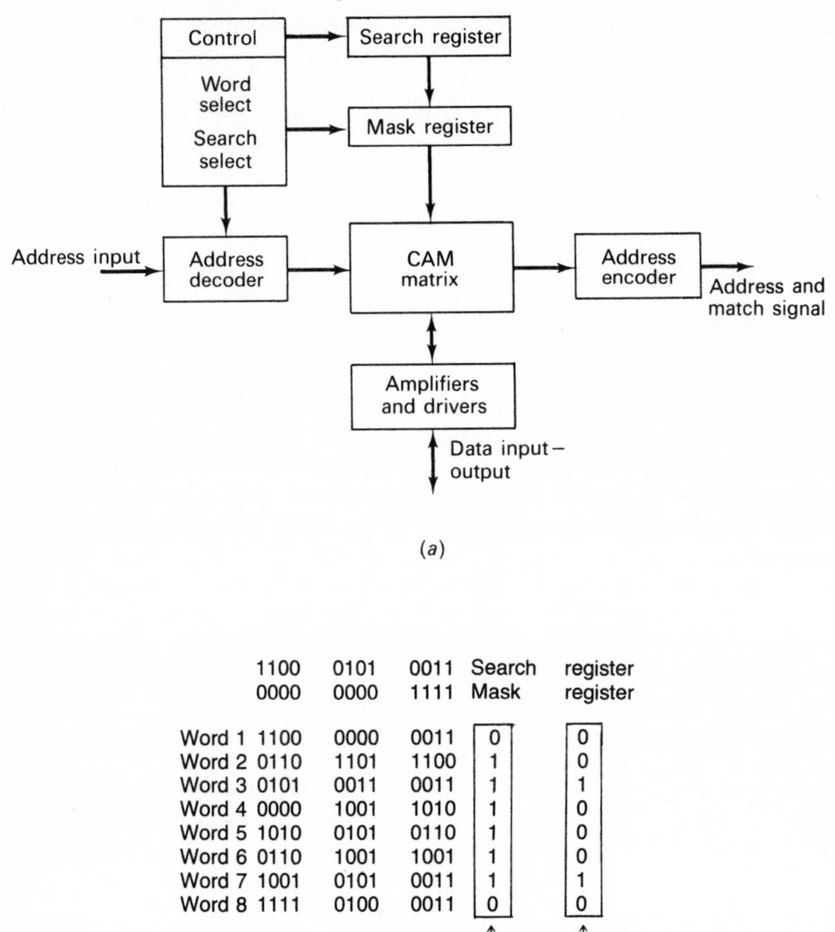

Fig. 5.31. CAM organization; (*a*) memory arrangement, (*b*) operational example.

be increased. In this respect area per bit is also important since the smaller the cell size the greater the density for any chip area. Other parameters may be more important in certain applications. For instance it may be essential to have a memory with non-destructive readout or non-volatility, and military or aerospace systems would require very high reliability. A further important consideration is the ease of usage, that is the amount of overhead circuitry in the form of level converters, refresh amplifiers and so on, required for the system. Furthermore, price is of prime importance in the bulk of commercial and industrial applications.

For a random access memory it is often necessary to choose between a bipolar and a unipolar technology. Both memory devices have approximately doubled in size every year in recent years although unipolar has maintained a two year lead and is likely to do so in the near future. The unipolar cell is generally smaller in size and is a simpler process, so that it is also lower in price. Any one technology is capable of a span of operating speeds, as shown in Fig. 5.38. PMOS represents an established low cost technology which is being overtaken by NMOS. This is capable of being made with larger bit capacities than PMOS so that the costs per bit tend to be lower. Bipolar devices will continue to be used in applications which demand speeds in excess of those available from NMOS. The growth of CMOS will be primarily in those areas in which power dissipation or noise immunity are important. Silicon on sapphire is expected to gain in popularity and will strongly challenge bipolar memories in speed.

At present ECL bipolar memories are unrivalled in speed. These are followed by the Schottky TTL process and then other systems, many of which are designed for high densities, such as Isoplanar, CDI, VATE, VIP and Injection Logic. Unipolar memories are available as static or dynamic cells. The latter generally has smaller cell sizes and lower power dissipation, but is not as easy to use. PMOS silicon gate is presently the most widely used memory technology although the higher carrier mobility of NMOS, which gives it higher speeds and lower threshold voltages, is enabling it to make inroads into the PMOS market. For very low stand-by power applications CMOS is presently unrivalled being capable of values less than 0.1 microwatts per bit.

The choice of a read only memory is very often determined by the application. Mask pro-

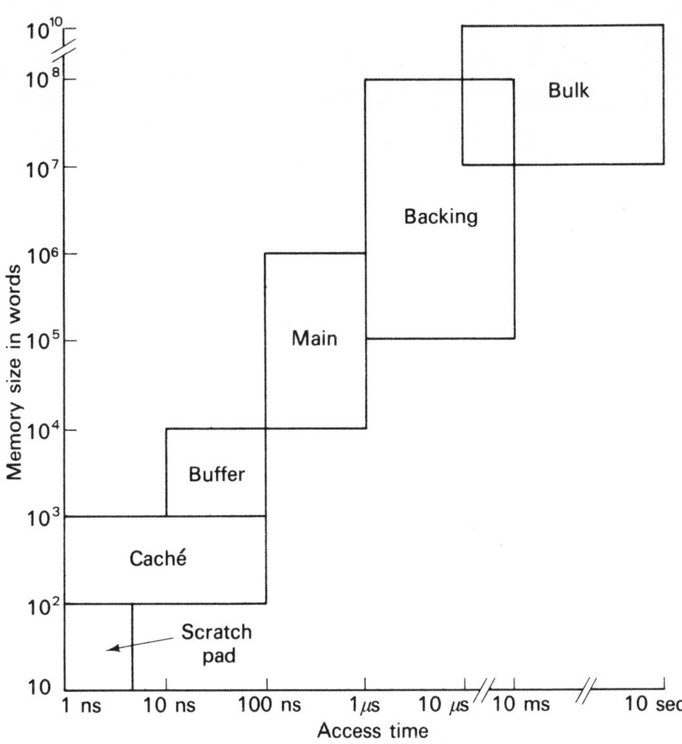

Fig. 5.32. Computer memory hierarchy.

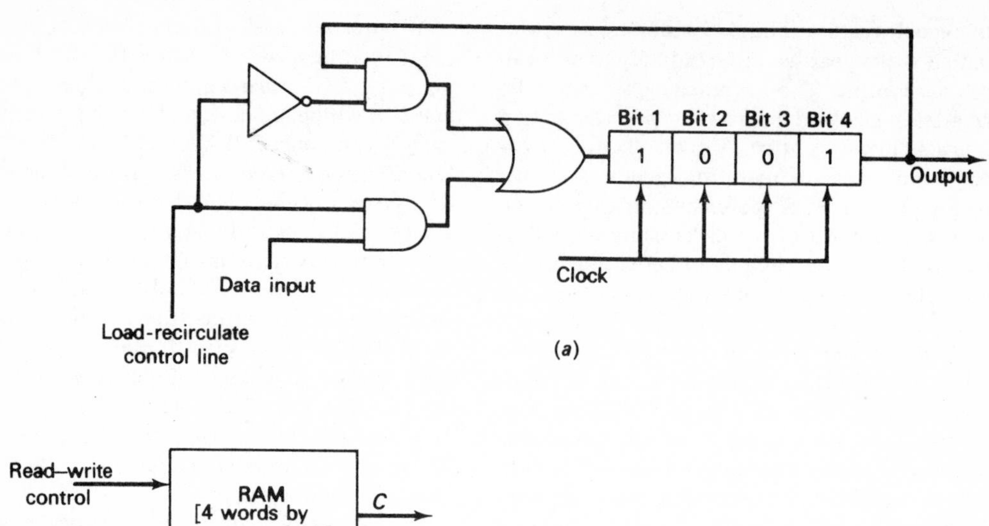

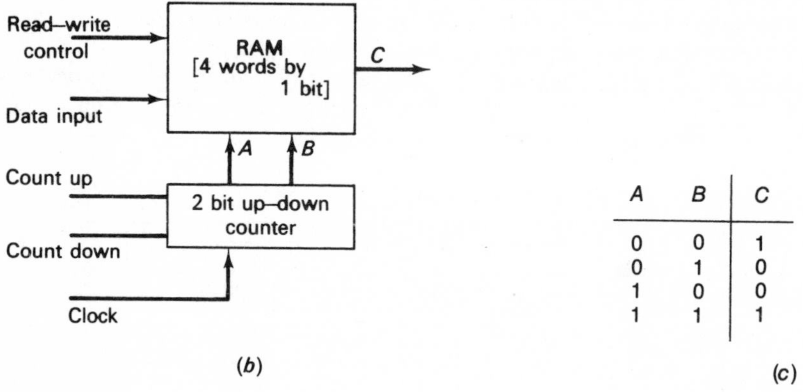

A	B	C
0	0	1
0	1	0
1	0	0
1	1	1

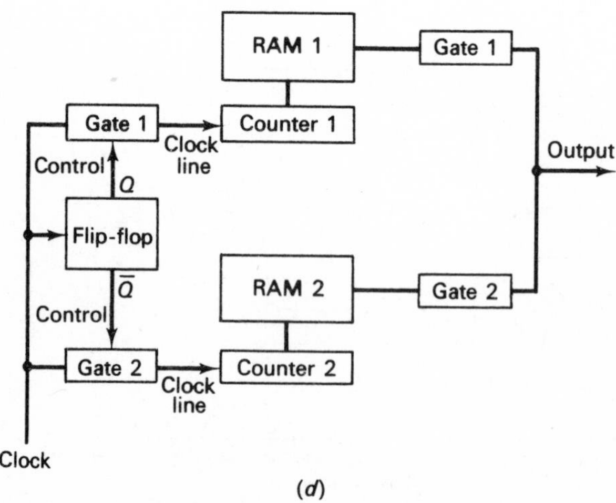

Fig. 5.33. Shift register simulation using a RAM; (a) circulating shift register, (b) RAM equivalent, (c) RAM coding, (d) multiplexed memories.

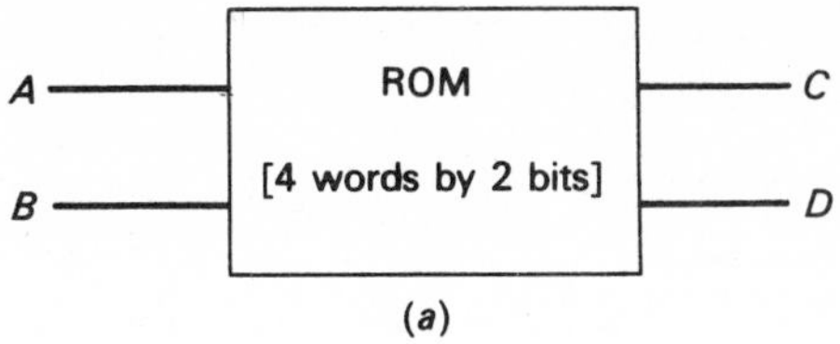

(a)

A	B	C
0	0	0
0	1	0
1	0	0
1	1	1

(b)

A	B	C	D
0	0	1	1
0	1	0	1
1	0	0	0
1	1	1	0

(c)

Fig. 5.34. ROM applications: (a) 4 by 2 ROM, (b) multiplier, (c) code conversion.

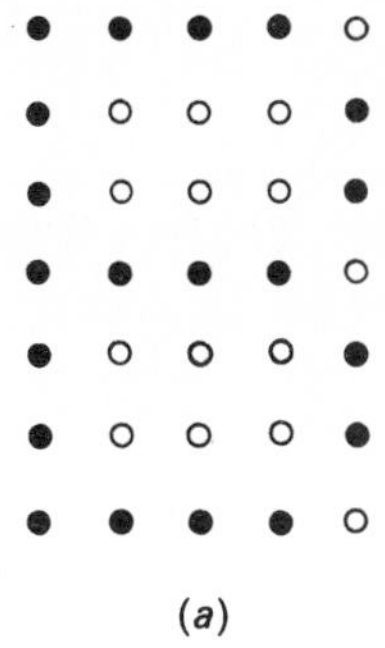

(a)

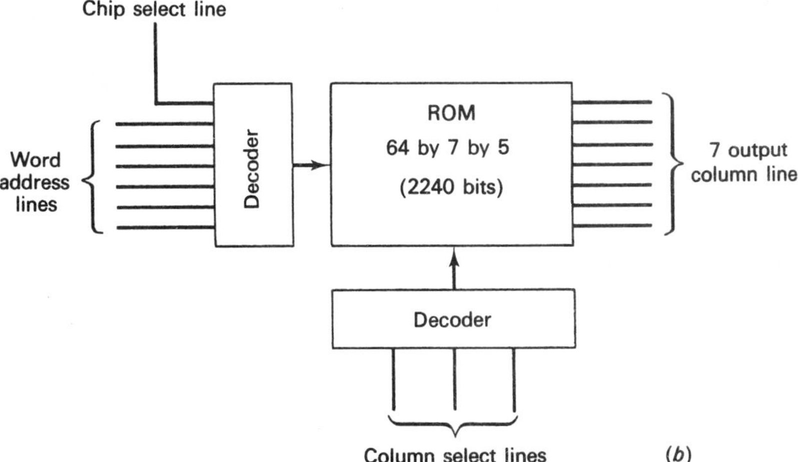

(b)

Fig. 5.35. A 7 by 5 dot matrix character generator; (a) dot matrix, (b) memory organization.

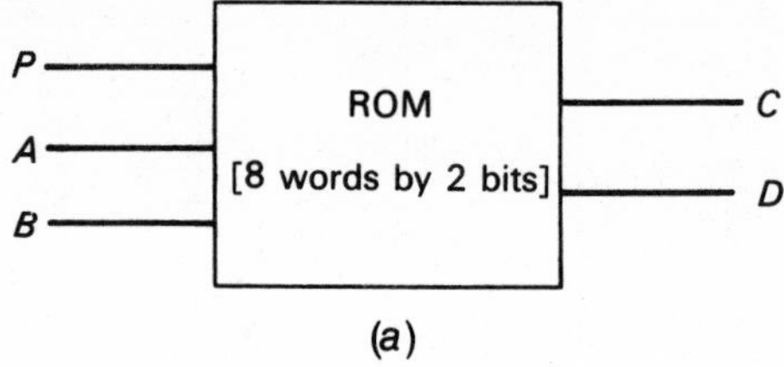

(a)

P	A	B	C	D
0	0	0	0	0
0	0	1	0	0
0	1	0	0	0
0	1	1	1	0
1	0	0	1	1
1	0	1	0	1
1	1	0	0	0
1	1	1	1	0

(b)

Fig. 5.36. Storing two codes in a common ROM; (a) memory arrangement, (b) coding table.

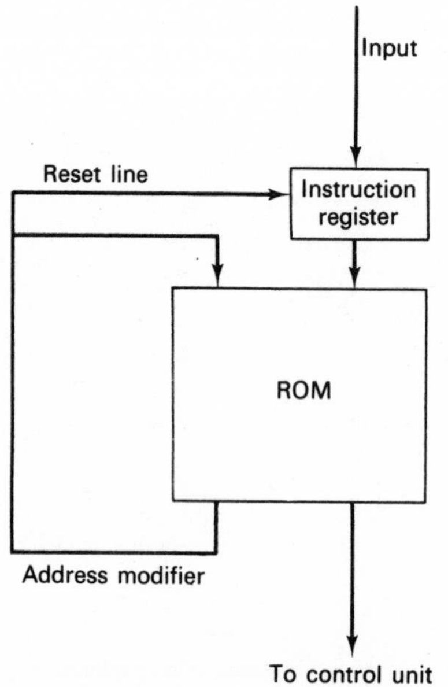

Fig. 5.37. Typical microprogram arrangement.

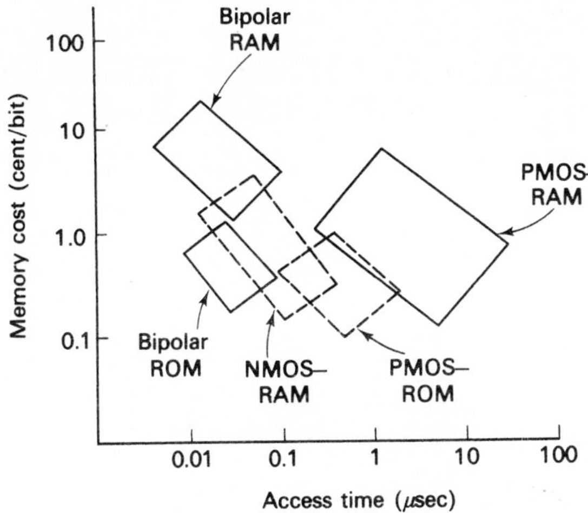

Fig. 5.38. Variation of memory cost with speed.

grammed memories are the cheapest in large quantities and unipolar devices are capable of bigger sizes and lower costs than bipolar, although the operating speeds are now also less. Field programmable ROMs are economic in low volumes since the masking charges are avoided, but their unit prices are generally two or three times that of mask programmed devices. They are also more complex and therefore not available in very large sizes. Where reprogramming is required it is possible to choose between FAMOS, MNOS and chalcogenide memories. The first device is presently the most popular but it is only capable of a limited number of erase operations. MNOS does not suffer from this problem but on the other hand it can only be accessed a limited number of times before requiring refreshing. Chalcogenide memories have relatively large cell sizes and are expensive. Hence they are not widely used at the present time.

6. Universal logic elements

6.1 Introduction

Universal logic elements endeavour to maintain the advantages of custom circuits while minimizing some of their disadvantages. They consist of large scale integrated components which are produced as standard devices but are so designed that they can be made to fit the user's special application with very little modification. This modification usually consists of a single custom design for the metallization mask used on some form of standard logic array. In this chapter these types of universal logic systems will be examined in greater detail. They are the uncommitted logic array, the read only memory, the programmable logic array and, perhaps the most important of them all, the microprocessor.

6.2 Uncommitted logic array

The uncommitted logic array (ULA) basically consists of an array of devices or gates in which the silicon wafers are processed up to the stage just before the final metallization, and then stored. When a customer application arises the metallization mask, which defines the interconnection pattern on the previously processed wafers, is designed and used to produce the custom chips. Since each new application requires only a single layer design both the cost and the time scales involved are much lower than in a more traditional custom design circuit.

There are several different types of ULAs on the market and no doubt the future will see the introduction of many more. Fig. 6.1 shows the structure of one form of cell which is made up of individual devices. These can be interconnected together in order to give a variety of digital and linear circuits, both types being on the same chip if required.

Fig 6.2 shows alternative forms of ULAs which are primarily gate arrays. Two levels of metallization are used. The first layer determines the gate structures as shown in Fig. 6.2, and following this the wafers are stored. When a custom application arises the wafers are recalled and the final specially designed metallization layer used to form the unique interconnection pattern for the customer. The advantage of this approach is that greater design automation can be used since the final layer requires cell routing only. Furthermore each cell can be designed for optimum performance as a gate. However this circuit is now capable of digital applications only. Fig. 6.2 (*a*)

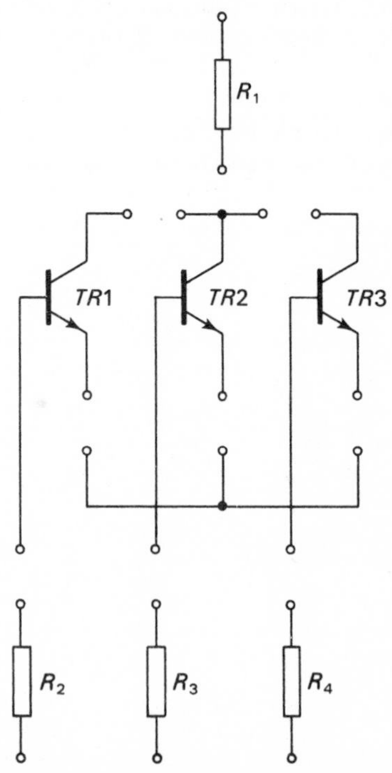

Fig. 6.1. An uncommitted logic array cell.

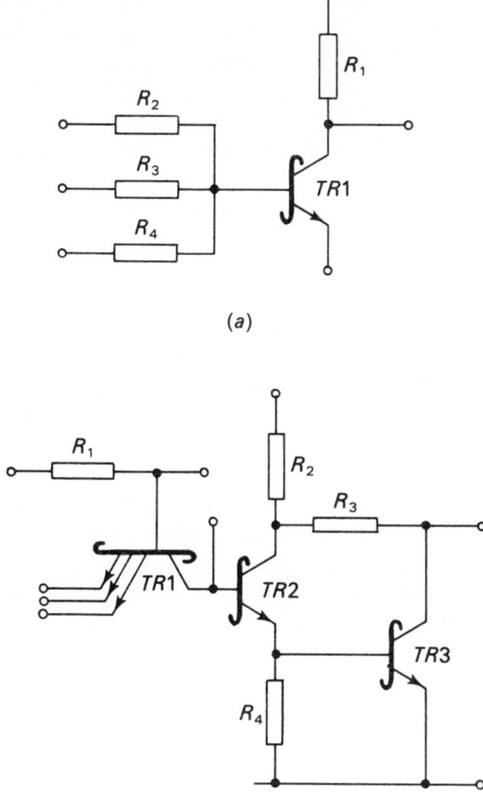

Fig. 6.2. Alternative forms of uncommitted logic arrays using gates. (*a*) RTL–Schottky, (*b*) TTL–Schottky.

shows a resistor transistor logic gate using Schottky transistors. It is capable of propagation speeds in the region of 25 nanoseconds with gate dissipation of about 3 milliwatts. For higher speeds TTL must be used as in Fig. 6.2 (*b*) which has a gate propagation delay of about 10 nanoseconds and a dissipation of 15 milliwatts. Alternative designs, using totem-pole output stages have delay times as low as 5 nanoseconds at the above dissipation value.

The advantages of the various types of ULAs in providing quick and cheap custom designs for low volume applications are evident from the above discussions. However these arrays are relatively inefficient in terms of silicon area used compared to custom designs since cells are generally not optimally placed for any particular application and there is often some redundancy of functions. Therefore the chip size is larger than that obtained from a more traditional custom design, and in large volume applications the cost is also likely to be greater for ULAs.

6.3 Read only memory

The construction and a few applications of read only memories were described in chapter 5. The present chapter will describe ways in which ROMs can be used to implement random logic. Although such a technique has several advantages over the more conventional logic systems it has not gained as widespread a use as ROMs for code conversion or character generation. Random logic circuits can be divided into two types, combinational and sequential. A combinational circuit has no store function so that the output is directly related to the inputs at the instance of time under consideration. The output of a sequential circuit, on the other hand, is determined by its present inputs as well as the previous state of the outputs from the circuit. The implementation of combinational and sequential circuits using ROMs is described in the following sections.

6.3.1 *Combinational logic*

The use of a ROM for combinational logic is best illustrated by an example. Consider a diode matrix ROM, of the type described with reference to Fig. 5.1. A 16 by 4 bit device of this type is shown in Fig. 6.3 in which the diodes at the required intersections of a matrix are represented by dots. This ROM has a four bit decoded input, giving a total of sixteen words, and four bits of output. Ignoring the dots on the matrix for the present let us suppose we wish the outputs to produce the following four equations:

$$O_1 = I_1 + I_2 + I_3 + I_4 \quad (6.1)$$

$$O_2 = I_1 \,.\, I_2 \,.\, I_3 \,.\, I_4 \quad (6.2)$$

$$O_3 = I_1 \oplus I_2 \oplus I_3 \oplus I_4 \quad (6.3)$$

$$O_4 = \overline{I_3} \,.\, I_4 + I_1 \,.\, \overline{I_2} + I_1 \,.\, \overline{I_3} \quad (6.4)$$

Output O_1 is clearly obtained by a diode at the intersection of this bit line and every word line save $\overline{I_1} . \overline{I_2} . \overline{I_3} . \overline{I_4}$. This is shown in Fig. 6.3 and represents a logical OR output. Similarly O_2 is obtained by a single diode at the intersection of the O_2 and $I_1 . I_2 . I_3 . I_4$ lines to give a logical AND output. Bit O_3 represents an EXCLUSIVE–OR function. It is obtained in Fig. 6.3 by a diode on the word lines containing a single '1'.

So far the implementation of the logic has been fairly easy and has not required any of the techniques discussed in chapter 3. However, Karnaugh mapping is very useful when the system becomes large and random. The maps for the four equations are shown in Fig. 6.4. Once these maps have been constructed the ROM programming is straightforward. For instance it would be difficult to go from equation (6.4) to the ROM circuit shown in Fig. 6.3. However, Fig. 6.4 (*d*) gives the Karnaugh map for equation (6.4). This shows clearly the locations which require diodes in the ROM implementation.

From the above discussions it is clear that since a ROM can produce a coded output of almost any combination of inputs it is a very useful element for implementing random logic. In fact the attraction of a ROM over conventional circuits becomes greater as the complexity of the circuit is increased. There are also several possible variants in implementing logic systems. For instance Fig. 6.5 shows a 64 by 1 bit ROM in which two of the inputs C_1 and C_2 are used as control lines. These lines can be decoded to give four blocks of 16 bits. Each of these blocks is capable of giving the patterns illustrated for O_1, O_2, O_3 and O_4 in Fig. 6.3. Therefore the single output can now be switched to give four different functions at different instances in time.

A ROM is generally much faster than a conventional logic system when complexities are high. It is now no longer necessary to total the individual gate propagation delays and the maximum delay of the ROM is equal to its access time.

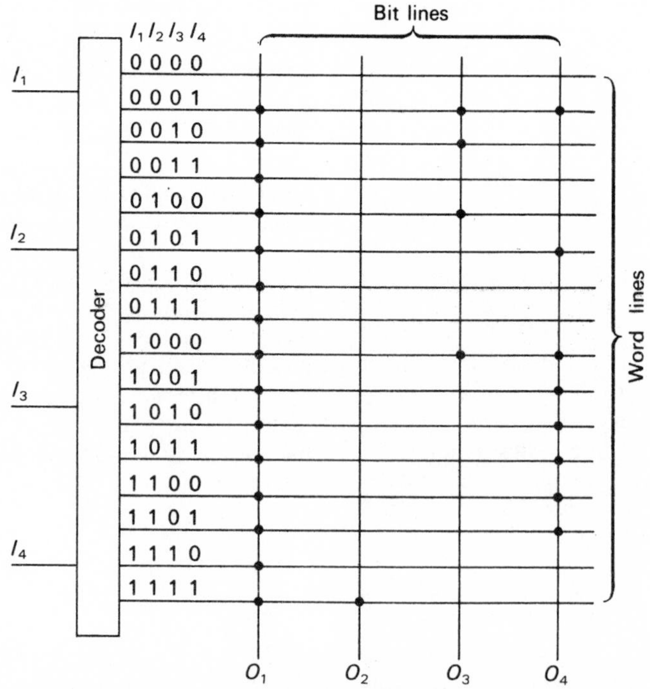

Fig. 6.3. A 16 by 4 bit ROM implementation of random logic.

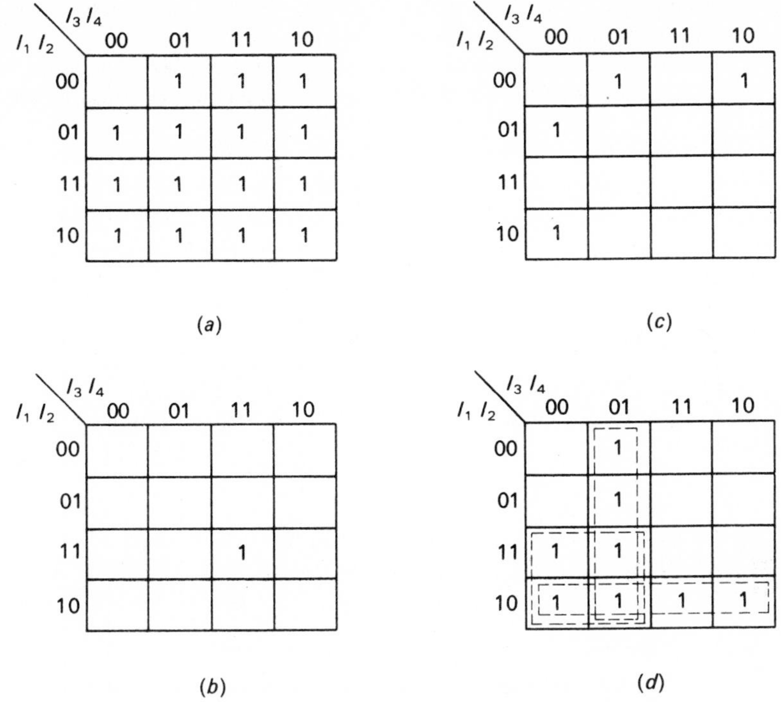

Fig. 6.4. Karnaugh maps for the ROM shown in Fig. 6.3; (a) $I_1 + I_2 + I_3 + I_4$ (b) $I_1 . I_2 . I_3 . I_4$ (c) $I_1 \oplus I_2 \oplus I_3 \oplus I_4$ (d) $\overline{I_3} . I_4 + I_1 . \overline{I_2} + I_1 . \overline{I_3}$

6.3.2 *Sequential logic*

For sequential logic some form of memory is required so that the output is affected by the previously stored state of the ROM. This can be accomplished in several ways, two of which are illustrated in this section. Fig. 6.6 uses external flip-flops for storage. It shows the use of a ROM to obtain a four bit binary counter and uses a 16 by 4 bit ROM feeding four *D* type flip-flops. The operation of the circuit can be followed by its truth table. Assuming initially that all flip-flops are cleared the ROM address input is 0000 and the ROM is programmed to produce an output of 0001. During the first clock pulse this is read into the flip-flops and represents the output of the counter. The ROM input is now changed to 0001 and the ROM is programmed to give an output of 0010. At the next clock pulse this is read into the flip-flops and gives the third stage of the counter. The ROM address is now also 0010 and the output of this word is 0011. The rest of the operation can be followed from the truth table. Since the ROM can be programmed to produce any form of output for a given input

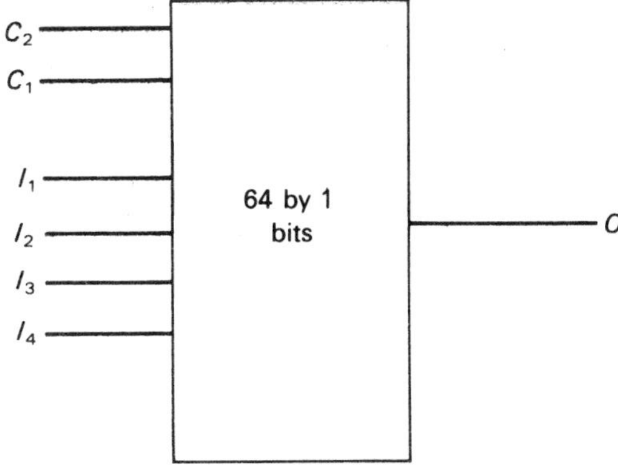

Fig. 6.5. A 64 by 1 bit ROM implementation of the logic illustrated in Fig. 6.4 with a single line output.

an identical arrangement can be used to give a *BCD* counter or a down counter. Alternatively by using two extra control bits, as was done in Fig. 6.5, it is possible to have a single counter which produces up–down counts in binary or in *BCD*.

The counter shown in Fig. 6.6 requires 16 words or locations for the 16 count operations.

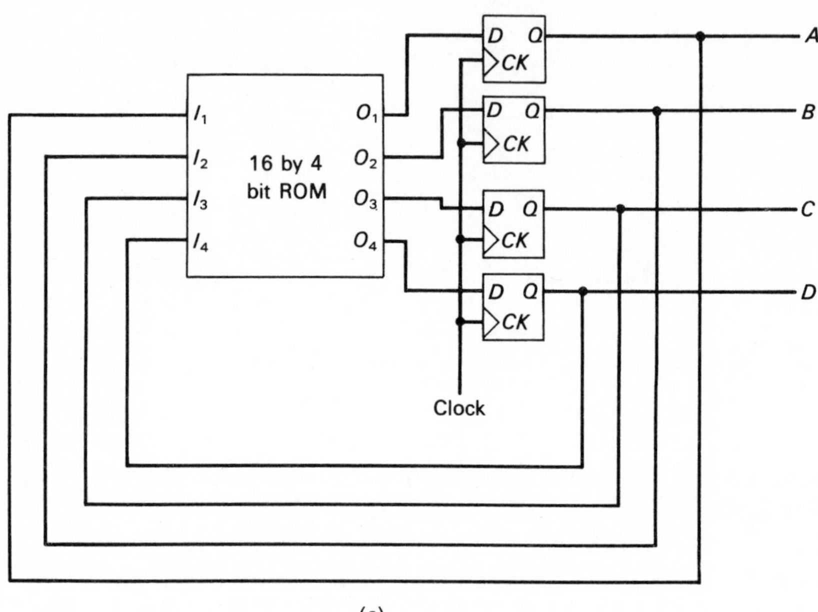

(*a*)

I_1	I_2	I_3	I_4	O_1	O_2	O_3	O_4	A	B	C	D		
0	0	0	0	0	0	0	1	0	0	0	0	←	Clock
0	0	0	1	0	0	1	0	0	0	0	1	←	
0	0	1	0	0	0	1	1	0	0	1	0	←	
0	0	1	1	0	1	0	0	0	0	1	1	←	
0	1	0	0	0	1	0	1	0	1	0	0	←	
0	1	0	1	0	1	1	0	0	1	0	1	←	
0	1	1	0	0	1	1	1	0	1	1	0	←	
0	1	1	1	1	0	0	0	0	1	1	1	←	
1	0	0	0	1	0	0	1	1	0	0	0	←	
1	0	0	1	1	0	1	0	1	0	0	1	←	
1	0	1	0	1	0	1	1	1	0	1	0	←	
1	0	1	1	1	1	0	0	1	0	1	1	←	
1	1	0	0	1	1	0	1	1	1	0	0	←	
1	1	0	1	1	1	1	0	1	1	0	1	←	
1	1	1	0	1	1	1	1	1	1	1	0	←	
1	1	1	1	0	0	0	0	1	1	1	1	←	
0	0	0	0	0	0	0	1	0	0	0	0	↓	Repeat

(*b*)

Fig. 6.6. A four bit synchronous counter using a ROM with external storage; (*a*) schematic, (*b*) truth table.

By doubling the size of the ROM it is possible to eliminate the use of the external flip-flops. This is shown in Fig. 6.7. An extra ROM input is required for the clock. Referring to the truth table assume that the clock is initially at 0 and the ROM address lines are also all at 0 giving an output from the ROM, which is also the counter output, of 0000. The clock now goes to 1 and addresses the ROM word 10000. This produces a ROM output of 0001 which is fed back to give a ROM address of 10001. If this memory word also contains the information 0001 then the counter is stable at this value. When the clock next changes to 0 the memory address is 00001 and the ROM output goes to 0010. This is fed back as a modified ROM address of 00010 which also contains the information 0010, so that the counter reading is unchanged. This sequence is repeated all the way through the various counts as illustrated in Fig. 6.7.

6.4 Programmable logic array

Although programmable logic arrays (PLA) have been available commercially for many years their popularity compared to read only memories, which they closely resemble, has been low. This is partly due to the fact that a PLA is best suited to random logic type of applications whereas the ROM has established itself as a memory element in applications such as computers. This very large potential market for ROMs has also encouraged many vendors to make and sell a variety of different ROM configurations so giving the user a much wider choice and boosting the sales of the memory.

In this section the characteristics of the PLA will first be described and compared with the more familiar ROM. This will then be followed by a description of PLA applications with emphasis on random logic uses.

6.4.1 *Construction of a PLA*

Fig. 6.8 (*a*) shows a simple 2 input 2 output ROM using a diode matrix of the type described in chapter 5. The input decoder gives the ROM four word lines so that with the two bit, or output, lines this is referred to as a 4 by 2 bit ROM. There are 8 cross over points between word and bit lines and at each point a diode may be connected or omitted as required by the program. Suppose the diodes are connected as indicated in Fig. 6.8 (*a*). Then as the word lines are addressed in turn, corresponding to the values of I_1 and I_2, the outputs on the bit lines are either zero volts or a positive voltage (logic 0 or logic 1) depending on the absence or presence of a programming diode. For instance for I_1 equal to 0 and I_2 equal to logic 1 the second word line is addressed, or raised to a logic 1. This has no effect on output line O_1 which is held at zero volts through resistor R_1. However diode D_2 conducts so that line O_2 is raised to a logic 1 giving an output O_1O_2 equal to 01. If I_1I_2 now changes to 10 then the third word line is addressed giving an output O_1O_2 equal to 11 since both their programming diodes D_3 and D_4 are present.

From the above description it is seen that the diodes cause the output lines to perform an OR function of the word lines since with the arrangement shown in Fig. 6.8 (*a*) line O_1 is a 1 when word lines 00 OR 10 are addressed and line O_2 is a 1 when word lines 01 OR 10 are addressed. It should also be noticed that the number of inputs is less than the number of word lines due to the presence of the input decoder. Fig. 6.8 (*b*) shows an alternative ROM matrix which performs an AND function in positive logic and in which the input decoder is absent. Input lines I_1 and I_2 are normally at zero volts so that all diodes connected to them are forward biased and the output lines are at logic 0. If I_2 goes to a positive voltage (logic 1) then D_2 and D_3 are reverse biased. Output O_2 therefore goes to a logic 1 but O_1 remains at logic 0 since D_1 is still forward biased. For O_1 to go to a logic 1 both I_1 AND I_2 must be at a logic 1. Therefore this is an AND ROM.

The AND and OR ROMs can be combined into a single device as shown in Fig. 6.8 (*c*). This is one form of a PLA. Partial decoding of inputs is illustrated since both the direct and inverse values are used. Therefore diodes in the above half of the matrix connect to one of these input lines. Clearly P_1 and P_2 represent AND terms of the inputs and they are referred to as product terms. These are then ORed in the second ROM to give outputs O_1 and O_2. The actual operation of the PLA will become clearer in the

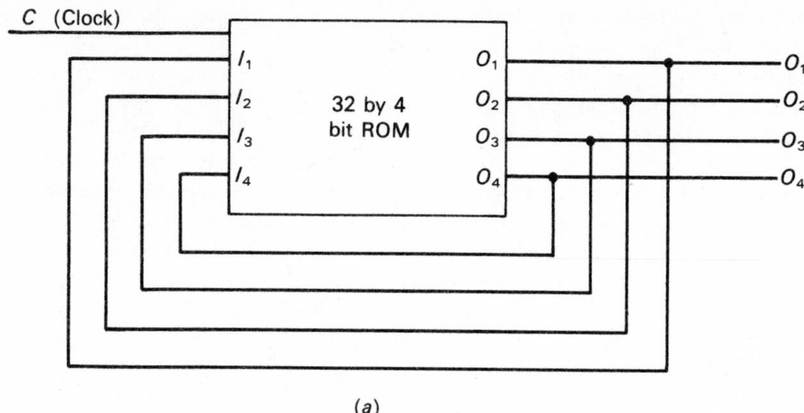

(a)

C	I_1	I_2	I_3	I_4	O_1	O_2	O_3	O_4
0	0	0	0	0	0	0	0	0
1	0	0	0	0	0	0	0	1
1	0	0	0	1	0	0	0	1
0	0	0	0	1	0	0	1	0
0	0	0	1	0	0	0	1	0
1	0	0	1	0	0	0	1	1
1	0	0	1	1	0	0	1	1
0	0	0	1	1	0	1	0	0
0	0	1	0	0	0	1	0	0
1	0	1	0	0	0	1	0	1
1	0	1	0	1	0	1	0	1
0	0	1	0	1	0	1	1	0
0	0	1	1	0	0	1	1	0
1	0	1	1	0	0	1	1	1
1	0	1	1	1	0	1	1	1
0	0	1	1	1	1	0	0	0
0	1	0	0	0	1	0	0	0
1	1	0	0	0	1	0	0	1
1	1	0	0	1	1	0	0	1
0	1	0	0	1	1	0	1	0
0	1	0	1	0	1	0	1	0
1	1	0	1	0	1	0	1	1
1	1	0	1	1	1	0	1	1
0	1	0	1	1	1	1	0	0
0	1	1	0	0	1	1	0	0
1	1	1	0	0	1	1	0	1
1	1	1	0	1	1	1	0	1
0	1	1	0	1	1	1	1	0
0	1	1	1	0	1	1	1	0
1	1	1	1	0	1	1	1	1
1	1	1	1	1	1	1	1	1
0	1	1	1	1	0	0	0	0
0	0	0	0	0	0	0	0	0 ↓ Repeat

(b)

Fig. 6.7. A four bit synchronous counter using a ROM with no external storage; (a) schematic, (b) truth table.

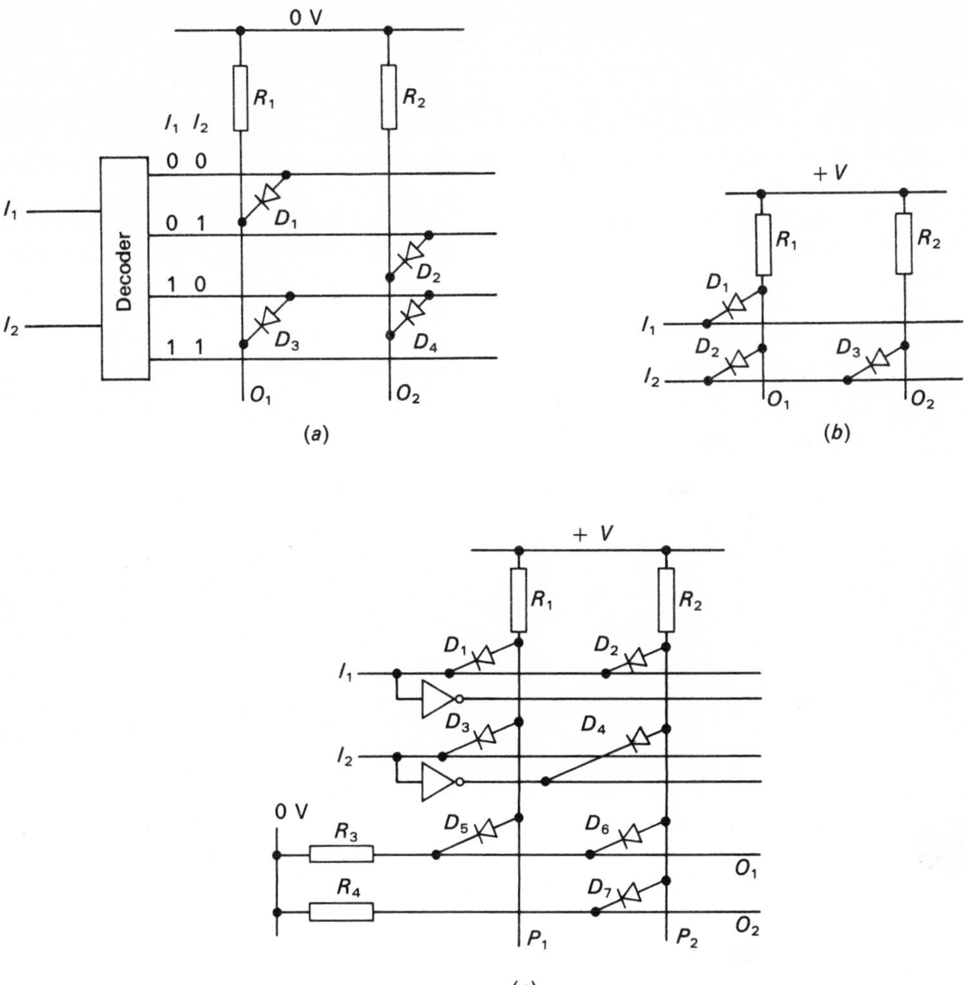

Fig. 6.8. Development of a PLA; (a) OR ROM, (b) AND ROM, (c) AND–OR ROM, also called a PLA.

next section when a few applications are described.

Fig. 6.9 illustrates more directly the differences between a ROM and a PLA. In a ROM all the inputs must generate a product term via the decoder, and this is the case whether these product terms, which are ROM words, are used or not. The product terms then go through an encoder, which is programmed by the absence or presence of a diode, to give outputs which are an OR function of selected product terms. In a PLA on the other hand there are two programmable encoders. The first produces the product terms, and the number of terms can be designed to be any selected value below a maximum equal to the number of equivalent ROM word lines. There is therefore no widespread redundancy in some applications as can happen in a ROM. The outputs are produced by an OR function of the product terms in the second programmable encoder. These differences between a ROM and a PLA will be illustrated further in the next section with reference to some applications.

Fig. 6.10 shows several representations of a PLA. As explained earlier it consists basically of an AND ROM which generates the product terms P_1 and P_2 from the inputs I_1 and I_2, and an OR ROM which produces outputs O_1 and O_2 from the product terms. The logic array of Fig. 6.10 (*a*) is closest in representation to the actual PLA layout in which the programming diodes are replaced by dots. The block diagram shows the functional operation much more clearly, whereas the gate logic diagram gives a more conventional representation. In this section the array logic representation will be used almost exclusively. It must be emphasized, however, that although only diode programming has been described for the actual arrays, most of the technologies described for ROMs in chapter 5 can also be used for PLAs. Therefore the present section is restricted to a functional rather than a technological description of the PLA.

The static form of PLA described so far is the most commonly available type and it can be used with external storage elements for sequential applications. Some PLAs also do not have the input inverters so that unless both the direct and inverse of the input is required this can

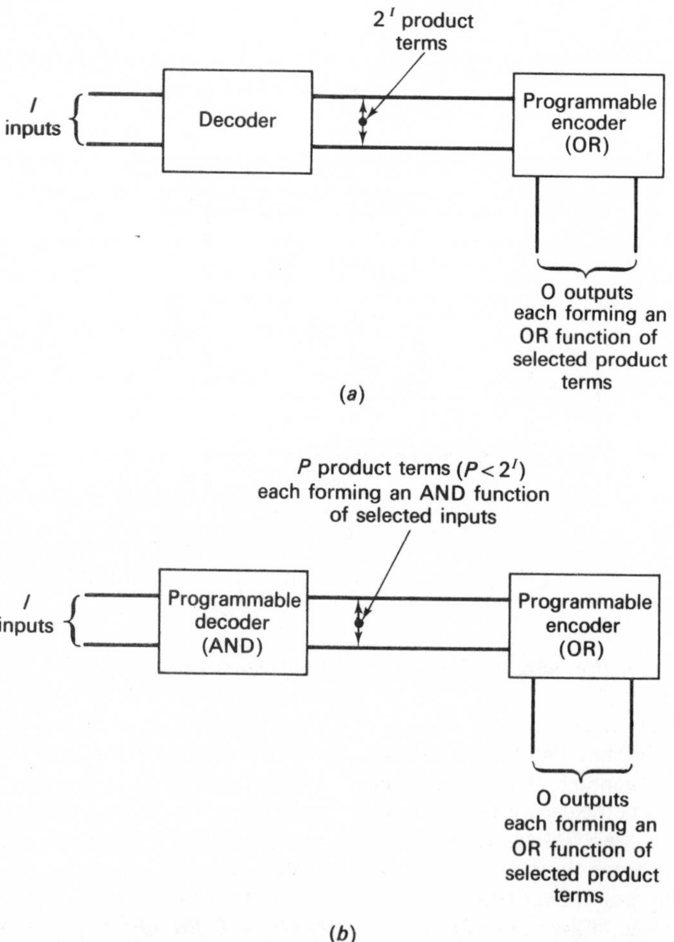

Fig. 6.9. Comparison between a ROM and a PLA; (*a*) ROM, (*b*) PLA.

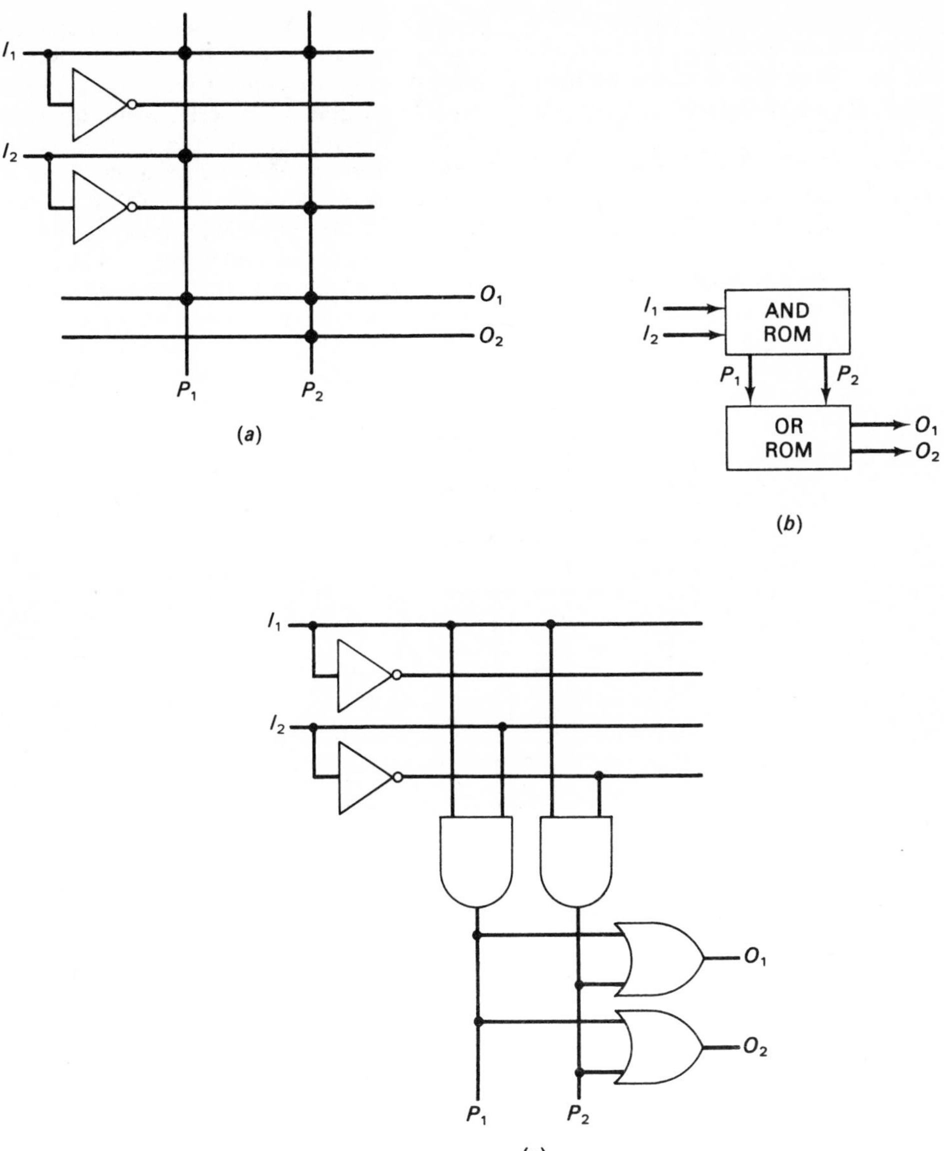

Fig. 6.10. PLA representations; (a) array logic, (b) block schematic, (c) gate logic.

result in savings. A further variation of commercially available PLAs include flip-flops on the silicon chip so that they are suitable for sequential applications. This is shown in Fig. 6.11. Both *D* type and *J–K* flip-flops may be used, and their use will become clear when applications are described as in the next section.

6.4.2 *PLA applications*

A PLA can be used to perform functions similar to a ROM, for instance code conversion, microprogramming and random logic operation. As an example in the use of a PLA, and its comparison with a ROM, consider an arrangement to solve the following equations:

$$O_1 = I_1 \,.\, I_2 \,.\, I_4 + \overline{I_1} \,.\, I_3 + I_2 \,.\, I_3 \qquad (6.5)$$

$$O_2 = \overline{I_1} \,.\, I_3 + \overline{I_2} \,.\, I_3 \,.\, I_4 \qquad (6.6)$$

$$O_3 = I_1 \,.\, \overline{I_2} \,.\, \overline{I_4} + \overline{I_1} \,.\, I_3 + I_2 \,.\, \overline{I_3} \qquad (6.7)$$

Fig. 6.12 shows the Karnaugh maps for the outputs, and the corresponding ROM implementation. Alternatively a PLA arrangement is given in Fig. 6.13. It is now necessary to generate only six product terms since one of these, $\overline{I_1} \,.\, I_3$ is common to all three outputs. Once again it is seen from these two figures that the PLA represents a much more direct implementation of the equations than the ROM. Furthermore the ROM is primarily a storage device in which the inputs address words to produce the corresponding bit outputs. Therefore for example, for a logic 1 on O_1 any one of seven words

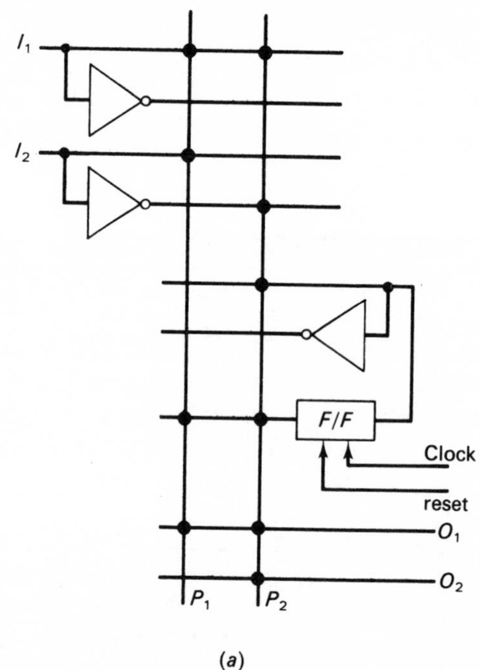

(a)

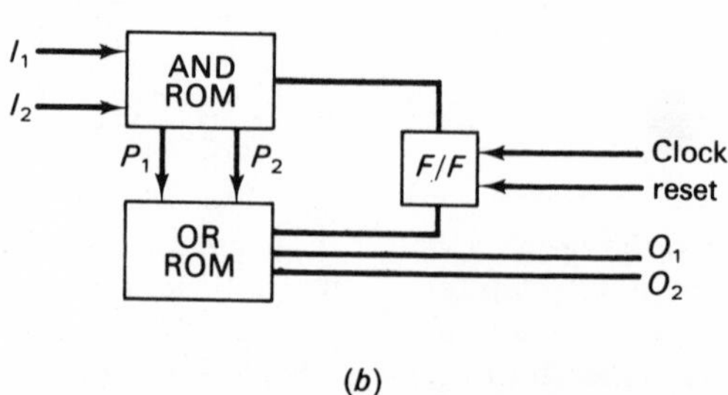

(b)

Fig. 6.11. PLA with internal memory; (a) array logic, (b) block schematic.

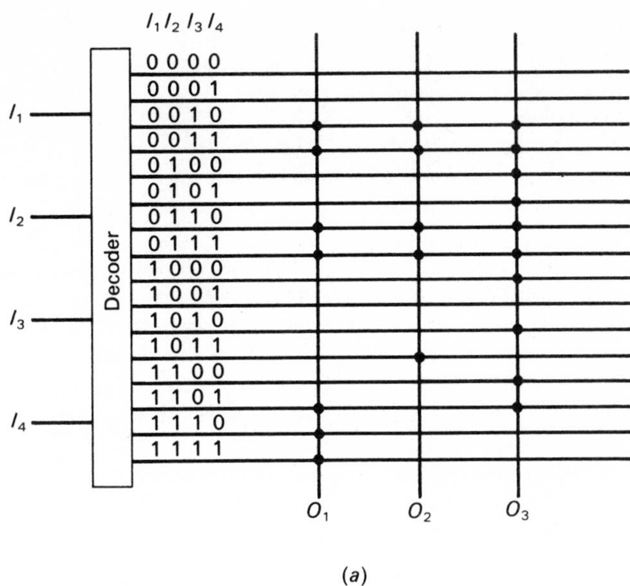

(a)

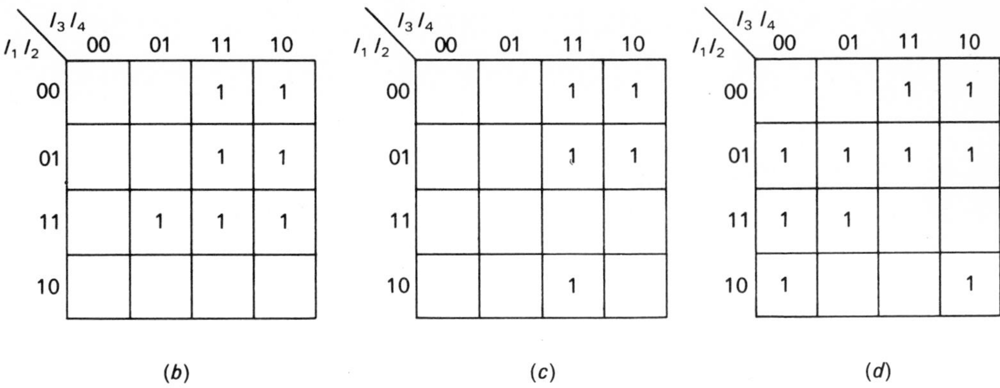

(b) (c) (d)

Fig. 6.12. Use of a ROM to implement random logic; (a) ROM arrangement, (b) map for $O_1 = I_1 . I_2 . I_4 + \overline{I}_1 . I_3 + I_2 . I_3$ (c) map for $O_2 = \overline{I}_1 . I_3 + \overline{I}_2 . I_3 . I_4$ (d) map for $O_3 = I_1 . \overline{I}_2 . \overline{I}_4 + \overline{I}_1 . I_3 + I_2 . \overline{I}_3$

can be addressed as shown in Fig. 6.12. A comparison of Fig. 6.12 and Fig. 6.13 indicates that the ROM uses a fewer number of array bits than the PLA. However this is not always the case. If I is the number of input variables, O are the number outputs and P the product terms required, then the ROM needs $2^1 \times O$ bits since it produces all the product terms. However the number of bits needed by the AND and OR parts of the PLA are given by $2 \times I \times P + O \times P$ assuming that the inverse of all inputs are generated. Therefore if $I = 4, O = 3$ and $P = 6$, as in the previous example, then the ROM will require 48 bits whereas the PLA needs 64 bits.

However if I was 14, O was 3 and P equalled 6 then the ROM would need to have 79142 bits whereas the PLA requires only 186 bits. Therefore the PLA scores where there are a large number of input variables and a relatively small number of product terms since, unlike the ROM, it does not need to produce all the possible product terms.

As mentioned earlier PLAs are also available with internal flip-flops for sequential applications. To illustrate their use consider the design of a BCD decade counter as shown in Fig. 6.14. The Karnaugh map for each bit is derived from the truth table. Therefore bit I_1 will be a logic 1 when the previous count state is $\overline{I_1} . I_2 . I_3 . I_4$ or $I_1 . \overline{I_2} . \overline{I_3} . \overline{I_4}$. There are, of course, a number of unobtainable states since the maximum count sequence is 10. From the Karnaugh maps the following equations can be derived for the counter:

$$(I_1)^{n-1} = I_2 . I_3 . I_4 + I_1 . \overline{I_4} \qquad (6.8)$$

$$(I_2)^{n-1} = I_2 . \overline{I_3} + I_2 . \overline{I_4} + \overline{I_2} . I_3 . I_4 \qquad (6.9)$$

$$(I_3)^{n-1} = \overline{I_1} . \overline{I_3} . I_4 + I_3 . \overline{I_4} \qquad (6.10)$$

$$(I_4)^{n-1} = \overline{I_4} \qquad (6.11)$$

Fig. 6.15 shows the implementation of the counter equations using a PLA. F_1 to F_4 are internal storage flip-flops and their outputs are fed back to the upper AND part of the PLA. Therefore the input to flip-flop F_1 is equal to the sum of product terms P_7 and P_8, that is: $I_2 . I_3 . I_4 + I_1 . \overline{I_4}$. This is equal to $(I_1)^{n-1}$ from equation (6.8) and after the next clock pulse it is fed back as I_1, as required. Similarly for the other terms of the counter.

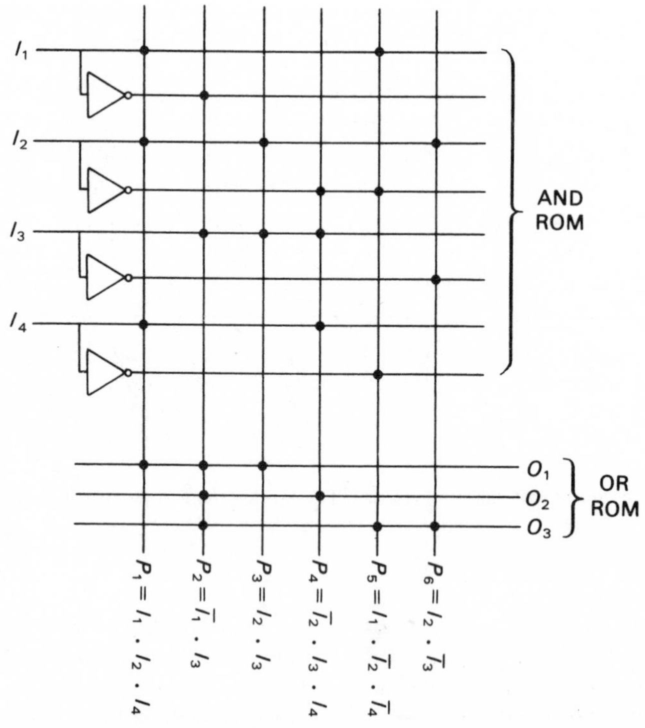

Fig. 6.13. Use of a PLA to implement random logic where; $O_1 = I_1 . I_2 . I_4 + \overline{I_1} . I_3 + I_2 . I_3$ $\quad O_2 = \overline{I_1} . I_3 + \overline{I_2} . I_3 . I_4$ $\quad O_3 = I_1 . \overline{I_2} . \overline{I_4} + \overline{I_1} . I_3 + I_2 . \overline{I_3}$

Time count	I_1	I_2	I_3	I_4
0	0	0	0	0
1	0	0	0	1
2	0	0	1	0
3	0	0	1	1
4	0	1	0	0
5	0	1	0	1
6	0	1	1	0
7	0	1	1	1
8	1	0	0	0
9	1	0	0	1

(a)

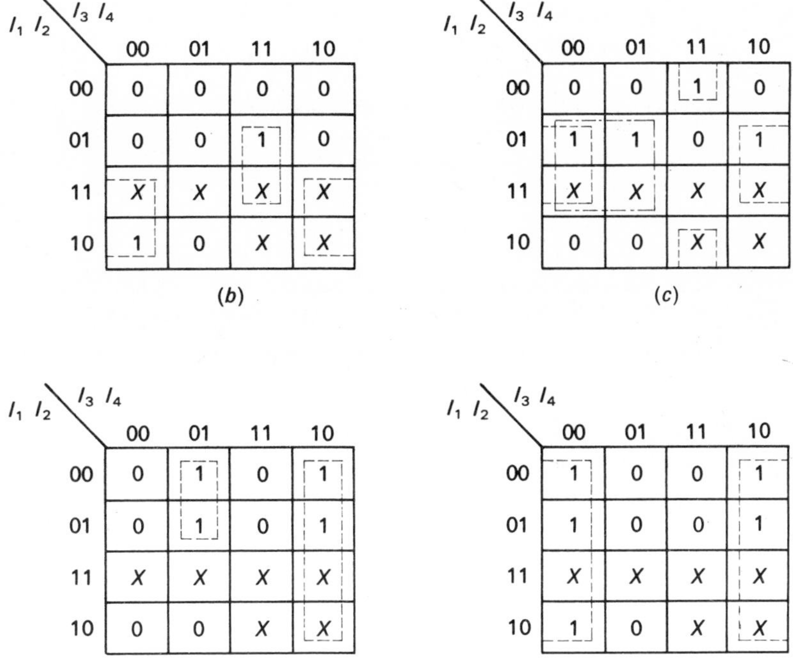

Fig. 6.14. Design of a BCD decade counter; (a) truth table, (b) map for $(I_1)^{n-1}$, (c) map for $((I_2)^{n-1}$, (d) map for $(I_3)^{n-1}$, (e) map for $(I_4)^{n-1}$.

There are several ways in which the output from the PLA counter may be taken to the outside world. Utilizing the flip-flop terminals is not recommended since these generally represent relatively high impedances. Furthermore it does not allow one to code the output if so desired. Fig. 6.16 (a) shows how the outputs from the counter may be accommodated if no code conversion is required. Alternatively code changes are easy to implement as Fig. 6.16 (b) illustrates. For example the decimal output 0_{10} is equal to BCD $I_1 . \overline{I_2} . \overline{I_3} . I_4$ and is so coded. Similarly for the other outputs.

6.5 Microprocessors

A microprocessor is akin to a ROM or a PLA in that it is a general purpose unit which can be programmed to meet any special application. In fact microprocessor programming is carried out within a ROM and (or) a PLA which are integral parts of the microprocessor system, so that the similarity between these devices is even closer. The prime distinguishing feature between a microprocessor and a ROM or PLA is that the microprocessor contains a logic and arithmetic processor in addition to the programmable array of the type used in the ROM or PLA. It is therefore a much more complex and flexible device and is capable of performing tasks which are beyond the capability of both the

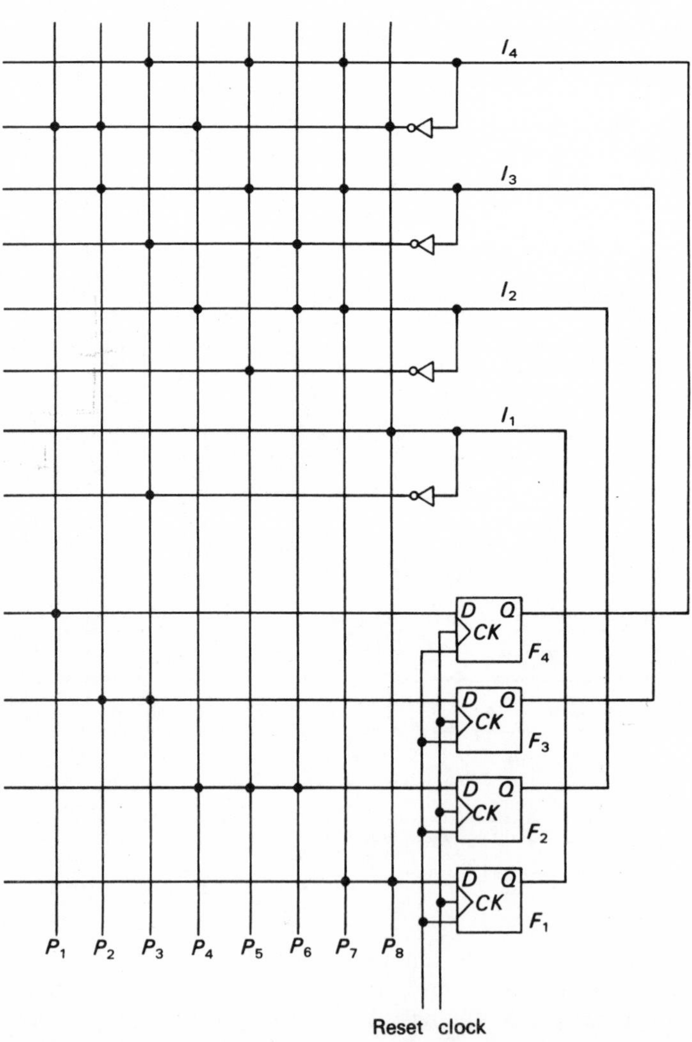

Fig. 6.15. PLA implementation of a BCD decade counter.

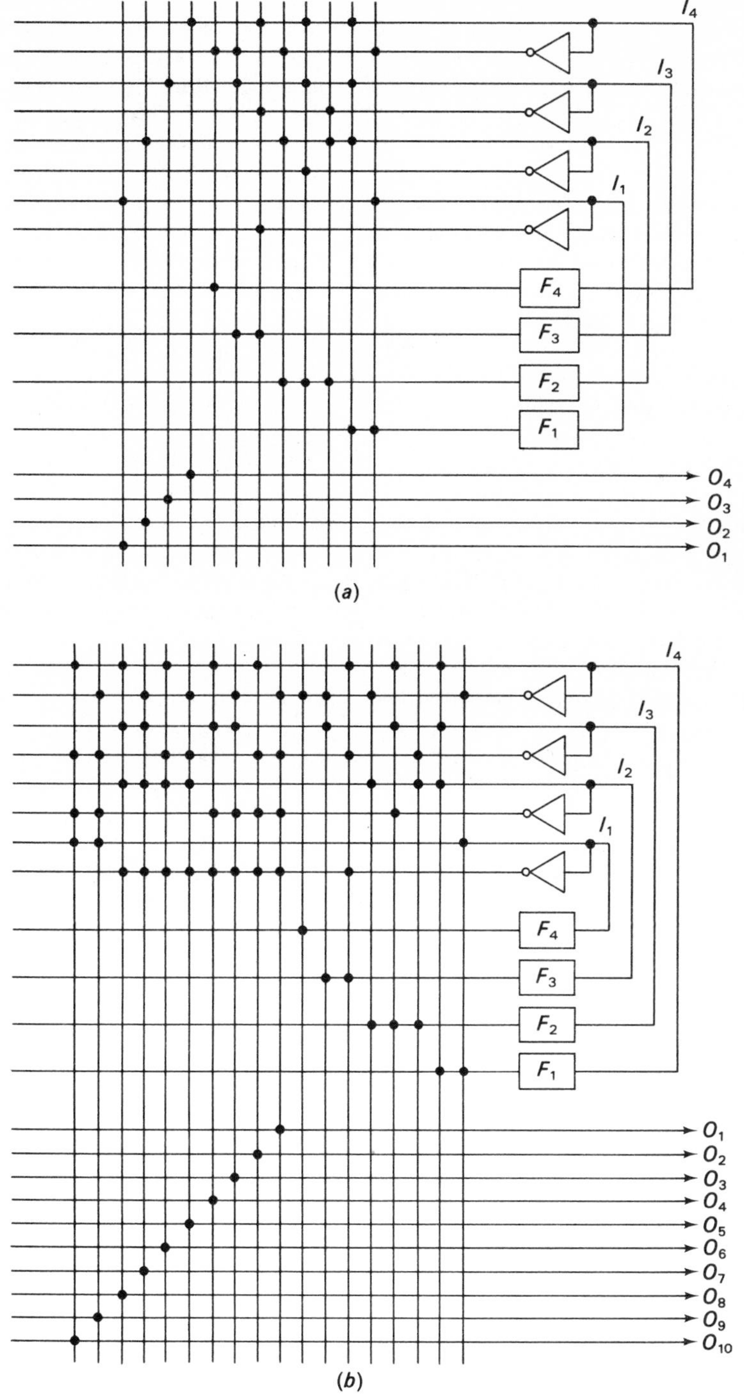

Fig. 6.16. Output lines for a PLA counter; (a) BCD output, (b) decimal output.

ROM and the PLA. As such the microprocessor is a stronger competitor for random logic systems which are implemented using small or medium scale integrated devices, and for those using custom large scale integrated chips.

Because of its greater complexity the microprocessor needs a more involved programming technique. This has resulted in the development of extensive software aids with the result that a microprocessor is often referred to as a software programmed system as opposed to the *hardware* programming used on the ROM and PLA. It is however essential to keep in mind that the end result is still the same i.e. hardware programming of an array in order to store and implement a sequence of events.

The first microprocessor was launched in December 1971 and since that time the market has been flooded with a variety of different devices covering a wide spectrum of sizes, organizations and technologies. In this chapter it is intended to largely ignore differences between commercial devices and to concentrate on the basic principles of a microprocessor. Both the software and hardware aspects will be discussed and these will be illustrated by means of examples of complete systems.

6.5.1 *Microprocessor hardware*

The basic organization of a microprocessor system is shown in block schematic form in Fig. 6.17. An examination of this figure will show that it is very similar to that of a computer. This is the reason why it is often called a microcomputer. The heart of the microcomputer is the central processing unit (CPU) which performs all the arithmetic and logic functions in the machine and also generates signals to control the correct performance of the rest of the blocks shown in Fig. 6.17. In a conventional computer the CPU may be made from many individual integrated circuit packages and it may occupy several printed circuit boards. For a microcontroller the CPU is normally constructed as a single LSI chip. Because of this the CPU of a microcontroller is called a microprocessor. The difference in these terminologies is important. The microprocessor is the CPU of the total system, and it is the total system which is called a microcontroller. Unfortunately, however, the total system is often referred to, mistakenly, as a microprocessor.

Although the microprocessor typically consists of a single LSI chip the total microcomputer can vary considerably in size depending on the application and the amount of read–write memory, *control memory* and input–outputs required. However there is another class of microcomputer which contain all the functions illustrated in Fig. 6.17 on a single LSI chip. These are therefore one-chip microcomputers. Although they are relatively inexpensive these devices are also limited in their processing capabilities. Their prime use is in control type applications which require a large number of input–outputs but a small amount of memory and low operating speeds. Because of this the one-chip microcomputer is sometimes also referred to as a microcontroller. The calculator chip is a special class of microcontroller which has been designed for a particular application. Irrespective of the overall size of a microcomputer its operating prin-

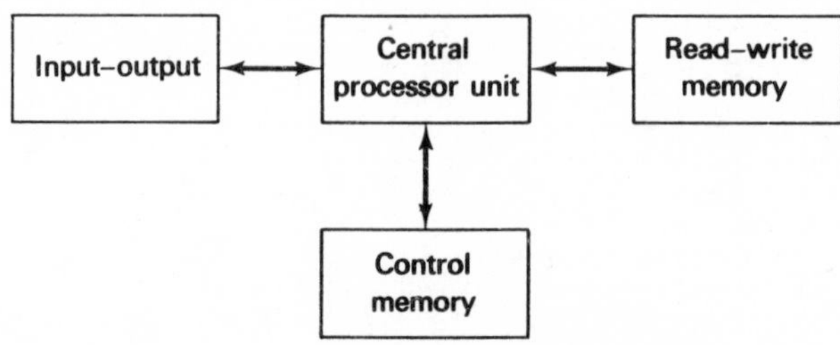

Fig. 6.17. Block diagram of a microcomputer.

ciples remain unchanged. The present discussion will therefore concentrate on the functional aspect of microcomputers and will ignore the finer differentials between the various types.

Returning to the block diagram for a microcontroller, as shown in Fig. 6.17, the CPU is the part which carries out all the arithmetic and critical control functions within the system. The operation of the CPU is controlled by a program which is stored in the control memory. This program causes the microcomputer, which is a general purpose device, to become a specific component for a user's application. Although the program may initially be written in a variety of languages, many of which resemble English words, it is translated into a series of logical 0s and logical 1s when it resides in the control memory, which is primarily a read only store.

In addition to a control memory the microcontroller requires some read–write memory or RAM to be able to store and recall intermediate data which is generated during the normal course of operation. Finally the CPU communicates with the outside world through input–output lines which can be arranged in a variety of ways to meet different requirements. In the sections which follow the various hardware aspects of the microcomputer will be considered further.

6.5.1.1 *Microprocessor architecture*

The microcomputer illustrated in Fig. 6.17 is referred to as a fixed instruction set device since the way that the CPU reacts to a given instruction from the control ROM is fixed by the manufacturer and is always the same for all users. An alternative is to have a microprogrammed microcomputer. The concept of microprogramming was introduced in chapter 5 and illustrated by use of a read only memory. A microprogrammed microcomputer is illustrated in Fig. 6.18. Whereas in a conventional microcomputer the user's program controls the flow of information in the CPU, for a microprogrammed system the user's program only represents a starting point in the microprogram memory for one or more microprogrammed steps. The microprogram memory generates its first instruction and feeds back the address within its own memory for subsequent instructions. Since both the microprogram memory and control memory can be programmed to the user's requirements such a system presents considerable flexibility. The internal microinstructions of the CPU can now be programmed to obtain a set which is tailor made to

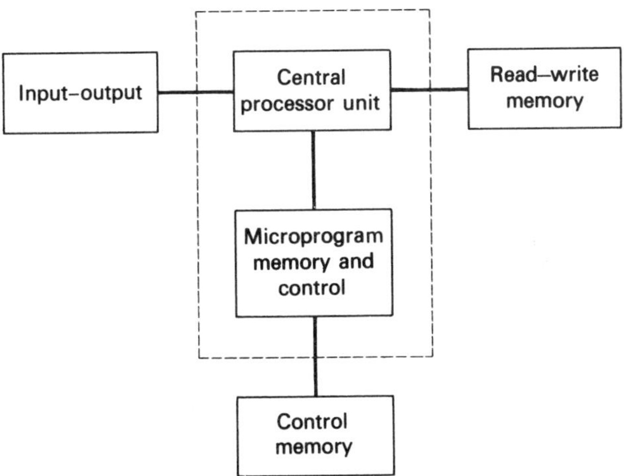

Fig. 6.18. Block schematic of a microprogrammed minicomputer.

a given application so that considerable savings in overall hardware are possible. Microprogramming is also very useful if one type of microcomputer is required to have a similar instruction set to another. This is known as emulation. A further advantage is that complex or critical routines, such as trigonometrical functions, can be programmed as microinstructions and not have to reside in the user's control memory.

In spite of its advantages microprogrammed microprocessors are not as common as ones with fixed instruction sets. The disadvantage is that considerable skill is required for microprogramming since one needs to deal with the internal *architecture* of the CPU. It is also not easy to transfer software from one microcomputer to another since each has a different microprogrammed memory. Finally currently available software design aids cannot be easily used since these are primarily geared to common fixed instruction sets.

Internally there are many possible arrangements for the CPU or microprocessor. Fig. 6.19 shows a simplified structure of one type. Basically it consists of an arithmetic logic unit (ALU), a collection of registers and counters, an input–output bus structure, and a timing and control section.

The arithmetic logic unit can vary in complexity from a powerful arithmetic device to a simple adder. Generally it has the ability to do hardware addition and subtraction, and to shift data any number of places in either direction. Associated with the ALU are flag bits, which are essentially flip-flops, which can be set to record certain conditions which arise during arithmetic operations. Examples of these conditions are overflow or carry, zero, negative, etc.

A register which is often linked to the ALU is the accumulator. This is capable of operations such as shift and rotation, and it stores one of the operands used by the ALU as well as acting as the destination for all its results. It is possible to have several registers within a microprocessor capable of performing the functions of an accumulator. The program counter is responsible for keeping track of where the CPU is in its program. All program memory locations are numbered and this number is called its address. The program counter keeps the address of the next instruction. The contents of the counter are automatically incremented by one every time an instruction is completed so that the microprocessor works down step by step through its program memory. However it is possible to force any number into the program counter so as to start from, or jump to, any required location within memory. Another register often used in the microprocessor is the address register. This temporarily stores the address of the memory location for the data. Sometimes one can alter the contents of this register so that a complete address can be built up before access is made to the memory.

All instructions from memory arrive eventually at the instruction register. This register and its associated decode circuitry respond to the positions of the logic 1 and logic 0 bits in the instruction and send signals to the control and timing circuitry in order to carry out the correct sequence of operations within the microprocessor to execute the instruction.

The microprocessor also usually has within it two sets of registers which connect to the data bus via multiplexers. The first set is primarily intended as a temporary store for data during manipulation and is a scratch pad memory. The second set of registers act as a last-in first-out (LIFO) system. It has several uses, the prime one being subroutine nesting. The operation of subroutines will be explained further in section 6.5.2. The prime requirement during this phase is to store address locations and then to recall them in reverse order. A LIFO memory is the obvious solution. Generally this store may be located on the actual microprocessor chip, as in Fig. 6.19, or else it can be located in a RAM which is connected to the microprocessor via its data bus. A pointer is now resident on the microprocessor chip to show the location in the external random access memory where storage is taking place. It is important not to confuse this pointer with a similar one used with an on-chip memory stack in order to create a LIFO memory, as shown in Fig. 6.19. The LIFO pointer is basically a bidirectional counter which counts one way during address stacking and the opposite way during address recall.

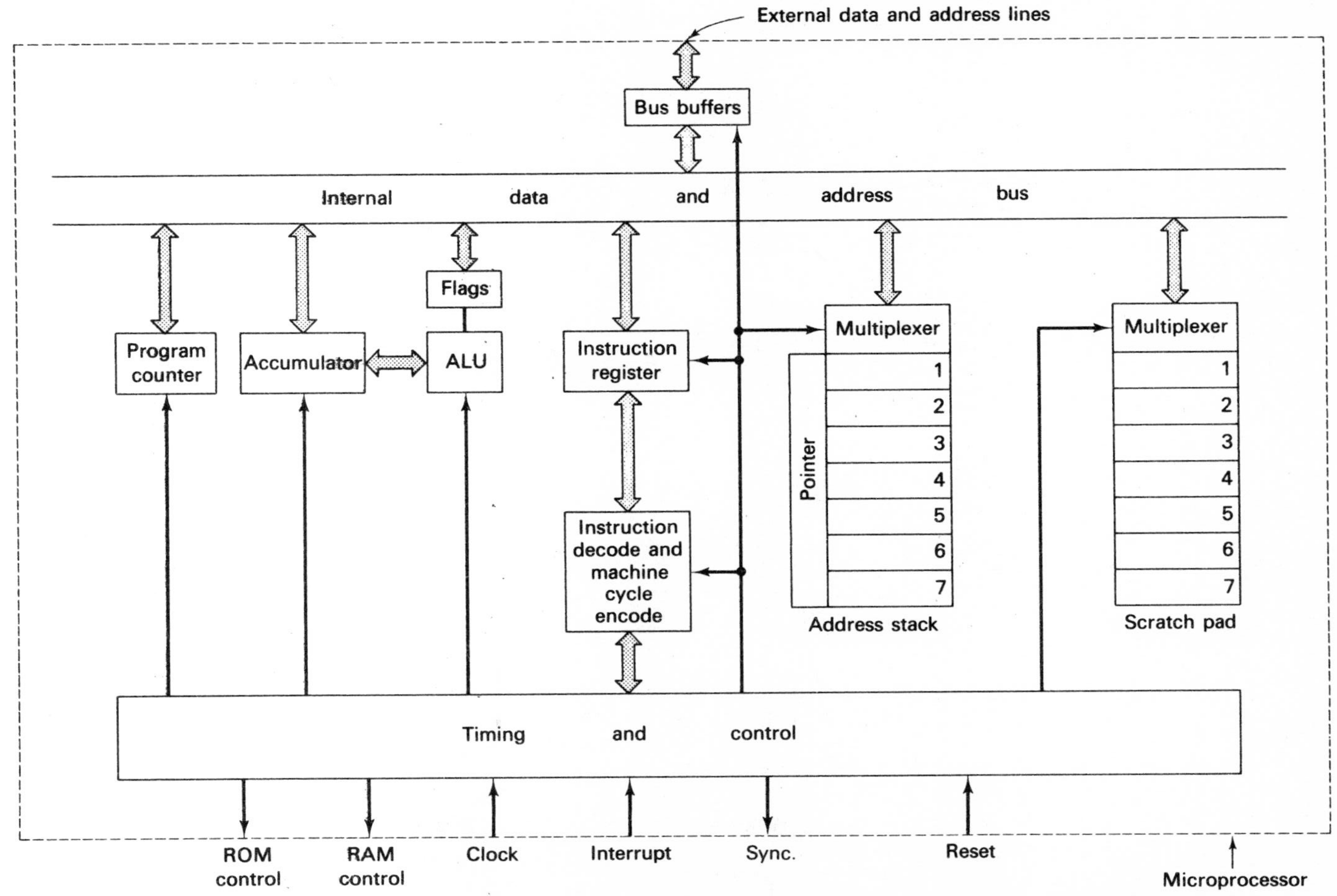

Fig. 6.19. Simplified internal architecture of a microprocessor (CPU) using a common data and address bus and an on-chip address stack.

When the address stack is on the microprocessor chip it is also sometimes referred to as a hardware stack whereas the off-chip arrangement is called a software stack.

6.5.1.2. *Microprocessor input–output*

The input–output structure of a microprocessor is determined by two considerations, the technique used to transfer information between the microprocessor and its peripherals, and the bus structure.

The commonest method of transferring information is by program transfer. This is controlled by the microprocessor and occurs at instances specified in the user's program. The microprocessor checks that the required device is ready to interchange information and then addresses the device and commences data transfer with it. The disadvantage of program transfer is in application where it is not possible to predict when a peripheral is ready for data transfer. The microprocessor must be constantly returned by the program to check the ready state so that valuable processing time is wasted. In these instances an interrupt mode of data transfer is preferred. It consists of the peripheral informing the microprocessor that it is ready for data transfer by signalling one of its input lines. When an interrupt occurs the microprocessor must first complete its present instruction and then store away the contents of all important registers such as the program counter and accumulator. After this is done it is ready to jump to another part of the control store which tells it what to do about the interrupting peripheral. On completion of this routine the microprocessor restores all its stored registers and continues in the main program at the point at which it was interrupted. The interrupt response time of a microprocessor is usually the time lag between it receiving the interrupt request and entering the subroutine to service the interrupt.

If several peripherals are connected to a microprocessor and each is capable of generating an interrupt, then on the receipt of an interrupt signal the microprocessor must first determine which device caused the interrupt. This can be done by polling each peripheral in turn and then jumping to the service routine associated with that peripheral. Alternatively a 'vectored' interrupt system may be used. In this a few input bits of the microprocessor are assigned to the peripherals and on generating an interrupt they also place their unique code on these vectored lines. The microprocessor now automatically jumps to the subroutine associated with this code. Some microprocessors have facilities for multiple interrupt levels in which one can assign priorities to each level. Now a peripheral can interrupt an interrupt caused by another peripheral provided it has been assigned a higher priority. It is also possible to disable and enable the interrupt facility under program control. In some devices there is the added feature that key registers are automatically stored when an interrupt occurs and are automatically recalled at the end of the interrupt routine. This saves programming time and space compared to microprocessors which require register storing under software control every time an interrupt occurs.

It is important to appreciate that an interrupt transfer is really a programmed transfer. The peripheral breaks into the normal running of the microprocessor to request an interrupt. But it is the microprocessor which is in control throughout via its program. If required, the interrupt can be ignored and even when data transfer does occur it is totally under program control. An alternative form of data transfer is known as direct memory access (DMA) or data break. In this a link is set up between the peripheral and the read–write memory of the microcomputer. Data transfer now occurs directly between the two under hardware control. The microprocessor takes no part in this transfer and is usually engaged on some other task, as determined by the program. Data transfers via DMA are extremely fast but are more expensive, in terms of control hardware, than programmed transfers.

There are several ways of connecting peripherals to the microprocessor. These are shown in Fig. 6.20. The radial method is the most direct and allows simultaneous access of all peripherals to the microprocessor. It however makes considerable demands on the microprocessor as regards number of output lines. The party line system requires that the

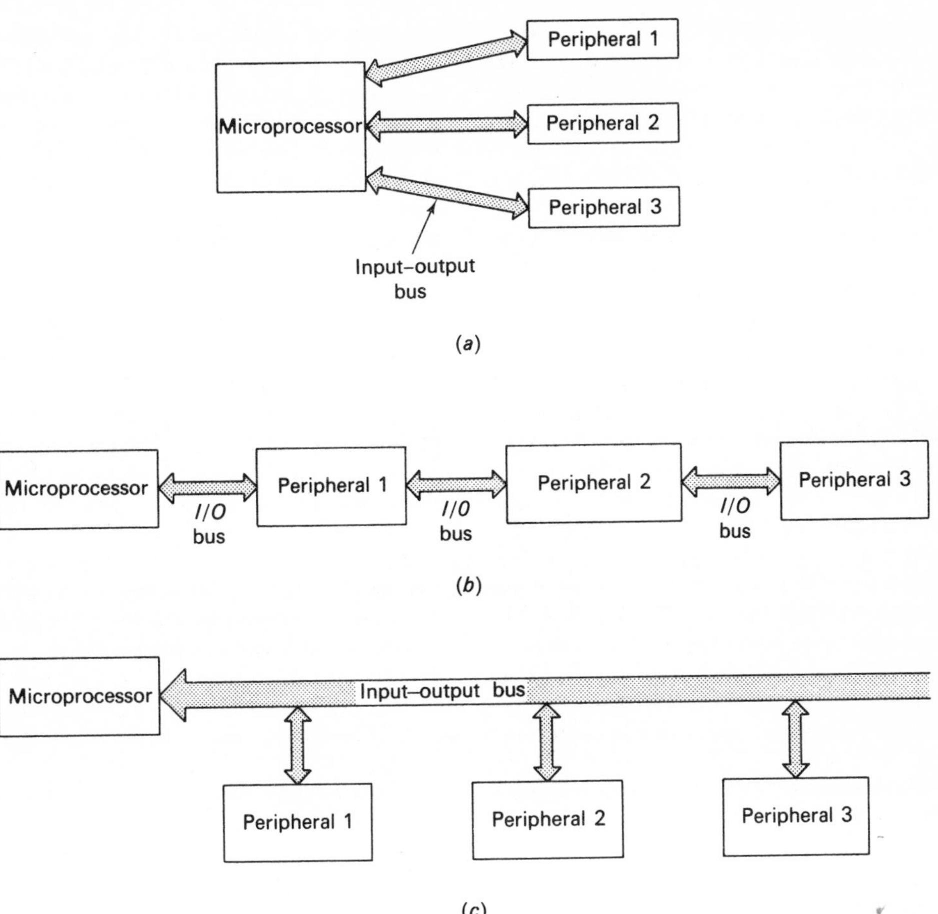

Fig. 6.20 Microprocessor-peripheral connection; (a) radial, (b) daisy-chain, (c) party line.

peripheral time share the input–output lines to the microprocessor since no two devices can engage in data transfers at the same time. It is now also necessary to be able to identify which peripheral is communicating with the microprocessor. Although this method is slower than the radial system it saves on the number of input–output lines. The daisy-chain arrangement is similar to the party line method with the exception that each peripheral passes on the input–output lines from the microprocessor to the following devices. As such, they have the facility to alter any data and to completely cut off the lines if they so wish. This represents an easy way to establish priorities of interrupt since the closer a peripheral is to the microprocessor the higher its priority.

The input–output bus from the microprocessor must be capable of carrying two different types of information. Firstly it must contain the address of a peripheral or a memory location. Secondly it must contain the data associated with this address. The address bus is usually unidirectional from microprocessor to memory. However there is data transfer in both directions into and out of the microprocessor depending on the operation required. It is possible to have three separate bus structures, one for the address and two for the data in either direction. However, this increases the number of output pads required on the microprocessor chip. A better solution is to use a two bus structure. The first carries the address and the second is a bidirectional bus which carries data. Such an arrangement is shown in Fig. 6.21 which should be compared to Fig. 6.19. It illustrates the separate data and address buses. Note that this is an eight bit machine (word lengths are discussed in section 6.5.1.4). The data bus is therefore 8 bits wide but the address bus has been made much larger. With the 16 bits it is now possible to address 2^{16} or 65536 memory locations directly. The arrangement shown in Fig. 6.19 uses a single bus for both data and address. It is more economic as regards microprocessor pins since for an 8 bit machine it would need only 8 outputs for the bus compared to 24 for the arrangement of Fig. 6.21. However, it is now necessary to time multiplex the data lines. For instance the timing clock associated with the input–output bus would be divided into three parts. On the first part the 8 most significant bits of the memory address would be put out into a latch. The 8 least significant bits of the address would follow on the second part of the clock, and the final part would produce the 8 data bits. When all address and data bits are available the instruction will be carried out. Clearly this system is slower than that using separate address and data bus lines since three clock periods are required as opposed to one. Note that the microprocessor in Fig. 6.21 uses an off-chip or software stack so that this can be made as large as required so long as there is adequate external RAM memory.

6.5.1.3 *Microprocessor timing and control*

A simplified timing diagram for a microprocessor is shown in Fig. 6.22. A master clock, which may be situated on the microprocessor chip or connected to it externally determines the timing intervals. The time between clock pulses is called the clock period. Several periods correspond to a microprocessor phase during which it carries out defined parts of its activity. In Fig. 6.22 three clock pulses are shown as being associated with each phase. There are generally three phases during the operation of a complete instruction, called the instruction cycle. During the first phase the microprocessor sends a read command to the control memory along with the contents of the program counter, which determines the memory address. Several clock pulses may be required for this if time multiplexing of the data bus is used, as explained in section 6.5.1.2. During the second phase the microprocessor receives the content of the memory location which has been addressed and it places this in the instruction register. Once again if the instruction has more than one word several clock pulses may be required to receive the complete instruction. Generally phases one and two are referred to as the 'fetch' stage of the machine timing. After the whole instruction is in the instruction register the program counter is incremented by one, ready for the next microprocessor cycle. The timing now enters the third phase where the instruction is decoded and obeyed or 'executed'. The instruction may have been for a memory read in which case the microprocessor will send a read command, and the specified address, to memory, and receive back the contents of that memory location. However this information is now data and not an instruction so it is placed in the accumulator, or one of the other general purpose registers, and not in the instruction register. If the original instruction was memory write, the microprocessor sends a write command to the RAM along with the address and then the data. For an input–output instruction the microprocessor needs to address the appropriate peripheral port and then send or receive the data.

In Fig. 6.22 each phase of the microprocessor operation always takes the same time as determined by the master clock. This is called synchronous operation. All phases, however, do not always require the same time period since some instructions can be performed in a shorter time than others. If one were to design the microprocessor with the longest phase in mind then there would be instances when the microprocessor was doing nothing and time would be wasted. An alternative system, called asynchronous, does not use a master clock. In this the execution of an instruction is commenced by a start pulse, but then each successive stage

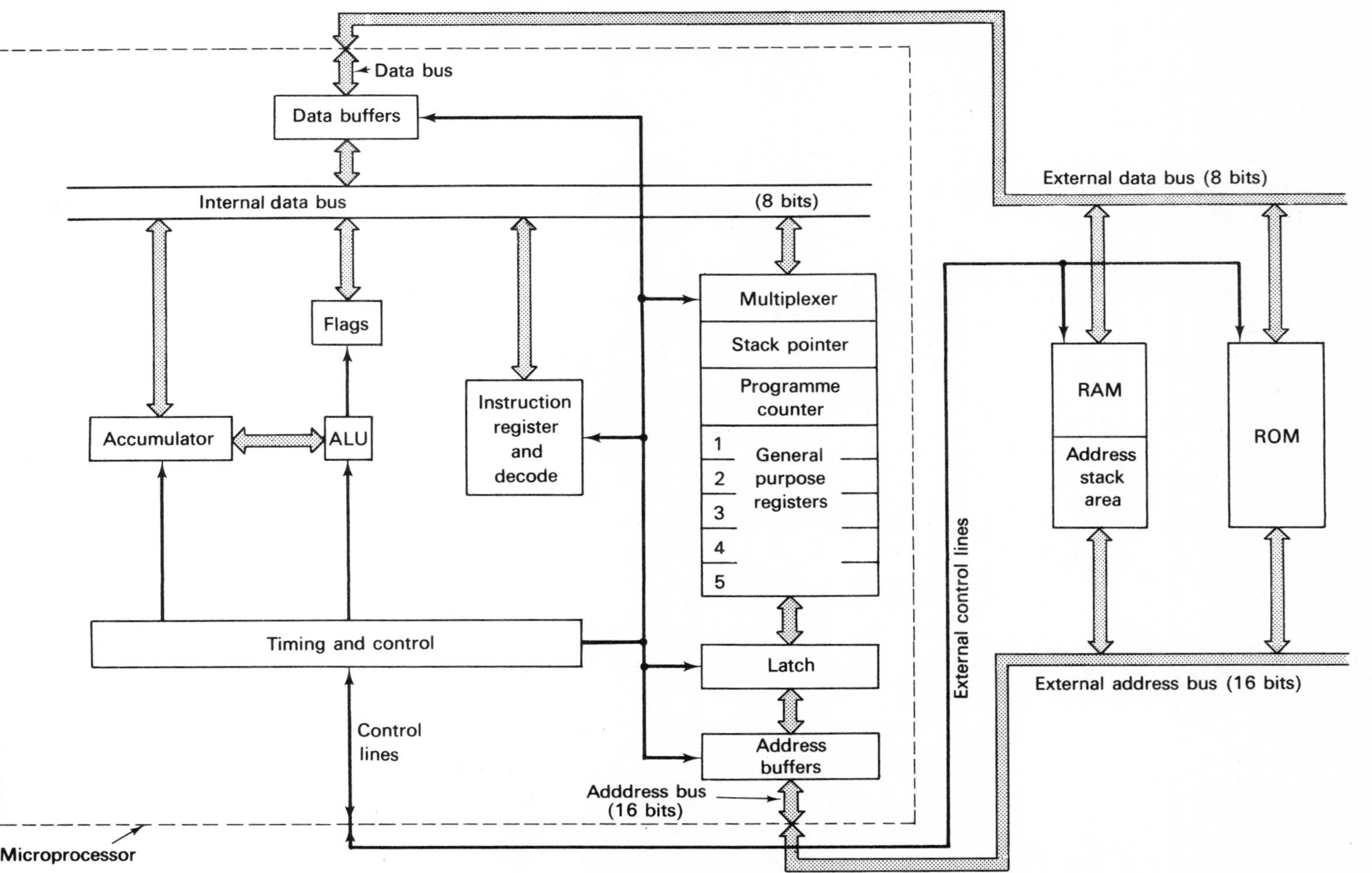

Fig. 6.21. Simplified architecture of a microprocessor using a separate data and address bus and an off-chip address stack.

follows on completion of the previous one. When the instruction is completed a new cycle begins with the fetching of another instruction from memory. Most commercial microprocessors adopt a system which is in between synchronous and asynchronous operation. A master clock is used but the microprocessor varies the number of pulses needed depending on the type of instruction being performed.

The detailed timing variations of a microprocessor are clearly linked very closely to its architecture and data paths. It is also usual to

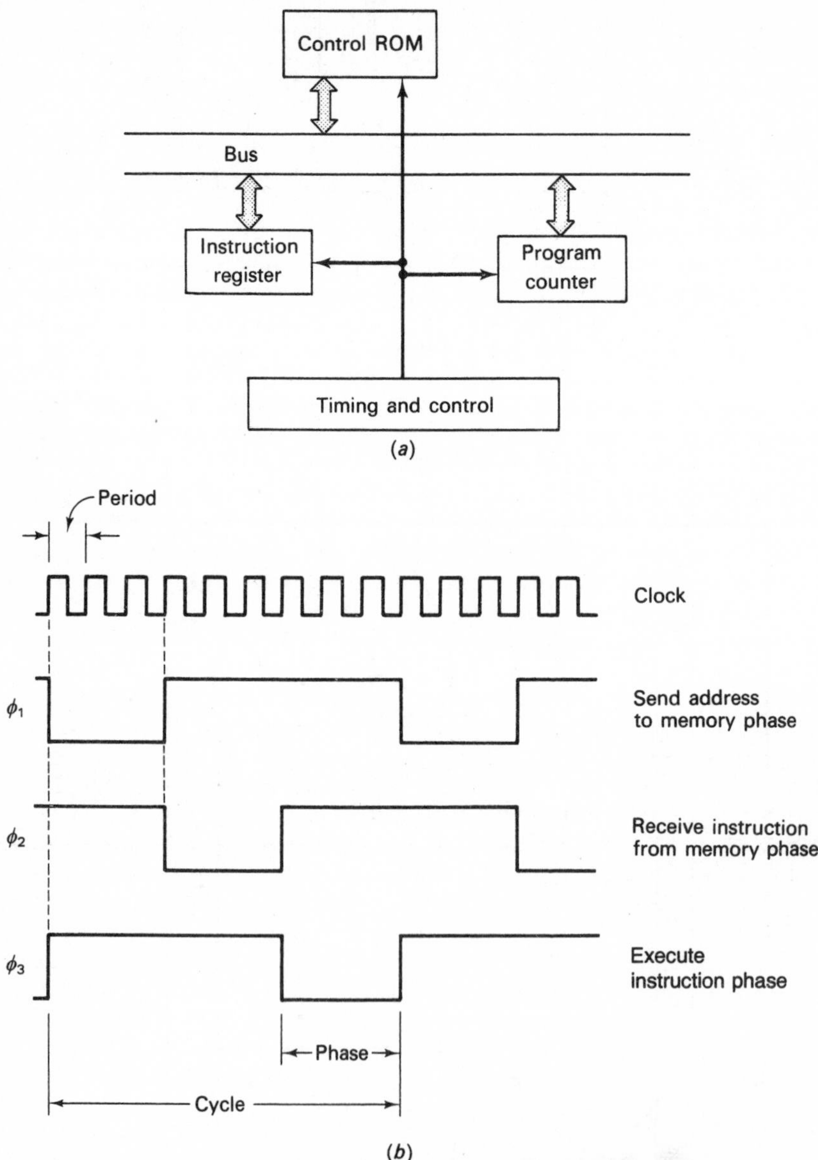

Fig. 6.22. Simplified microprocessor timing cycle; (a) part of microprocessor architecture, (b) microprocessor timing diagram.

send out a synchronization pulse (Fig. 6.19) at the start of each microprocessor cycle to synchronize all peripherals to its operation. The control section is also capable of accepting a reset pulse which clears all internal flip-flops (flags) in the microprocessor and forces it to an instruction start location (usually zero) within its memory.

6.5.1.4 *Microprocessor selection*

In the previous sections the basic hardware concepts behind a microprocessor have been described. The present section summarizes many of these and indicates the bewildering choice which faces the would-be user in selecting between different commercial devices.

Perhaps one of the key parameters to consider is word length. Instruction and data are transferred and manipulated within the microprocessor as sets of numbers called words. The larger each set, the faster the data rate, and the higher the accuracy of any mathematical operation. Similarly larger words can generate a greater variety of codes for handling character sets. For instance BCD needs a 4 bit word, but USASCII code used in data transmission, requires 7 bits. The larger word size also permits direct access to a greater number of memory locations, and since instructions are simply a collection of different bit patterns the repertoire of the microprocessor is also increased with its word length. The disadvantage of larger word sizes is the increased complexity of both the internal and external circuitry that is required to handle this. Commercial microprocessors generally vary in size between 4 bits and 16 bits. Although most devices are designed to operate as fixed bit machines several are available with bit slice architecture. This means that the microprocessor is structured so as to be able to expand in size. Therefore a 4 bit slice processor can be combined with four other devices to give an effective 16 bit microprocessor.

The speed at which a microprocessor is capable of operating is a very important parameter for many applications. However, it is also very difficult to define microprocessor speed. For instance, should it be based on the clock period, cycle time, phase time, minimum instruction time, or the time to add two numbers? In fact the overall operating speed is closely linked to the power and flexibility of the available instruction set. Therefore the best way to compare competing devices for a given application is by a benchmark, i.e. building the systems and seeing which runs faster.

Microprocessor bus structure and architecture are also important considerations. Are there enough input–output lines? Is data handled in serial or parallel form? Serial operation is less demanding on internal hardware and output pins but is slower. Are there adequate facilities for interrupts and DMA, and for nesting subroutines by a push-down stack? Generally the number of general purpose registers available is less important than their quality. Registers should be able to operate off memory, rather than always via the accumulator. It is also useful if they can be tested for zero, so that they can be used for counting, and if they can act as a source or destination for arithmetic or logic operations. Another consideration in microprocessors is their interface capabilities. It should be possible to connect the outputs of the microprocessor to other MOS or TTL systems, and it should be capable of operating with a variety of RAM, ROM and PROM devices.

The software capability of a microprocessor is an important consideration, and this is discussed in section 6.5.2. Other aspects include the amount of peripheral circuitry required to build a minimum microcomputer system and the ease with which this can be expanded with future needs. The design aids, both for hardware and software which are available for use with the different devices, are also important. Since microprocessors are complex and powerful devices they require considerable prototyping effort, and any aid which minimizes this time is likely to be very useful. In fact one of the advantages of microprocessors over the more conventional random logic is that such aids are readily obtainable so that the overall prototyping time is much reduced.

6.5.2 *Microprocessor software*

The hardware architecture described in the previous section represents a general purpose universal logic device which can be made to

Memory address	*Memory content*
0 0 0 0 0 1 0 0 0 0 0 0	1 1 1 0 1 0 0 0 0 0 0 0
0 0 0 0 0 1 0 0 0 0 0 1	0 0 1 0 0 1 0 0 0 1 0 1
0 0 0 0 0 1 0 0 0 0 1 0	0 0 1 0 0 1 0 0 0 1 1 0
0 0 0 0 0 1 0 0 0 0 1 1	0 1 1 0 0 1 0 0 0 1 1 1
0 0 0 0 0 1 0 0 0 1 0 0	1 1 1 1 0 0 0 0 0 0 1 0
0 0 0 0 0 1 0 0 0 1 0 1	0 0 0 0 0 0 0 0 1 1 0 1
0 0 0 0 0 1 0 0 0 1 1 0	0 0 0 0 0 0 0 1 0 0 0 1
0 0 0 0 0 1 0 0 0 1 1 1	0 0 1 0 0 0 0 1 0 0 1 0

Fig. 6.23. Portion of a typical program.

perform a specific application. The part of a microprocessor system which does this, that is, which converts a general purpose device into one which is made for a specific application, is called the program, or microprocessor software. The architecture of this software and a few typical systems are illustrated in this section.

6.5.2.1 *Software architecture*

The program for a microprocessor is stored in its control memory. The microprocessor hardware may be designed to operate as a 4, 8, 12, or 16 bit device and the memory system will usually be compatible. In the present discussions a 12 bit structure will be assumed although the same considerations apply for other word sizes.

Fig. 6.23 shows a portion of a typical program. Eight sequential memory locations are shown. The program in effect consists of a string of logic 0's and logic 1's and it is the task of the microprocessor hardware to respond to the position of these logic bits and carry out the coded instruction. The detailed mechanics of how this is done will not be considered here, since it consists of relatively involved circuitry, but the principles involved should be clear at this stage. Note also that the program contains instructions and data but that they both look identical. They are a series of binary numbers which are interpreted as instructions or data primarily by the context in which they appear in the program. For instance suppose that we wish to add the number 001001000011 which is stored in memory location 000001000011 to the current contents of the accumulator and then to increment the result by one. It will be seen in the next section that, for the hypothetical microprocessor used for these discussions, the code for 'add contents of location 000001000011 to the accumulator' is given by 001001000011, and the code for 'increment the accumulator' is 111000000001. Therefore the program will have the form shown in Fig. 6.24. Two locations now have an identical content. However, if the microprocessor starts at location 000001000000 it will assume that the first word is an instruction and will place this in its instruction register. This will inform the device that the word in location 000001000011 is data and is to be added to the contents of the accumulator. The microprocessor then proceeds to the next memory location which is 000001000001 and it assumes this to be an instruction since it has not been told otherwise. This increments the accumulator. Next location 000001000010 is examined as if it was an instruction. This would, in our example, halt the microprocessor. If it did not do so then

Memory address	*Memory content*	
0 0 0 0 0 1 0 0 0 0 0 0	0 0 1 0 0 1 0 0 0 0 1 1	←Instruction
0 0 0 0 0 1 0 0 0 0 0 1	1 1 1 0 0 0 0 0 0 0 0 1	
0 0 0 0 0 1 0 0 0 0 1 0	—	
0 0 0 0 0 1 0 0 0 0 1 1	0 0 1 0 0 1 0 0 0 0 1 1	←Data

Fig. 6.24. A program illustrating the similarity between an instruction and a data word.

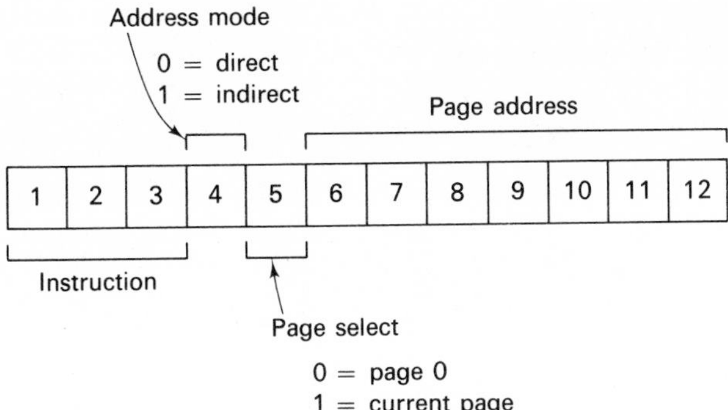

Fig. 6.25. Format for a memory reference instruction.

during the next cycle, location 000001000011 would be examined as if it was an instruction and would result in its contents being added to the accumulator again! Therefore the memory word is acting as data and as an instruction!

The next section will describe a few typical instructions which are used in microprocessors. They will be described in relation to a hypothetical 12 bit device although for convenience, the instruction set has been kept close to that of a PDP8 minicomputer. Six of these instructions will be used in conjunction with some address in memory. Therefore three bits are required for this, leaving 9 bits for memory address. However, 9 bits can only access 2^9 or 512 locations in memory directly. To increase this amount we will use what is known as indirect addressing as illustrated in Fig. 6.26. For direct address the instruction word contains the address of the memory being referred to. For indirect address the word being referred to by the instruction contains the full address of the location in which the data is stored. Since this second word does not contain any instructions all 12 bits may be used for memory address so that 2^{12} or 4096 locations can be accessed. Fig. 6.25 shows that one bit of the instruction word is being used to indicate the address mode. A zero in bit 4 will indicate a direct address and a 1 will indicate an indirect address. The remaining eight bits can address 256 locations. These will be assumed to be split into two 'pages' of 128 locations each. The first page will be designated page zero and will be addressed when bit 5 is a logic 0. The second page is the current page on which the instruction resides and is applicable when bit 5 is a logic 1. Therefore the data word XXX000001111 will be a direct address referring to location 0001111 on page 0. However, the word XXX010001111 is a direct address referring to current location plus 0001111, i.e. decimal 15. If it is situated on the page starting at 129, its location is 129 + 15 or 144. Note that although it has been convenient to think of the memory as split into pages of 128 locations each, no such physical division exists in practice.

In the next section some actual instructions are described. In all these discussions the octal code will be used, rather than the binary format, since it is more convenient. Therefore the program illustrated in Fig. 6.24 may be re-written as in Fig. 6.27.

6.5.2.2 *Instructions*

The first three bits in the twelve bit word will be used to indicate the type of instruction. This allows eight combinations. Six of these reference the memory, the seventh is used for data manipulation within the accumulator and the eight is an input–output instruction.

(1) AND instruction. This is programmed by a 0 in all three instruction bits and it performs the logical AND between the accumulator and the word addressed by the remaining bits. This is illustrated in Fig. 6.28. Note that the program refers to contents of location 0203 (in octal) which is a direct address on the page current to the instruction. Therefore the address mode bit

4 is a 0 and bit 5 is a 1. This gives the instruction word in location 0200 as 000010000011 or 0203. As seen from Fig. 6.28 (a) the accumulator content after the AND instruction has altered although that of the memory location which was addressed remains unchanged.

(2) TAD instruction. This is represented by a '1' in bit 3 and it adds the contents of the word being addressed to the accumulator contents. Fig. 6.29 illustrates the use of a TAD instruction in which the addressed word is on page zero so that bits 4 and 5 both contain a '0'.

(3) ISZ instruction. This represents an instruction to add one to the contents of the memory location being addressed (i.e. to increment it) and then to skip the next instruction if the contents of the location goes to zero due to this addition. ISZ therefore stands for 'increment and skip if zero'. It is represented by a logic 1 in bit 2 of the instruction word.

(4) DCA instruction. This is represented by a 1 in bits 2 and 3. It deposits the contents of the accumulator into the memory location addressed and then clears the accumulator. DCA therefore stands for 'deposit and clear accumulator'. It is illustrated in Fig. 6.30. The instruction word resides at location 0200. Indi-

	Address	*Contents*
	30	–
Instruction → location	31	Add contents of location 100 to accumulator.
	32	–
	–	–
	–	–
	99	–
Data location →	100	5
	101	–
	–	–
	–	–

Accumulator contents after above operation = 5

(*a*)

	Address	*Contents*
	30	–
Instruction → location	31	Indirectly add contents of location 100 to accumulator.
	32	–
	–	–
	–	–
Pointer to	99	–
data location →	100	201
	101	–
	–	–
	–	–
Data	200	–
location →	201	12
	202	–
	–	–
	–	–

Accumulator content after above operation = 12

(*b*)

	Address	*Contents*
Instruction and	30	–
data location →	31	Add the number 18 to the accumulator
	32	–
	–	–
	–	–

Accumulator content after above operation = 18

(*c*)

Fig. 6.26. Memory address modes; (*a*) direct, (*b*) indirect, (*c*) immediate.

Memory address	*Memory content*
0 1 0 0	1 1 0 3
0 1 0 1	7 0 0 1
0 1 0 2	–
0 1 0 3	1 1 0 3

Fig. 6.27. Octal format for the program given in Fig. 6.24.

Memory address	*Memory contents*
0 2 0 0	0 2 0 3
0 2 0 1	–
0 2 0 2	–
0 2 0 3	1 1 0 1

(*a*)

001001000001	Memory location content
111010101001	Accumulator before AND
001000000001	Accumulator after AND

(*b*)

Fig. 6.28. Illustration of the AND instruction; (*a*) program, (*b*) accumulator and word contents.

	Memory address	Memory contents
	0 0 1 0	–
	0 0 1 1	0 0 1 1
	–	–
	–	–
	–	–
Starting → location	0 2 0 0	1 0 1 1
	0 2 0 1	–
	–	–

(a)

000000001001	Memory location content
000000000101	Accumulator content before TAD
000000001110	Accumulator content after TAD

(b)

Fig. 6.29. Illustration of the TAD instruction; (a) program, (b) accumulator and word contents.

rect addressing is now used to first reference location 0011 and then the main word at location 4101. The content of the accumulator is deposited into this location so that its original contents of 1321 is lost.

(5) RMS instruction. This is represented by a logic 1 in bit 1. It is a jump to subroutine instruction. It allows the programmer to insert pieces of auxiliary program into his main program at several locations. The address stack is used to ensure that return is made to the correct point in the main program after the auxiliary programs, called subroutines, have been completed. This return is made by a RES instruction donated by 7002. This is not a memory reference instruction and so is complete in the form shown.

Fig. 6.31 illustrates how subroutines are nested. The instruction in location 0201 calls the subroutine starting at location 0210. However, before this number is forced into the pro-

	Memory address	Memory content
	0 0 1 0	–
	0 0 1 1	4 1 0 1
	—	–
	—	–
Starting → location	0 2 0 0	3 4 1 1
	0 2 0 1	–
	—	–
	—	–
	4 1 0 0	–
	4 1 0 1	1 3 2 1

(a)

Accumulator content	Location 4101 content	
1 0 1 0	1 3 2 1	Before DCA
0 0 0 0	1 0 1 0	After DCA

(b)

Fig. 6.30. Illustration of the DCA instruction; (a) program, (b) accumulator and word contents.

Memory address	Memory content	Program counter	Stack
0200	–	0200	–
0201	4210	0201 → 0210	0202
0202	–	0202	–
–	–		
–	–		
0210	–	0210	0202
0211	4220	0211 → 0220	0212, 0202
0212	–	0212	0202
0213	7002	0213 → 0202	–
–	–		
–	–		
0220	–	0220	0212, 0202
0221	–	0221	0212, 0202
0222	7002	0222 → 0212	0202

Fig. 6.31. Illustration of the JMS and RES instructions.

gram counter its contents, which was increment when the last instruction was fetched, is stored in the push-down stack. When a second subroutine is called at location 0211 the contents of the stack are pushed down to store two return addresses. On the first RES instruction at location 0222 the stack is 'popped' and its contents of 0212 is forced into the program counter. Similarly the second 'pop' at location 0213 causes return to the main program at the correct instruction.

(6) JMP instruction. This is an unconditioned jump and transfers control of the program to the location indicated. It is represented by a logic 1 in bits 1 and 3. Fig. 6.32 illustrates its use in jumping to location 0207 from 0202, so that the intermediate instructions are in effect missed out.

Address	*Contents*
0200	–
0201	–
0202	5207
0203	–
–	–
–	–
0206	–
0207	–
–	–

Fig. 6.32. Illustration of the JMP instruction.

(7) Input–output instructions. The format for this instruction is shown in Fig. 6.33. A logic 1 in bits 1 and 2 tells the microprocessor that this is an input–output instruction. It then looks at bits 4 to 9 to determine which device is to be addressed. As explained in section 6.5.1. several peripherals can be connected to the data bus each having its own unique code to which it responds. The last three bits specify the operations needed on this peripheral. A 1 in position 10 will mean transfer information between microprocessor and peripheral. A 1 in position 11 indicates that the ready status flag of the peripheral should be checked and the next instruction skipped if it is set. Finally a 1 in position 12 clears the flag. The three instructions would be used in a programmed data transfer mode as shown in Fig. 6.34. The flag is initially cleared to indicate that the peripheral, which is assumed to be slower than the microprocessor, is busy. The next instruction at location 0202 tests the flag and as it is cleared it does the next instruction which loops it back to 0202. Therefore the microprocessor sits waiting for the peripheral to become free to transfer data. When this happens the peripheral sets its flag so that instruction 0203 is skipped and 0204 is performed.

Data transfer using the interrupt mode can occur by jumping to a service routine when the interrupt line is energized, as described in section 6.5.1. This routine can be located at a fixed area in memory, designated by the hardware design. The routine determines which device has caused the interrupt by checking all their flags and then jumping to the appropriate service routine. Alternatively a vectored interrupt system can be used, as described in section 6.5.1, which automatically sends the program to the correct service routine. Most microprocessors have the ability to turn the interrupt facility on or off by software control. For our present microprocessor these instructions will be designated by 6001 and 6002 respectively.

(8) Operate instructions. These instructions have a logic 1 in all the three code bits. There are two groups of instructions determined by a logic 0 or a logic 1 in bit 4. A 1 in any of the other

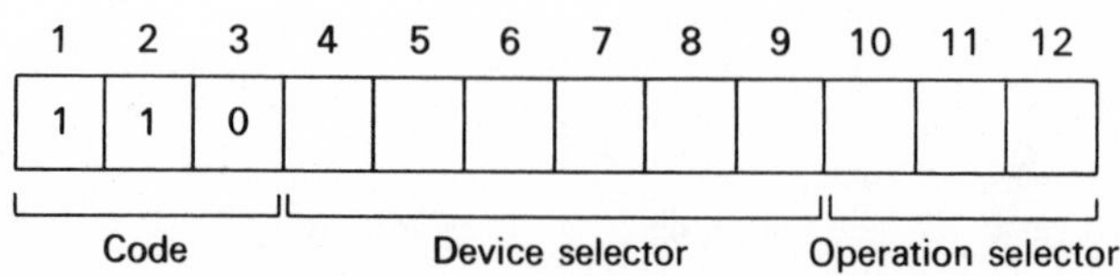

Fig. 6.33. Input–output instruction format.

Address	*Contents*
0200	–
0201	Clear flag
0202	Skip if flag set
0203	Jump to 0202
0204	Transfer information
0205	–
0206	–
–	–

Fig. 6.34. Programmed data transfer.

bit positions indicated in Fig. 6.35 will perform the instructions, as follows

CLA– clear accumulator
CMA – complement accumulator
RAR – rotate accumulator one place to the right
RAL – rotate accumulator one place to the left
RES – return from subroutine (i.e. load latest entry address stack into program counter)
IAC – increment accumulator
SMA – skip on minus accumulator
SPA – skip on positive accumulator
SZA – skip on zero accumulator
SNA – skip on non-zero accumulator
HLT – halt

The choice between the alternatives in bits 6 and 7 of Fig. 6.35 (*b*) is made depending on whether bit 9 is a 0 or a 1. Fig. 6.35 indicates very forcefully the effect of the instruction word in presenting logic 1s at various locations in order to activate internal circuitry and to carry out specific routines by the microprocessor. Places left blank can be used for other commands which are not specified at this stage. It is, of course, possible to combine several instructions within a single word. For instance a 1 in bits 5 and 7 and a 0 in bit 4 will mean that the accumulator will be cleared and then complemented leaving it with all ones.

6.5.3 *Building a microprocessor based system*

In developing a microprocessor based system it is necessary to go through several stages. Clearly there is the problem of choosing a microprocessor which meets the technical and economic requirements of the application. To do this one needs to spend a considerable amount of time defining the problem and drawing extensive system flow charts. This is probably the single most important part of the whole operation and should not be rushed. Fig. 6.36 shows, by means of a flow chart, the alternative routes which may be followed once the problem has been defined.

In writing the software the programmer has the choice of three different languages. Machine language is the most basic and consists of strings of numbers of the type described in the last section. This is the only language which the computer can understand since a computer requires the logic 0 and 1 signals at the correct spaces to cause it to go through its various operations. However machine language coding can be time consuming when large programs need to be written. One can use

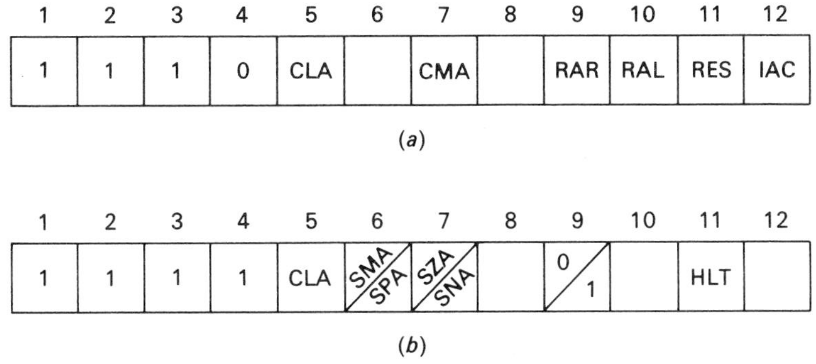

Fig. 6.35. Operate instruction formats; (*a*) group 1, (*b*) group 2.

English letters to represent the numbers. For instance CLA was used to indicate the clear accumulator instruction which was 111010000000 or 7200 in octal code. This is called mnemonic coding. The computer cannot understand the term CLA so that it is necessary to convert this into its machine language equivalent of 111010000000. This is done by a program called an assembler. The assembler basically gives a direct correspondence between the mnemonic and machine language codes. It does however have several other facilities such as allowing memory addresses to be 'labelled' by names instead of absolute addresses. This is illustrated in Fig. 6.37. Supposing we wish to do two different routines depending on whether or not the number stored in a particular location is positive. The instructions first clear the accumulator and then bring this test word into the accumulator. The word has been labelled X so that it is written simply as TAD X and the programmer is freed from the drudgery of remembering absolute addresses. Note however that one cannot completely ignore the locations in memory since it is still only possible to directly address two pages, that is page 0 and the current page. If the test word was stored on a different page to these two then indirect addressing would be required. For this an *I* is used and the TAD *X* instruction changes to TAD I X. Location X will now contain the address of the test word instead of the word itself.

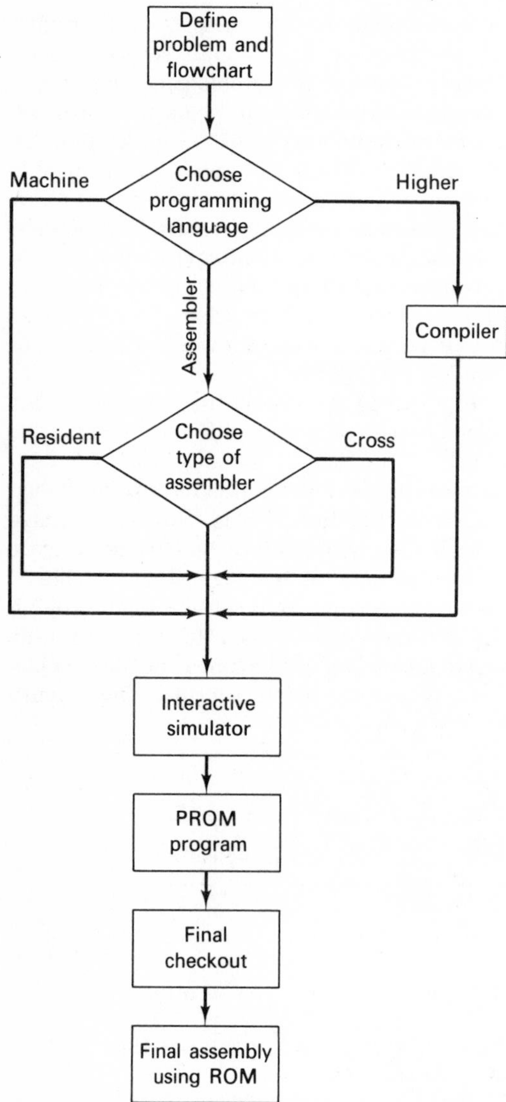

Fig. 6.36. Alternative routes in developing a microprocessor based system.

Once the word has been loaded into the accumulator, in Fig. 6.37, it is tested by the SPA instruction. If it is positive then the next instruction is skipped and routine *A* completed. If the word is not positive then the program jumps to location *L* and does routine *B*.

It is usual when developing a microprocessor based system to use a prototyping aid. This is similar to a minicomputer and has facilities such as expanded memory, interface for teletype or visual display units, power supplies, control panel, programming system for PROMs, and various software systems. The user's program can be loaded into a RAM inside the prototyping aids using the teletype or VDU and it is usually translated into machine language by a resident assembler. The prototyping aid has an on board microprocessor similar to the type the user wishes to employ. It can therefore be plugged into place within the total system and used to drive the machine. Errors will be indicated and these can be corrected by various de-bug and edit packages. When the user is satisfied with his program he can transfer it onto a PROM and finally into a ROM if the production volumes are large enough.

Sometimes it is possible to use a cross-assembler instead of a resident assembler. This resides in a much larger computer than the prototyping aid and as such it can be made more complex with added facilities. The cross-assembler will accept the mnemonic text, and output a paper tape which is suitable for loading into the prototyping aid.

Assembler or mnemonic coding represent a degree of sophistication over machine language. It is possible to use higher languages which make coding even simpler. The user can now largely forget about the details of the microprocessor hardware and concentrate on his program. However the end result must still be a machine language program. To obtain this conversion requires a complex software translator called a compiler which is used in conjunction with a larger computer.

Fig. 6.38 illustrates the difference between an assembler and higher level language. It would be a useful exercise for the reader to follow through this program using the flow diagram. Although the higher level language is easier to use it is less efficient in terms of operating speed and program space in memory. This is primarily due to the fact that a compiler is not as good as the human programmer when it comes to translating between higher level language and machine code. A trade-off occurs between ease and cost of programming, and the amount of hardware used. This choice is usually governed by the total number of systems which are to be built.

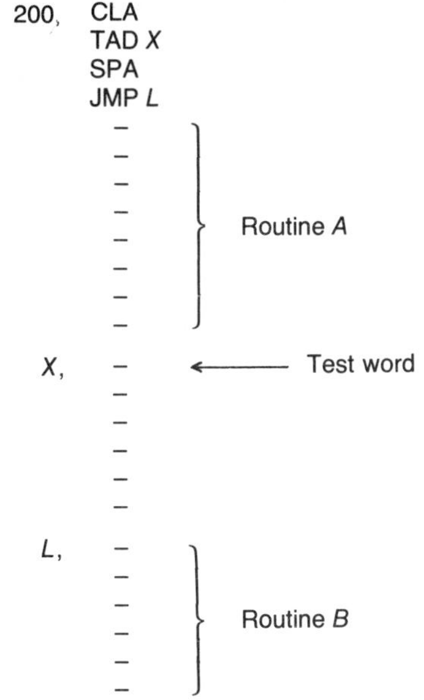

Fig. 6.37. Illustration of the use of labels.

Returning to the flow chart of Fig. 6.36, once the program has been written it can be loaded into the prototyping aid, which acts as a hardware simulator, or else it can be used with a software simulator. This runs the program as if it was actually controlling a machine and allows the user to observe and correct errors. Software simulation cannot check out a system completely and the final test is whether the program works when it is on PROM and controlling its own microprocessor.

6.5.4 *A comparison of systems*

Four types of universal logic elements have been described in this chapter. These are the ULA, ROM, PLA and microprocessor. In addition one can use random logic or a minicomputer for logic implementation. How does the user decide between these various types?

Generally a ROM or a PLA can be used for a relatively uncomplicated combinational or sequential system only. A logic gate can typically be replaced by between 5 and 20 bits of ROM. Therefore if one assumes that an average integrated circuit package, used in a random logic system has about 10 gates, then between 50 and 200 bits of ROM are equivalent to a package. This means that a 16K bit ROM can replace from 800 to 3200 gates or 80 to 320 integrated circuit packages. A 16K ROM is available as a single package so that the advantage of such a replacement is very attractive. Not only is reliability considerably increased, but the cost is also likely to be much less. This is specially so when one considers the overheads associated with an I.C. package. Some of these are listed in Fig. 6.39 from which it is seen that a dollar I.C. costs over three dollars by the time it is ready for use in a production assembly.

The choice between a ROM and PLA is usu-

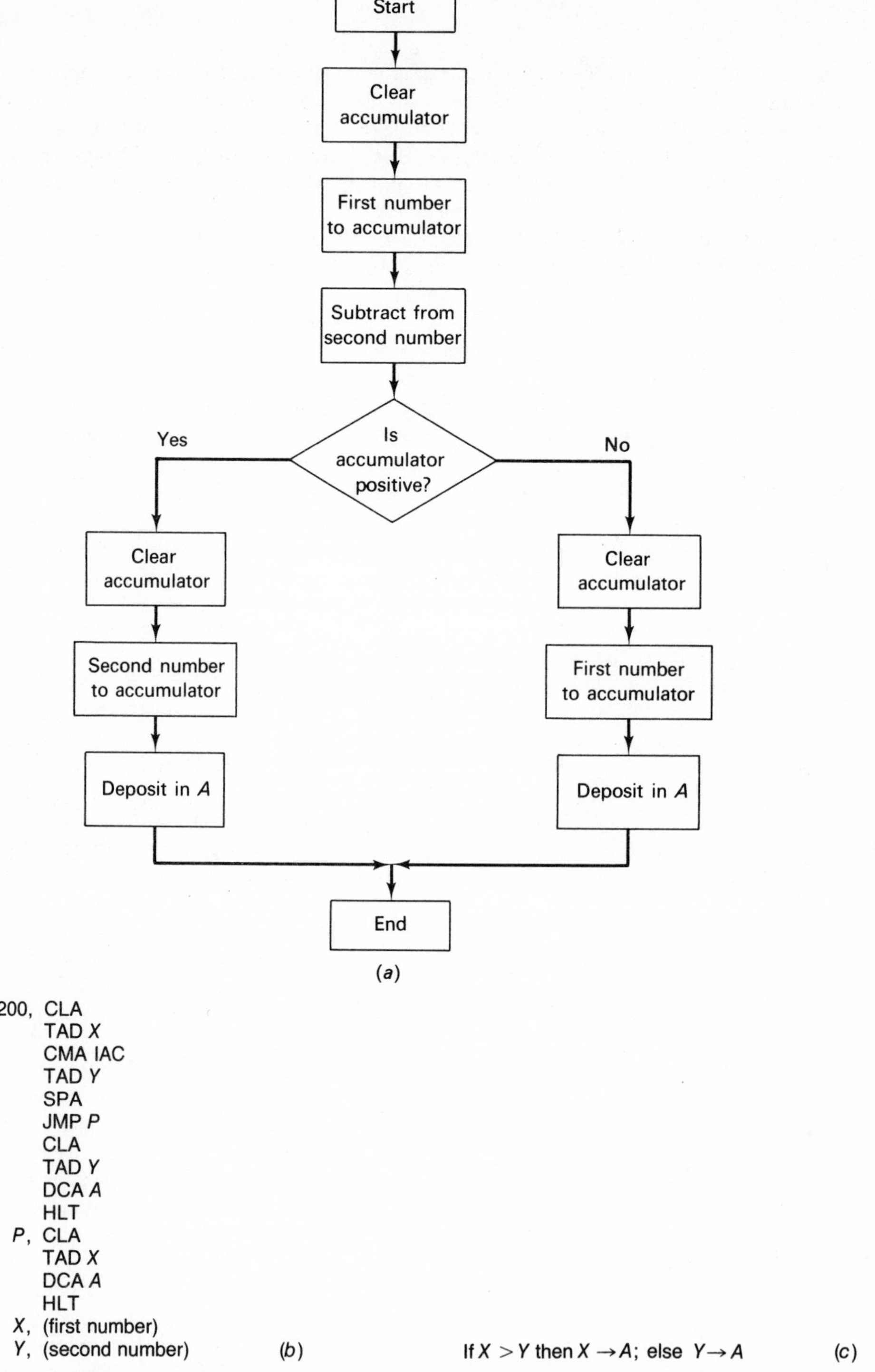

Fig. 6.38. An example of a program for placing the larger of two numbers, which are stored in locations X and Y, into location A; (a) flow diagram, (b) mnemonic coding, (c) high level language coding.

Item	Cost (percentage of total)
Basic device	30
Printed circuit board	20
Assembly	10
Test and rework	5
Connector	5
Wiring	5
Power	5
Cabinet, etc.	5
Inventory and spares	5
Inspection and burn in	10
	100

Fig. 6.39. Costs involved in using an integrated circuit.

ally determined by the number of input variables and the number of different product terms generated. It was seen in section 6.4 that these factors are primarily responsible for the size differences between the two devices. However, both these devices can only be used effectively for combinational logic and are not as flexible as a random logic or microprocessor system.

The microprocessor represents a development of the ROM and PLA and often incorporate both devices. It can vary in complexity from a single chip to a large assembly resembling a minicomputer. The microprocessor can perform all the functions available from random logic and has the added flexibility of being software programmed so that changes can be made relatively easily. However a microprocessor is fairly complex and it is this complexity which gives it the ability to adapt to a range of applications. Therefore a microprocessor is more expensive than a random logic system if the application requires only a few logic functions. A microprocessor is also slower than random logic. The decision tree for choosing between microprocessors and random logic is shown in Fig. 6.40 and it highlights the points raised in the earlier discussion. Other advantages of microprocessors are the reduction in cost and increase in reliability due to the use of fewer packages. The microprocessor itself replaces about 50 to 200 general purpose I.C. packages and, as seen earlier, each 1K bits of ROM used with the microprocessor will replace between 5 and 20 additional packages. Microprocessor systems can also be developed faster once the initial software and hardware learning phase has been completed. This means that products can be put on the market ahead of competitors and modifications can be readily made to adapt to changed circumstances. The flexibility of the microprocessor also allows one to put a standard product on the market and then to adapt it to meet the specialized requirements of the individual customers.

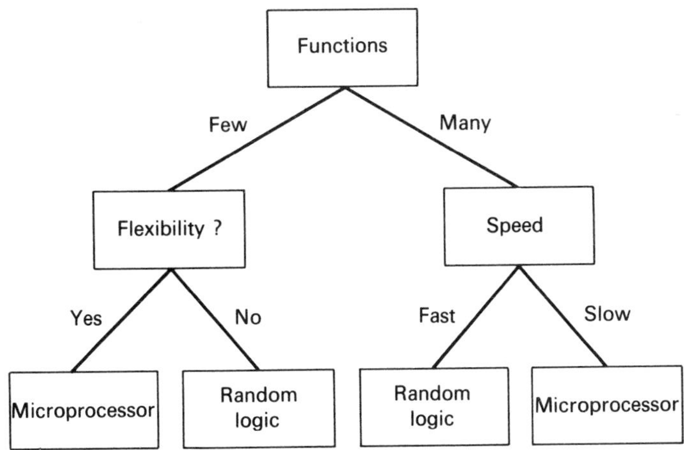

Fig. 6.40. Decision tree for choosing between a microprocessor and random logic.

Fig. 6.41 compares the cost per function differences between a minicomputer, microprocessor and random logic implementation of a system. The complexity and speed of the system are considered to be the variables. The cost per function of a random logic system is relatively unchanged with complexity since packages can be added or removed as needed to meet the system requirements. However, the minicomputer carries a substantial overhead in terms of power supplies, minimum size CPU and memory, so that it is very expensive when the system complexity is low but the cost per function falls rapidly when complexity increases. The microprocessor carries some overhead but less than that of a minicomputer. Its curve therefore lies between that of the minicomputer and random logic. Between points *A* and *B* in Fig. 6.41 (*a*) the microprocessor represents the most economic system. With the trend towards inexpensive one-chip microcontrollers and also towards fast complex systems the microprocessor is clearly expanding its area of competitiveness.

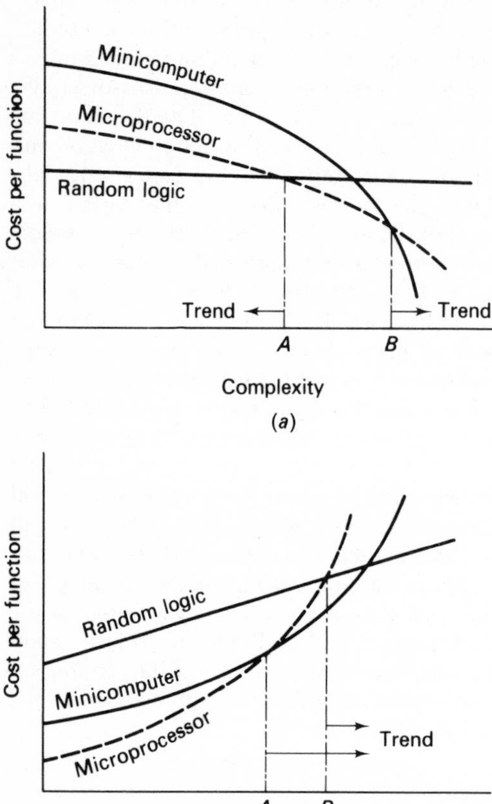

Fig. 6.41. Comparison between random logic, microprocessors, and minicomputers; (*a*) complexity considerations, (*b*) speed considerations.

If a system of medium complexity is considered, then for low speeds the microprocessor gives the lowest cost per function and random logic gives the highest. However, random logic can work at a higher absolute speed than either of the other two and its rate of increase in cost per function with speed is less than the others. Once again the trend is clearly towards faster microprocessors which can operate economically at higher speeds.

The ULA is generally a competitor to random logic only since if ROMs, PLAs or microprocessors are usable they are likely to be more economical. For mixed analogue and digital systems the ULA or random logic is the usual choice. Generally the ULA will be cheaper than random logic provided the quantities are large enough to amortise the initial development cost.

7. Linear integrated circuits

7.1 Introduction

Integrated circuits established themselves first in digital applications. It is only relatively recently that linear integrated circuits have begun to achieve the same sort of popularity as digital circuits. There are two prime reasons for this. Firstly a digital circuit is much less exacting than a linear circuit and it is capable of operating with much wider process tolerances. It was therefore only after considerable experience and control had been obtained over integrated circuit processes that they began to be applied to linear systems. The second reason for the late development of linear circuits was the market size. Digital integrated circuits found a vast and ready market in computers and associated peripherals, which required many identical circuits per system. It is only recently that market for the linear circuits has begun to approach this sort of potential with its entrance into consumer fields.

It was understandable that the first linear circuits to be available commercially were general purpose operational amplifiers which were intended to cover a wide range of applications. These early devices spawned a large family of amplifier circuits and related systems such as voltage comparators. At present, operational amplifiers still dominate the linear integrated circuit field. However, they have now been joined by other classes of devices, which are described in this chapter. These include voltage regulators, analogue to digital and digital to analogue converters, analogue storage devices, analogue switches and arrays, phase locked loops, and finally consumer circuits for radio, television and other consumer equipment.

7.2 Linear integrated circuit processes

Unipolar processes are not suited to the fabrication of high performance integrated circuits so that bipolar technology is the one most often used. There are several reasons for this. Bipolar devices, for instance, have a much lower 'on' resistance than unipolar components. This enables higher output currents to be obtained and the power dissipation to be minimized. Both these are important considerations since linear circuits, due to their operating mode, dissipate considerable power. Furthermore they interface directly with external systems and need to be capable of providing output power.

Bipolar transistors can be built with closely controlled turn on voltages and this parameter is matched to within a few millivolts for devices on the same silicon chip. For unipolar transistors the turn on voltage is less predictable and close matching is difficult to achieve. Therefore unipolar amplifiers designed with conventional differential circuits are not successful.

The advantage of a unipolar device over a bipolar device is its much higher input impedance. This characteristic is desirable in operational amplifiers since ideally these should take zero input current. Unipolar technology is therefore employed in the input stages of some amplifiers. CMOS is also used fairly often. It gives a *load-driver* arrangement which has high voltage gain, low power dissipation, good linearity and good temperature stability.

The characteristics of unipolar devices are ideally suited to two applications. These are analogue storage and analogue switching and multiplexing. In the former case the low input current and process simplicity of the unipolar device is used, whereas in the second case its bidirectional properties are important. These and other applications are described in the sections which follow.

7.3 Linear arrays

There are a variety of transistor arrays which are available in monolithic form. Some of these were described in chapter 3 since they are more frequently operated in a power switching mode. Other arrays contain a combination of matched Darlington pairs and these are clearly intended to be used in linear applications.

Another important class of linear arrays is the analogue switch. In concept its operation is very simple and is illustrated in Fig. 7.1 (*a*). The d.c. control line operates the switch which is capable of handling analogue signals. A variety of switches are available integrated on a single chip. Fig. 7.1 (*b*) shows a change over version and a multiplexer is illustrated in Fig. 7.1 (*c*). These integrated circuit switches have the advantage over conventional relay types in being

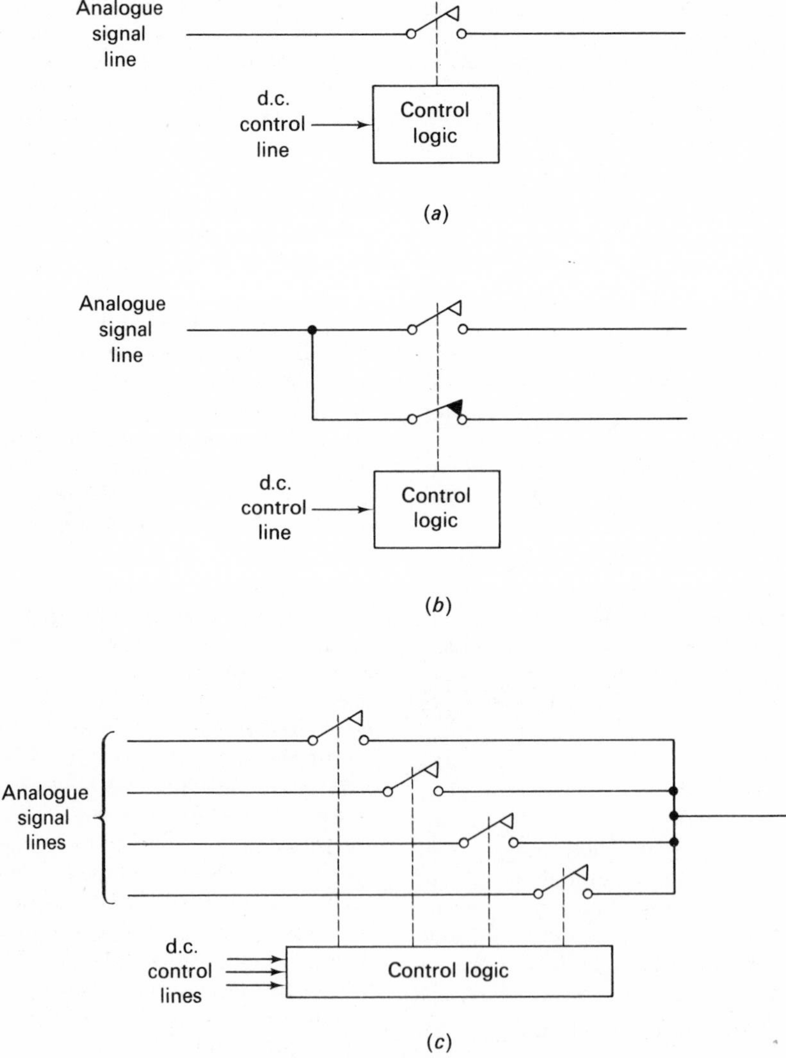

Fig. 7.1. Switches and multiplexers; (*a*) single pole, normally open switch, (*b*) single pole, change over switch, (*c*) four line multiplexer.

faster, operating at lower power dissipation, having logic compatibility, and attaining lower costs per switch since several devices can be integrated on a single chip. The disadvantages are that the switch resistance is higher than that of a mechanical contact and its voltage rating is lower.

Unipolar technology is used more commonly than bipolar in switches. It is capable of higher voltages, lower offsets and bidirectional operation. Both junction field effect transistors (J–FET) and metal oxide semiconductor field effect transistors (MOS–FET) are common. Fig. 7.2 illustrates the differences between these two. For the J–FET a channel exists between source and drain until it is depleted by a negative gate voltage. Therefore in operation the system is very similar to the depletion mode MOS–FET shown in Fig. 7.2 (*c*). The enhancement mode device requires a positive gate voltage to turn it on. The turn off voltage for a J–FET and a depletion mode MOS–FET is called the pinch off voltage (V_P) and the turn on voltage for an enhancement mode MOS–FET is the threshold voltage (V_T).

A unipolar technology which is very commonly used for analogue switches is CMOS. It has many advantages, as described later, but also some problems. One of the commonest is the possibility of latch up. Fig. 7.3 (*a*) shows the arrangement of a CMOS switch with guard bands. Under normal operating conditions the source of the p-channel transistor is connected to the positive supply terminal and the source of the n-channel transistor to the negative terminal. However, due to parasitic action a pnpn structure now exists across the supply, as shown in Fig. 7.3 (*b*). This means that a low impedance path can occur across the supply under fault conditions and this will destroy the switch unless external current limiting is employed. Several techniques have been used to reduce this parasitic pnpn action. These include floating the body of the n-channel device by omitting the negative connection; using a buried p^+ layer under the n-channel body so as to reduce the gain of the parasitic pnpn diode to less than unity and so avoid the latching action; and using dielectric isolation, as in Fig. 7.3 (*c*). These solutions are offered by

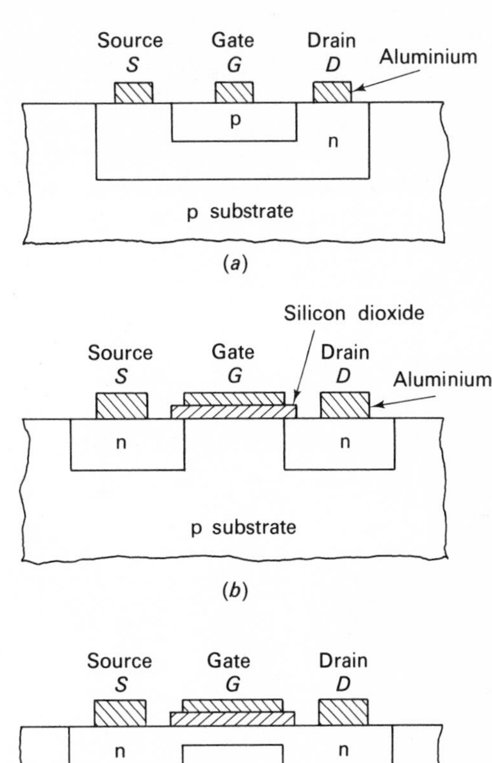

Fig. 7.2. Structural differences between J–FET and MOS–FET; (*a*) N channel J–FET, (*b*) N channel enhancement mode MOS–FET, (*c*) N channel depletion mode MOS–FET.

competing manufacturers and they all introduce some advantages and disadvantages.

In using analogue switches several parameters are of importance. The on resistance should be as low as possible. Furthermore this resistance should not vary appreciably with the signal level or else distortion will occur. Unfortunately the on resistance of a unipolar transistor is very dependent on the magnitude of the gate to source (signal) voltage.

The conductance of a unipolar transistor is given by the equation

$$g_{DS} = K \frac{W T}{L} \mu N_c \qquad (7.1)$$

Where L, W and T are the length, width and thickness of the channel, μ is the carrier mobility, N_c the mobile carrier concentration and K is a constant. For a fixed channel dimension a J–FET transistor has a higher conductance and a lower change in g_{DS} with the signal than a MOS–FET. However it was seen from Fig. 7.2 that the former are more difficult to integrate on a silicon chip. n-channel transistors have greater mobility and therefore higher conductance than p-channel. It is possible to increase the channel dimensions to give higher

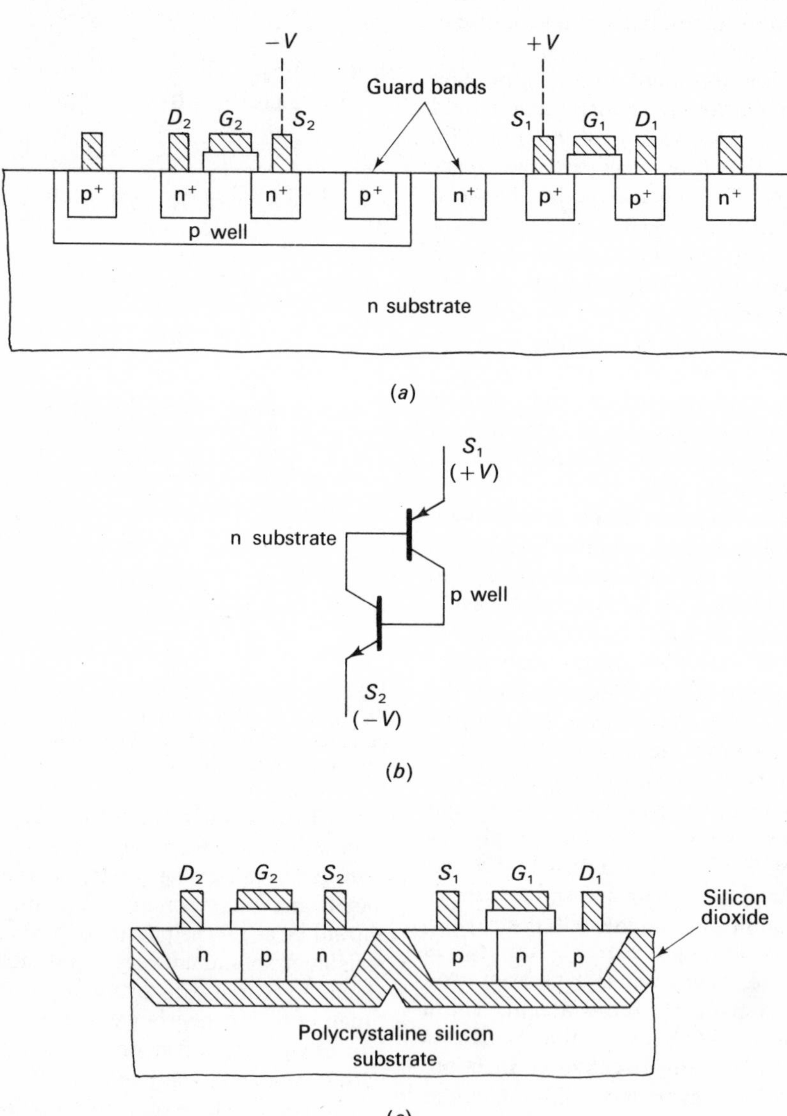

Fig. 7.3. Latch up phenomenon in CMOS switches; (a) conventional switch arrangement, (b) equivalent circuit of the parasitic four layer pnpn diode, (c) avoiding latch up by dielectric isolation.

values of conductance but this increases the leakage current, parasitic capacitance and cost. Since the CMOS switch is made from a combination of p- and n-channel devices its resistance variation with the signal level is much lower, although in practice it is still greater than a J–FET. A further advantage of the CMOS switch is that it has no pinch off or threshold voltage so that the analogue voltage can swing almost to the limits of the supply voltage.

The parasitic capacitance of the switch must be low to avoid offset voltages and to give increased speed and better high frequency isolation. A further effect of parasitic capacitance is to introduce signal coupling, called cross talk, between switches which are integrated on the same chip. Switches made from double diffused MOS technology (DMOS) are available and these have low parasitic capacitance so that they can operate at high speeds with very little interference between switches.

7.4 Charge transfer devices

There are two types of components which operate by transferring charge from one location to another on a silicon chip. These are known as bucket brigade, and charge coupled devices. Fig. 7.4 (*a*) shows a discrete component representation of a bucket brigade device. Signals are stored on capacitors and they can be moved from one location to the next by clock pulses which turn on the relevant transistors. Two facts should be noted from this figure. First, since capacitors are used for storage it is possible to handle digital or analogue information. For digital storage a fixed charge represents logic 1 and the absence of this charge is logic 0. For analogue storage the quantity of the charge

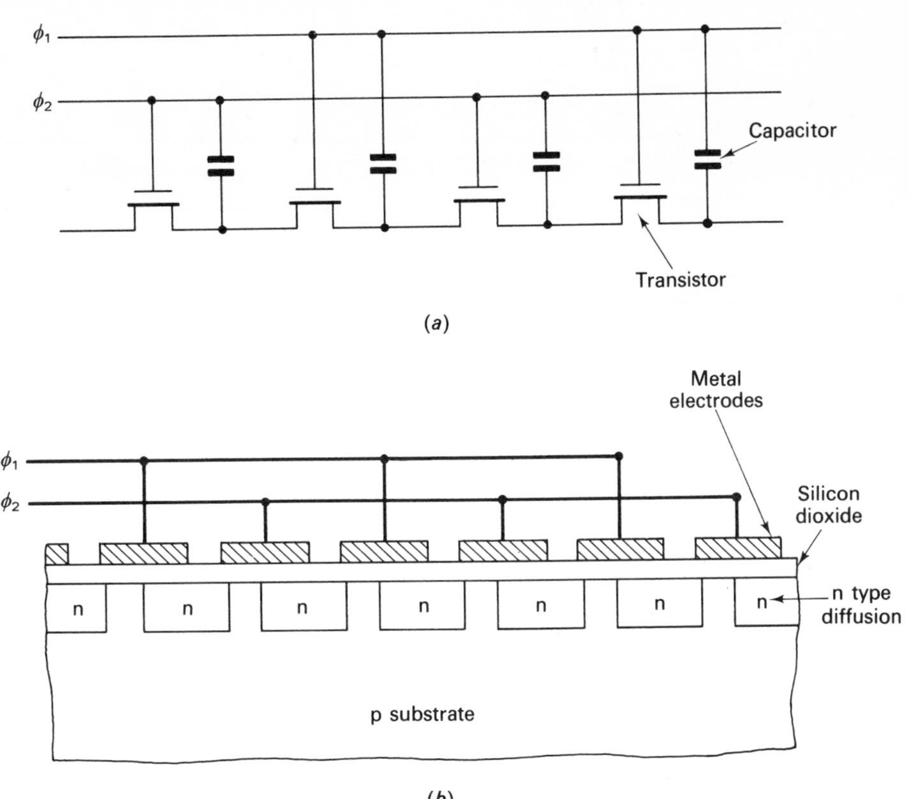

Fig. 7.4. Bucket brigade device; (*a*) discrete version, (*b*) integrated circuit version.

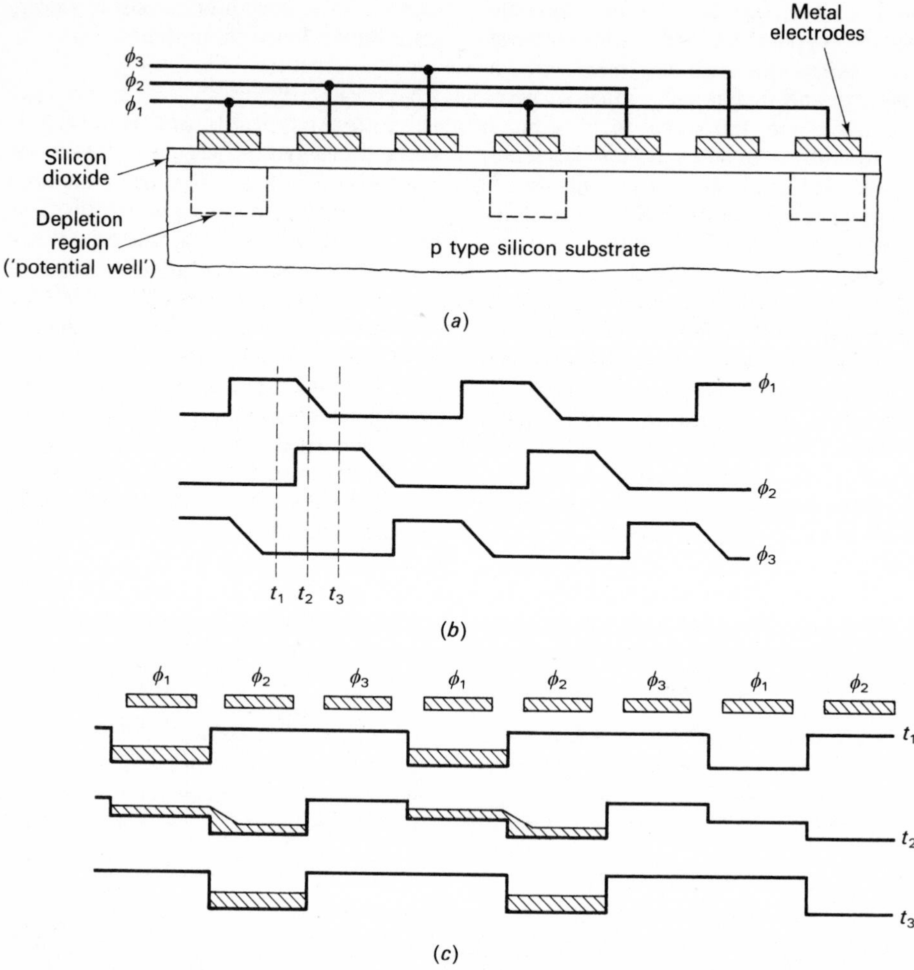

Fig. 7.5. Basic n type three phase CCD; (a) construction, (b) timing waveforms, (c) potential well profile.

is determined by the level of the analogue signal. The second point to note is that the transistors have finite leakage when off so that this memory must be operated in a dynamic mode.

The bucket brigade device is most frequently used in integrated circuit form and this is shown in Fig. 7.4 (b). In principle it operates very similarly to the discrete version except that capacitors are formed by the overlap of the metal electrodes on the n type diffusions so that the construction is greatly simplified.

Fig. 7.5 shows the construction of one form of charge coupled device (CCD). It is seen to be simpler than the bucket brigade device (BBD) since no diffusions are used. To explain the operation of the CCD it is necessary to examine the silicon area under each electrode. Normally this contains majority and minority carriers which, in the case of the p type substrate, are holes and electrons respectively. If a positive voltage is now applied to any electrode it will deplete the region underneath of holes and create a potential well. If the positive voltage is high enough it will even cause the silicon, at its interface with the oxide surface, to become inverted and store electrons. In any event the potential well will not last indefinitely since it will soon attract minority charge (electrons) which are thermally generated within the silicon.

The CCD works by storing and moving charge from one potential well to another. In practice the charge is stored at the silicon–silicon dioxide interface but it is convenient to think in terms of charge flowing in and out of a potential well when describing CCD operation. The depth of this well, and therefore the charge handling capability of the CCD, is given by

$$Q \propto \frac{V_G - V_T}{t} \tag{7.2}$$

where V_G is magnitude of the applied electrode voltage, V_T is the threshold voltage and t the thickness of the oxide layer under the electrode.

Fig. 7.5 illustrates a three phase clocking system for the CCD. This clock arrangement is required to ensure that charge is moved unidirectionally along the length of the device. Therefore at time t_1 the electrodes connected to φ_1 are energized and the charge, if present, is stored under these. At t_2 clock φ_1 has reduced so that its potential well is not as deep as that under φ_2. Any charge which was under φ_1 now starts to spill over to the well under φ_2 until at t_3 the transfer is complete. Similar considerations apply for the transfer from φ_2 to φ_3 and from φ_3 back to φ_1. It is only after these three transfers that the charge is moved one complete cycle. Therefore three electrodes are required to define one storage location. From this discussion it is seen that the CCD is very similar to the BBD in that it moves data in serial fashion, it can store analogue or digital information in its wells, and it must be run in a dynamic mode to prevent an empty well filling up due to thermally generated electrons. However, the CCD is simpler in construction and is consequently much more widely used than the BBD.

7.4.1 *CCD construction*

Although the CCD arrangement shown in Fig. 7.5 is widely used it has several disadvantages. The potential wells under the electrodes need to overlap for maximum transfer of charge, and this requires narrow gaps and leads to processing difficulties. The doped polysilicon arrangement shown in Fig. 7.6 (*a*) affords some advantages since the polysilicon layer provides protection to the CCD during subsequent process steps. Fig. 7.6 (*b*) shows an overlapping electrode arrangement in which the gap problem is largely avoided. The top electrode may be made from aluminium or polysilicon. This structure is more commonly used with two phase or four phase clocking arrangements. Fig. 7.7 shows one form of two phase clocking. Since directionality of charge transfer cannot now be obtained from the clock waveform alone, as for three phase systems, transfer barriers are made by implanting regions under the electrodes as shown. From the timing and potential well diagrams it is seen that the shift of charge from under a φ_1 electrode to a φ_2 electrode occurs on a φ_2 positive pulse and vice versa. The two phase arrangement has the advantage of a simpler clocking requirement and less space per storage location, but it also generally has a lower charge carrying capability than a three phase structure.

A further consideration in the construction of a CCD is to confine the charge transfer channel to a small area in the active region of the silicon and to prevent it spreading along the length of the electrodes. This is illustrated in Fig. 7.8 (*a*) in which the transfer channel is seen to be only a small region in the centre of the transfer electrodes. Fig. 7.8 (*b*) shows one method by which this may be achieved, the transfer of charge now occurring into the paper. A thick oxide is used for all areas except where the transfer channel is to occur. MOS action is prevented under this thick oxide so that no potential wells can be formed. The thick oxide is usually deposited by chemical vapour deposition techniques. The disadvantage of the thick oxide method of channel confinement is that an extra process step is required, and the electrodes, which are very thin since they need to be closely spaced, have to traverse the large step in the oxide. This makes them susceptible to cracking and reduces efficiency. An alternative method for channel confinement is shown in Fig. 7.8 (*c*). A heavily doped diffused layer is used in the areas

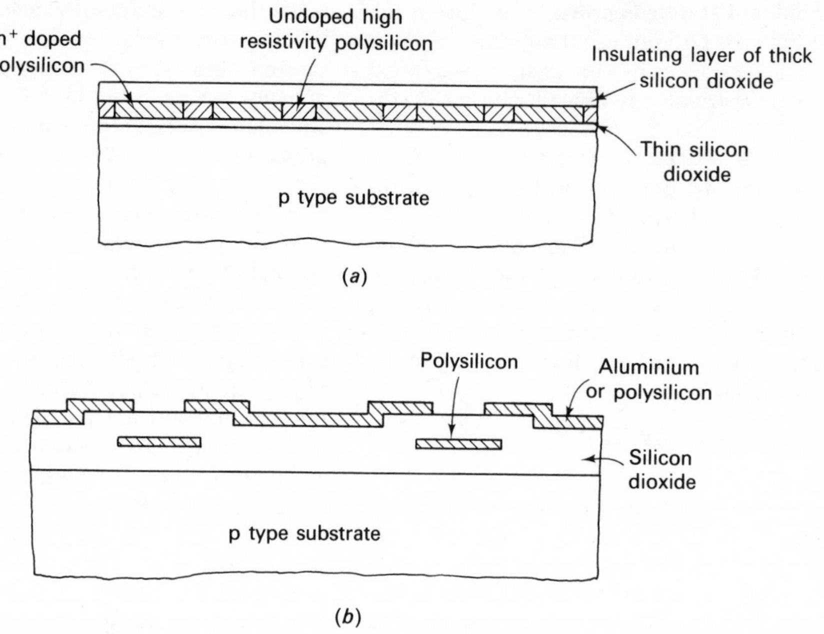

Fig. 7.6. CCD construction variations; (a) doped polysilicon gate, (b) two level gate.

where the channel is not allowed to spread since the depletion voltage required for this is now much greater. Although this method again requires an extra process step it does not result in a step in the conductors.

7.4.2 *CCD performance*

The most important parameter to be considered in the performance of CCD devices is its charge transfer efficiency (η). The charge in a CCD resides at its silicon–silicon dioxide interface where the potential for minority carriers is minimum. Loss of this charge due to recombination is negligible since the same potential which stores the minority carriers also repels majority carriers. However some charge is left behind after each transfer operation and this not only distorts the analogue signal but repeated shifts can so reduce the value of the charge that it may be falsely recorded in a digital system. There are two reasons for this residual charge at each transfer. These are the time needed for transfer and the trapping effects which occur at the silicon–silicon dioxide surface. The speed at which charge moves from one potential well to another is determined by the charge repulsion effect, by thermal diffusion and by the drift induced by the fringing effects of the externally applied fields. The repulsion effect becomes unimportant after about 99 per cent of the charge has been transferred and thereafter the remaining two effects cause a roughly exponential decay of charge with time. These effects fix the maximum operating speed for the CCD.

Charge loss by trapping occurs in the 'fast states' which exist at the silicon–silicon dioxide interface. The effect can be minimized by continuously circulating a small background charge, of magnitude equal to about 10 per cent to 20 per cent of the full signal level charge. This is called a 'fat zero' and it keeps the fast states full so that no further trapping occurs in them when the signal arrives. However since the edges of the potential well are sloping, as shown in Fig. 7.9, charge trapping still occurs along these edges. The trapping effect can be considerably reduced by using a buried channel arrangement, as shown in Fig. 7.10. It differs from the surface channel devices described so far in having a layer beneath the oxide which is doped with opposite polarity to the substrate. The potential well now attains a minimum at a location remote from the silicon dioxide surface so no trapping occurs. Therefore the buried channel CCD is capable of operating without a fat zero at much higher frequencies. However it also has lower signal handling capabilities and is more complex than a surface channel CCD.

Another important parameter for a CCD is its dark current. This comes from the thermally generated hole–electron pairs, which tend to fill the potential wells and distort the signal. Its magnitude should clearly be as low as possible. The buried channel CCD has a higher dark current than a surface channel device. Power dissipation should also be low in the CCD. Power loss occurs during a transfer operation due to the fall of the carriers through a potential drop. For n phases, and a voltage swing of V_G on the control electrodes, the power loss at a clock frequency f when transferring a charge Q is given by

$$P \propto nfVQ \qquad (7.3)$$

7.4.3 *CCD applications*

So far we have only considered the transfer mechanism within a CCD and have not described any system for putting charge in or reading it out. Several methods may be used, one is shown in Fig. 7.11. The input diffusion acts as a source, the G_1 electrode as a gate and G_2 as the drain. A strobe voltage is applied to G_1

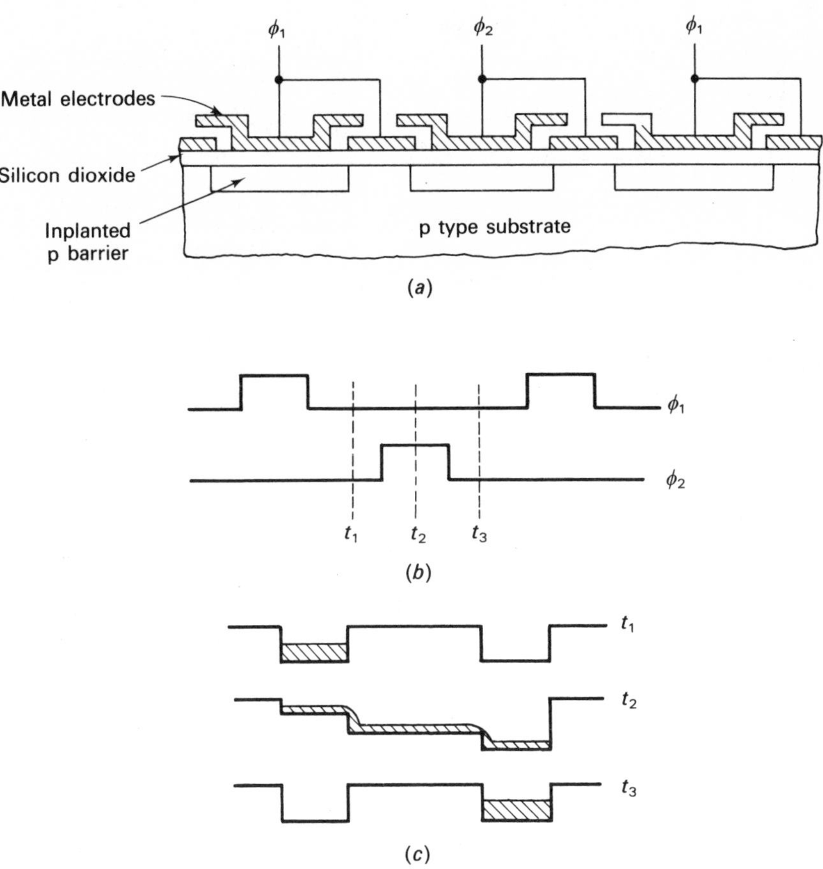

Fig. 7.7. Two phase non-overlapping clock operation of a CCD; (*a*) CCD construction, (*b*) timing waveforms, (*c*) potential well profile.

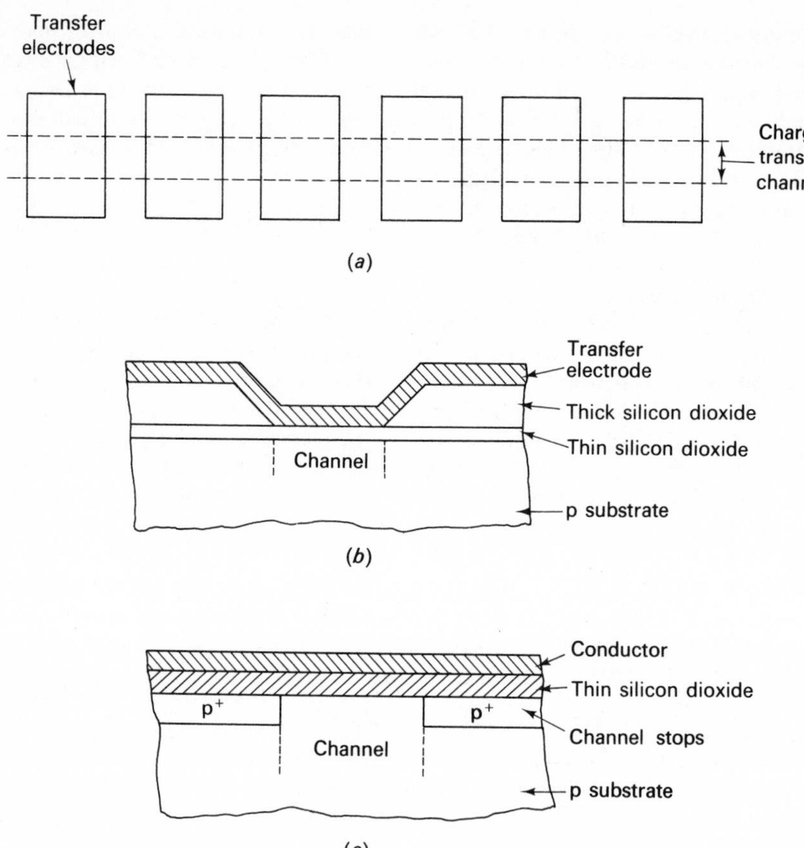

Fig. 7.8. Channel confinement in a CCD; (a) plan view of channel, (b) edge view of thick oxide confinement, (c) edge view of diffusion confinement.

and the signal is fed to the input diffusion. The magnitude of the source charge is therefore determined by the signal level and when the strobe occurs, the first potential well under G_2 will fill to this level. Charge also fills the well under G_1 but when the strobe voltage V_1 is removed this flows back to the source. In the output stage the n diffusion acts as a drain. d.c. bias is applied to G_5 in order to decouple the clock pulse from the drain diffusion. The charge from G_4 crosses over this potential barrier and is detected as an output at the drain diffusion. It is common practice to use amplifiers integrated onto the CCD chip in order to increase the output signal level. These can also be used to give non-destructive readout along the length of the CCD. However, very often the overall speed of the CCD is limited by that of the amplifier.

A CCD may be used for analogue and digital storage or for recording and storing an optical signal. If an analogue signal is applied to the input then it is automatically sampled and shifted along at clock frequency so that the output appears as a delayed pulse amplitude modulated version of the input. This may be amplified and filtered to give the original signal. If f is the frequency of the clock then it can be shown from sampling theory that the maximum bandwidth of any signal which can be sampled is $f/2$. Therefore a low pass filter of this bandwidth is required. Since a delay of $1/f$ occurs at each CCD shift operation, then for an n element CCD the delay is n/f. Therefore the delay–bandwidth product of the CCD is $n/2$. This means that one can design the CCD for any delay or bandwidth so that the device is a variable delay line.

The digital CCD is a special case of the analogue version in which zero and maximum charge are used for the logic 0 and 1 states. The charge, if required, is injected by pulsing the first control electrode at the same time as the input diode. The charge can then be moved along as in a dynamic shift register. The advantage of the CCD for this application is its simplicity so that it can be built in very large memory sizes to give a low cost per bit. However for these long memories a relatively low transfer efficiency could lead to a corruption of the data, i.e. the inability to differentiate between the logic 0 and 1 states. To prevent this regeneration amplifiers are used at periodic intervals along the length of the CCD. These are simple threshold devices which convert the signals to their original levels. Fig. 7.12 shows several ways in which these amplifiers can be connected into the memory system. The serial–parallel–serial arrangement is limited to relatively small memory sizes. It is capable of very high circuit densities since only one amplifier is required, for re-circulating in a dynamic mode. The serpentine arrangement is only limited in size by that of the overall chip. It has a lower circuit density and consumes more power than the serial–parallel–serial arrangement since many more regenerative amplifiers are used. The addressable arrangement is clearly capable of the fastest access times but is also uneconomical in chip area due to the requirement for additional chip peripherals. It also needs many more package pins.

In the imaging application of a CCD the input diode diffusion is not used. Instead the light is allowed to fall on the CCD array. One of the electrodes, say φ_1, is held at a positive voltage to create a potential well. This collects the minority charge from the electron–hole pairs that are generated by the incident light. The magnitude of the charge is proportional to the light intensity and the duration for which it falls on the CCD. After a fixed time the signal is shifted along and read out in serial fashion from the CCD. The light integration period should be much longer than the shift period to avoid undue smears in the signal. Larger area arrays are possible. However the maximum number in the array through which transfer can occur is determined by the charge transfer efficiency. Multiplexing can be used on the chip to increase this number. Fig. 7.13 (*a*) shows an arrangement consisting of two CCD areas. Light imaging occurs in one half and when this is completed the information is rapidly shifted in parallel to a second storage area from which light is excluded. The information can now be read out slowly in serial form without the danger of further smear. An alternative system is shown in Fig. 7.13 (*b*) in which the imaging and storage areas are interleaved in columns. The information from the imaging columns are transferred in parallel to the storage column and then read out in serial form as the video signal.

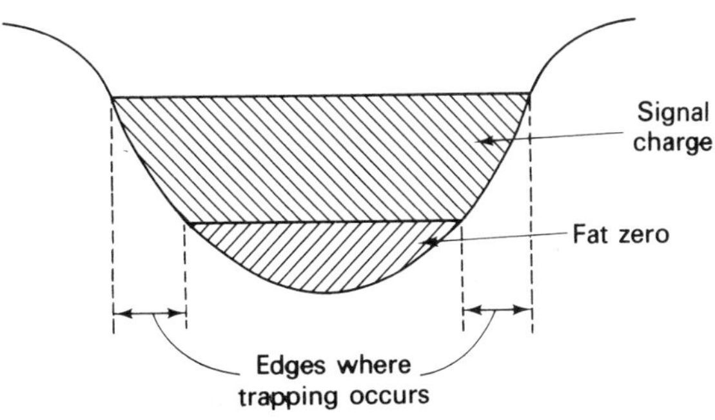

Fig. 7.9. Effect of fat zero in a potential well.

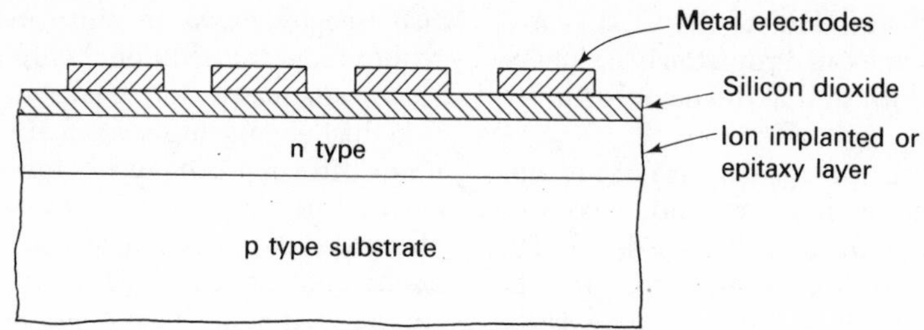

Fig. 7.10. Buried channel CCD.

7.5. The operational amplifier

The operational amplifier was one of the earliest of linear devices to be available as an integrated circuit and it still dominates this field. It would be impossible to cover all the characteristics and applications of the amplifier in this section. All that is attempted here is an introduction to the subject so as to give the reader a basic understanding of the capabilities of this component.

The operational amplifier is intended to be used as a general purpose device whose overall performance characteristics can be altered, by external components, to suit a specific application. However the internal design of the amplifier can also be varied during manufacture, so as to obtain certain special characteristics, and to make it more suitable for different types of applications. These variations will be described later in this section. Most amplifiers, however, tend to use a differential input configuration in preference to single ended inputs. Such an arrangement is shown in Fig. 7.14. It has the advantage that, since its operation is determined by the matching of the transistors in the differential pair, it is not very sensitive to fluctuations of the supply voltage and can be designed for good long term performance by ensuring that the components track each other. Although a differential input single ended output amplifier has been illustrated in Fig. 7.14, and this is the commonest arrangement, in practice it is possible to obtain amplifiers with single inputs and differential outputs.

7.5.1 *Amplifier characteristics*

Before discussing some of the amplifier characteristics two terms need to be defined. The amplifier of Fig. 7.14 has two inputs so that a

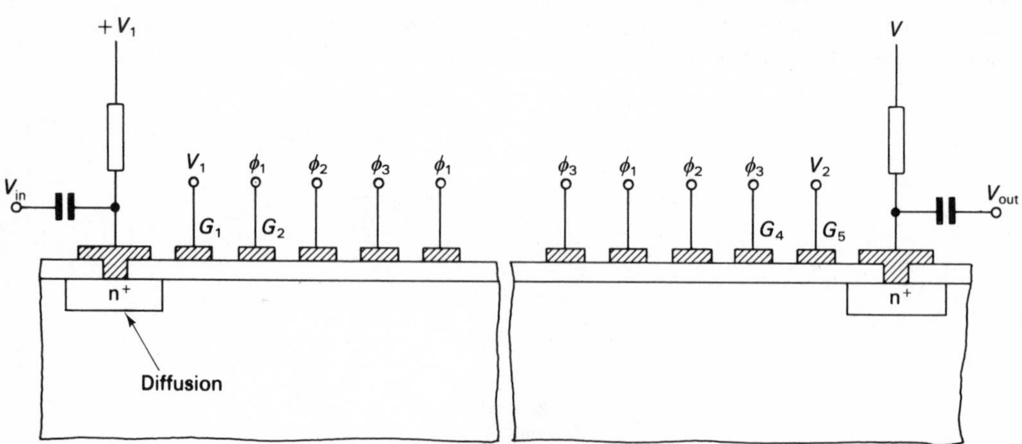

Fig. 7.11. CCD input–output arrangement.

signal can be applied to both of them. Suppose that two signals of 2.0 and 2.1 volts respectively are applied then the differential input to the amplifier is (2.1 − 2.0) volt, or 0.1 volt, whereas the common mode inputs are 2.0 and 2.1 volts with respect to ground. Ideally the amplifier should have infinite gain for a differential signal and zero gain for common mode inputs. In fact the ratio of the differential gain to common mode gain is not infinite and is called the common mode rejection ration (CMRR) of the amplifier and this is one of its most important parameters. The CMRR should be as large as possible and is stated in decibels. It is generally dependent on temperature, frequency and the magnitude of the common mode voltage.

There is a limit to the maximum value of differential and common mode voltages which can be applied to the amplifier. For large differential voltages the amplifier output will run

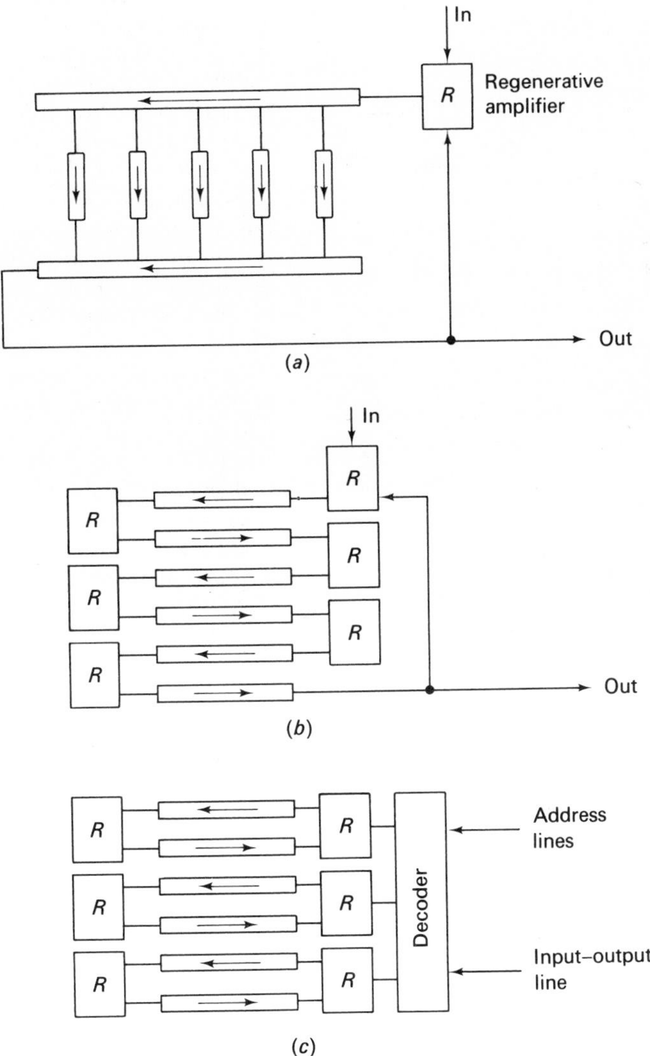

Fig. 7.12. Memory organization; (a) serial parallel serial, (b) serpentine, (c) addressable.

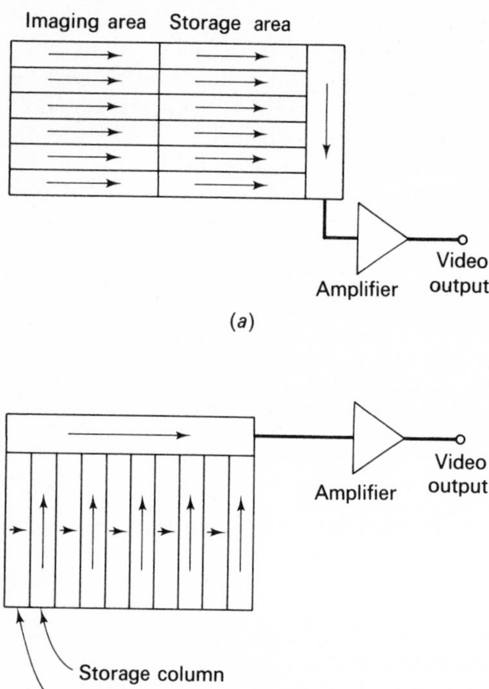

Fig. 7.13. CCD area imaging array; (a) frame transfer, (b) column transfer.

into saturation so that it will not respond to any further change in the input. This is called clipping. Large common mode inputs make the output signal progressively more non-linear, so that the peak value of common mode voltage is usually determined by the distortion allowable by the application.

All operational amplifiers have an error in that the output is not zero for a zero differential input. This error is usually stated in terms of the input offset voltage, which is the value of the input voltage required to force the output to zero. Related to the offset voltage is an offset current and it is this current flowing through the source impedance which forms the offset voltage. The input offset current is therefore the difference of the two input currents to the amplifier when its output voltage is zero. The input bias current is the average of these two currents.

The various amplifier characteristics tend to change in value with ageing and variation in long term conditions such as temperature and supply voltage fluctuations. Temperature effects are generally the most important.

Up to now we have been considering the static characteristics of the amplifier. The dynamic characteristics are equally important and are generally associated with its operating frequency. The gain of an amplifier falls with frequency and a phase shift is introduced between the output and input signals. At low frequencies the gain of the amplifier is equal to its d.c. value. At high frequencies the gain falls off ('rolls' off) at the rate of 20 decibels per decade, which is the same as 6 decibels per octave. This means that for every doubling in frequency the gain falls by 6 decibels, or every time the frequency is increased by a factor of ten the gain falls by 20 decibels. The break point frequency f_1 is the frequency at which the gain has fallen 3 decibels below its d.c. value and at which there is a phase shift of 45°. One can design the amplifier to have several break points. At each of these the roll off increases by a further 20 decibels per decade. However the phase shift also increases by 45° at each point so that it will eventually approach 180° and result in positive feedback. For stability at this point the gain must now be less than one.

Another important consideration is the response of an amplifier to a step input of voltage. After a short recovery or dead time the output voltage commences to change and rises relatively linearly towards its final voltage of $+V$. The time to change from 10 per cent to 90 per cent of this value is called the slew time and the rate at which it achieves this is measured in volts per microsecond, and is called the amplifier slew rate. This is an important parameter as it determines the speed at which the amplifier can operate without undue distortion. Thereafter the output of the amplifier will oscillate about the final value before gradually attaining it. The time required for the amplitude of these oscillations to fall within an acceptable band is called the settling time of the amplifier.

7.5.2 *Amplifier configurations*

Amplifiers are available in a variety of configurations and technologies to meet many dif-

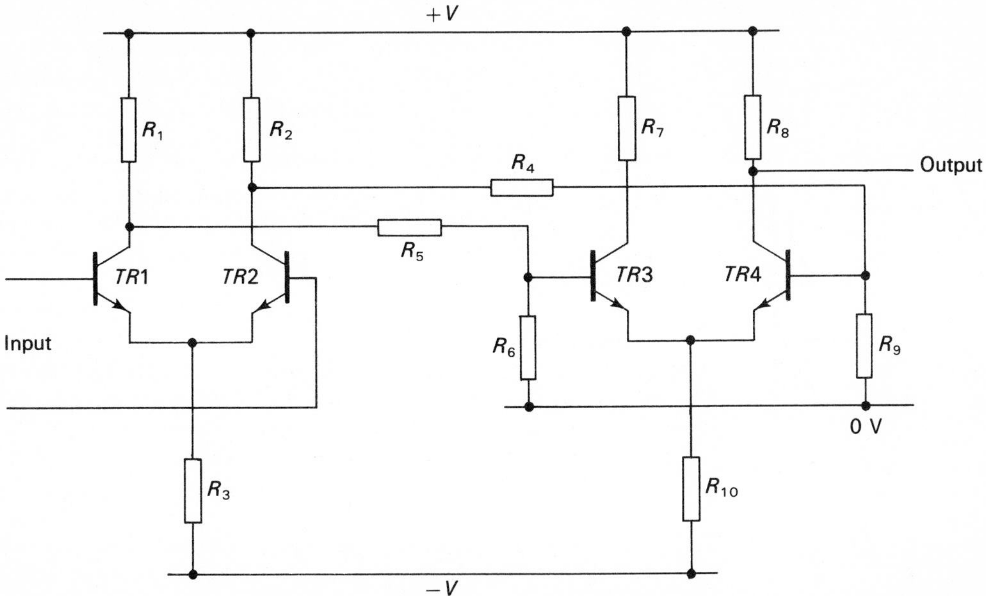

Fig. 7.14. Two stage differential amplifier circuit.

ferent types of applications. Very often the technologies are mixed so as to optimise on the characteristics which are available with each type. For instance an operational amplifier is currently on the market with a MOS–FET input stage, a bipolar intermediate stage, and a CMOS output stage. The FET stage provides a high input impedance and low currents, the bipolar stage is used to give high bandwidth, gain and slew rate, and the CMOS output stage provides large output currents with voltage swings close to that of the supply rails. Other amplifiers are available which use bipolar technology throughout except for the input which consists of a matched FET stage. The advantages are very low input bias and offset current and high input impedance. The FET stage can be built as a monolithic circuit along with the rest of the chip or an external FET chip may be used and the device packaged in hybrid form. The disadvantage of FET inputs is that the bias current doubles for every 10°C increase in temperature so that for a 100°C rise the current will increase by over 1000 times. Although bipolar stages have a much higher initial bias current this increases by less than 1 per cent per °C temperature rise.

Amplifiers are often required in many types of general purpose instruments and this has given rise to the instrument amplifier which is a development of the general purpose operational amplifier. Internal components are now used so as to optimize on several parameters such as input impedance, CMRR and drift. It is also now easy to modify the gain of the amplifier by changing, say, a single external component without upsetting any of the other parameters of the amplifier. Generally the input to the instrument amplifier comes as a low level signal from measuring transducers and the amplifier is especially designed for this application.

In many applications it is desirable to vary certain parameters of the amplifier or to select separate input lines by means of a remote signal. Such a need arises in analogue multiplexing or in programmable power supplies. The programmable operational amplifier has been designed for these applications and is shown in Fig. 7.15 (*a*). It is seen to be equivalent to four conventional operational amplifiers all of which feed into a common output amplifier. The individual input amplifiers can be selected by means of the decode–control circuit which

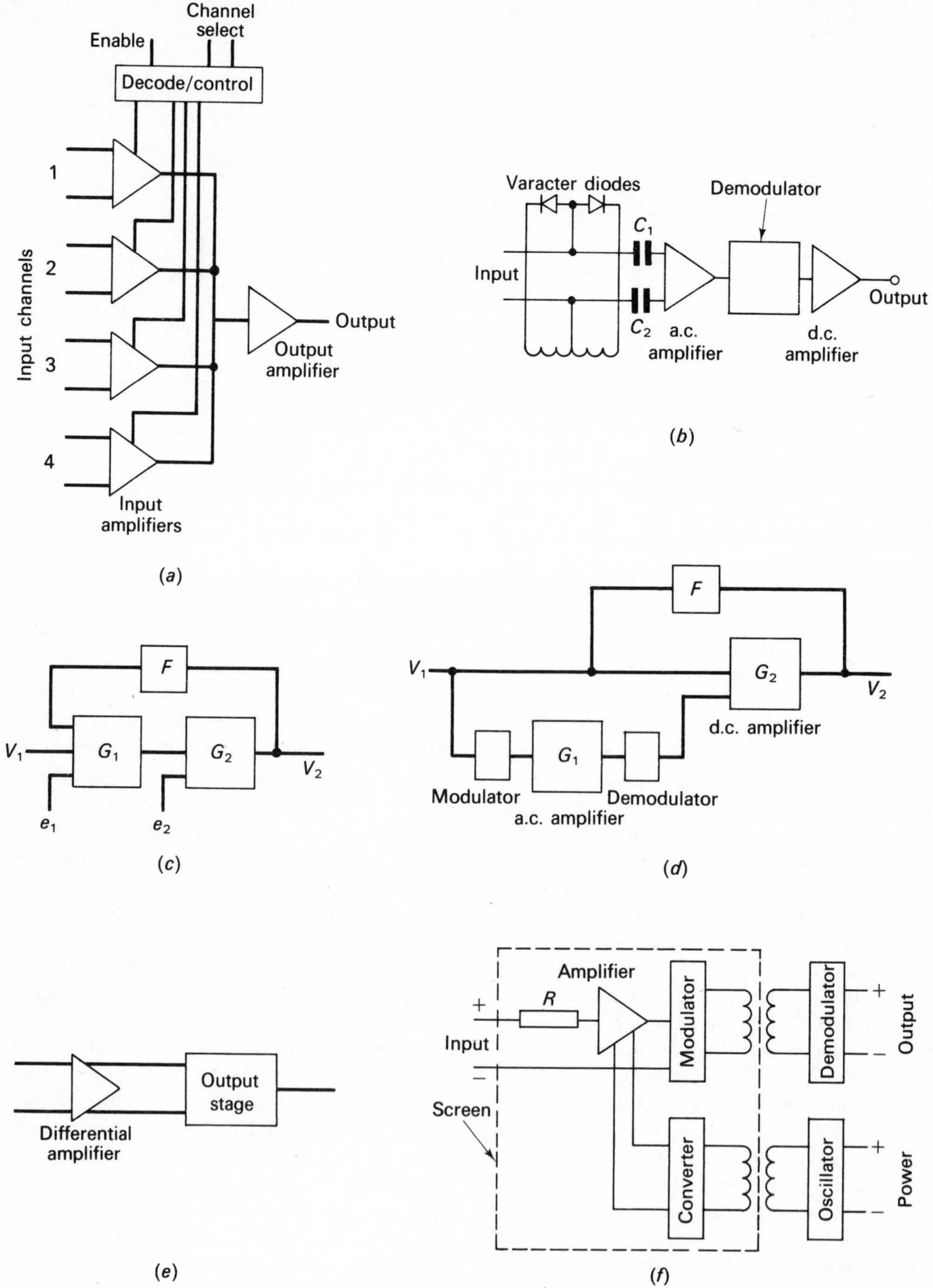

Fig. 7.15. Amplifier configurations; (*a*) programmable, (*b*) parametric, (*c*) two stage, (*d*) chopper, (*e*) comparator, (*f*) isolating.

is fed from the channel select lines. The enable line disables all amplifiers and so allows several identical circuits to be connected in parallel when more than four input channels are required. Parameters such as gain, input impedance and so on can be set in advance on the individual input channels and these can be switched in as required by the remote digital signal.

Amplifier current drift is often a problem in many applications and this can be overcome by using the parametric amplifier as shown in Fig. 7.15 (*b*). The input stage consists of a balanced bridge made from a pair of matched varactor diodes. This feeds an a.c. amplifier and then the output d.c. amplifier stage via a demodulator. The input signal varies the capacitance of the diodes and so feeds an a.c. signal to the a.c. amplifier stage. This is amplified, demodulated and then fed to the d.c. amplifier. Although the parametric amplifier has very low current drift it has large voltage drift, poor bandwidth and high noise.

Voltage drift which occurs in the first stage of a multistage amplifier is the most significant and it should be minimized. This is seen by considering the two stage configuration shown in Fig. 7.15 (*c*). If G_1 and G_2 represent the gain of the two stages, e_1 and e_2 their drift voltages and F the proportion of output voltage fed back to the input then:

$$V_2 = G_1 G_2 (V_1 + e_1 + FV_2) + e_2 G_2 \qquad (7.4)$$

If V_2 is zero then V_1 represents the offset voltage and it is given by:

$$V_1 = -\left[e_1 + \frac{e_2}{G_1}\right] \qquad (7.5)$$

This shows that the drift in the first stage is numerically the most important in determining the drift in offset voltage since the drift in subsequent stages is divided by the gain of preceding stages. Voltage drift in an a.c. amplifier is very small so that low drift amplifiers can be designed with a modulator, to convert d.c. signals to a.c., followed by an a.c. amplifier and demodulator, and then subsequent d.c. amplifier stages. Such an arrangement is shown in Fig. 7.15 (*d*). Note that the a.c. and d.c. stages are not in series as in Fig. 7.15 (*c*) since in this case the overall bandwidth would be limited by that of the a.c. stage. For Fig. 7.15 (*d*) the offset voltage is given by

$$V_1 = \frac{e_2}{1 + G_1} \qquad (7.6)$$

where e_2 is the drift in the d.c. stage, the drift in the a.c. amplifier being assumed to be negligibly small. Semiconductor switches are used in the modulator and demodulator circuits so that the amplifier is small and reliable. The disadvantage of this type of amplifier, which is called chopper stabilized, is that it has considerable internal noise due to the action of its modulator and demodulator switches.

It is very often required to compare the level of two analogue signals and to feed a logic 0 or 1 signal to the rest of the system depending on which of these two signals is greater. An operational amplifier may be used for this and if given sufficient gain it will be able to detect very small differentials. However a new class of component has been developed for this application and it has several advantages. It is called a comparator and is shown in Fig. 7.15 (*e*). The circuit has been optimized for high speed and the output stage is designed to be compatible with digital circuits. Since this comparator is often used in digital systems it is specified in different terms to those used for operational amplifiers. For example the speed is measured as a propagation delay and the offset voltage is now the minimum differential input voltage which makes the output non-zero.

The last type of amplifier configuration to be considered here is the isolating amplifier, as shown in Fig. 7.15 (*f*). It is used in applications which require large isolation, in the order of 2 kilovolts, between input and output. This isolation is achieved by using transformers to couple the signal and power supply lines. The CMRR for the amplifier is very high and it is relatively constant being dependent only on stray capacitances to ground and to the output. This effect can be reduced by shielding the input circuit by a screened box. The isolating amplifier has poor bandwidth, low gain, relatively large drift and is physically large. It is generally built in modular form from discrete components and not as an integrated circuit.

7.5.3 *Amplifier applications*

The multiplier is probably the next most popular linear circuit after the operational amplifier. Multipliers which are most commonly available in integrated circuit form are the variable transconductance and pulse width modulated devices. A simple circuit for the variable transconductance amplifier is shown in Fig. 7.16 (*a*). It is based on the principle that a linear relationship exists between the collector current and the transconductance of a bipolar transistor. This transconductance is controlled by changing the constant current through the long tailed pair by means of input *Y* whereas

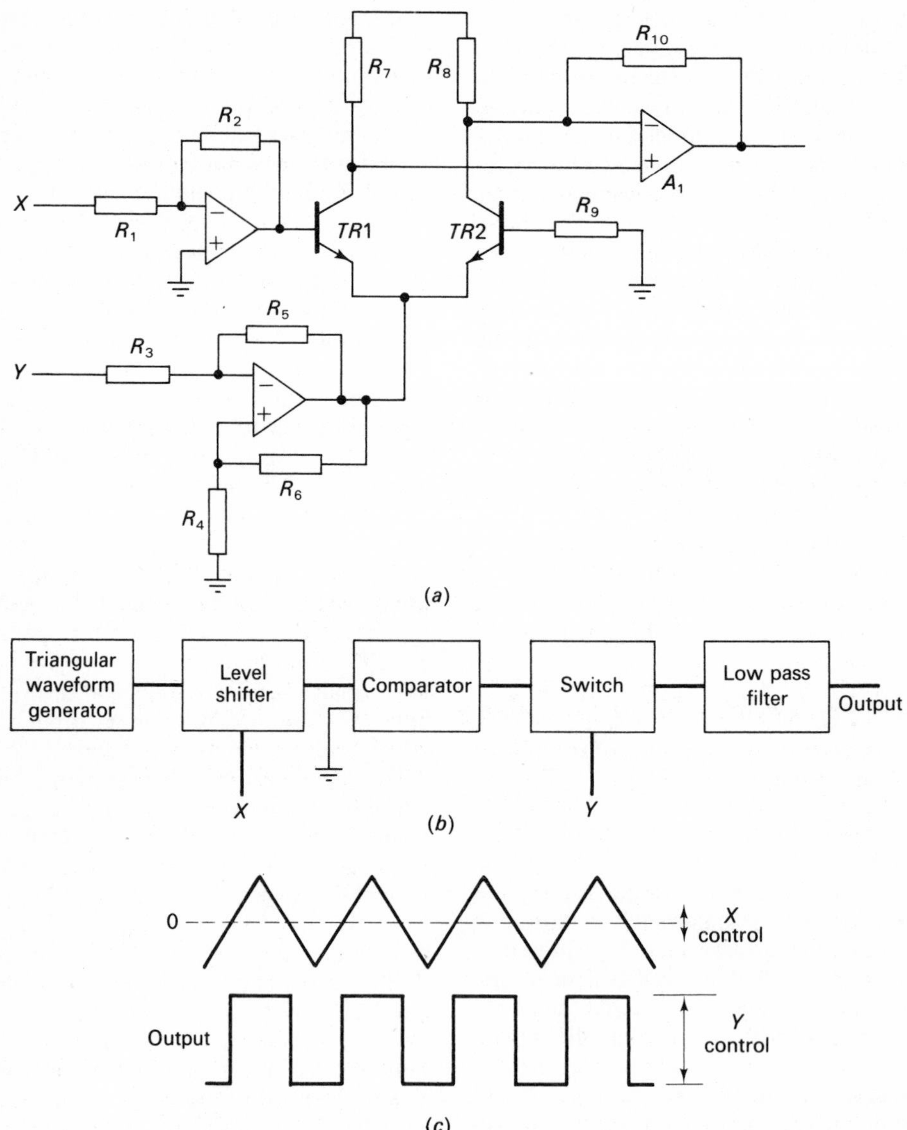

Fig. 7.16. Direct multipliers; (*a*) transconductance, (*b*) pulse width modulated, schematic, (*c*) pulse width modulated, waveforms.

input X is applied to the base of $TR1$. The output voltage is proportional to the product of X and Y and this is buffered by amplifier A_1.

The transconductance multiplier is cheap but has a limited termperature range. The pulse width modulated multiplier of Fig. 7.16 (*b*) works on the principle that input X changes the width of the output pulse by shifting the level of the triangular waveform, whereas input Y controls the pulse height. Therefore the area under the output pulse is proportional to the product $X \cdot Y$ and it can be filtered to give a d.c. signal. These multipliers are more complex than the variable transconductance types but they have good accuracy and stability and a wide operating temperature range.

Several parameters are important for a multiplier and these are usually specified in its data sheets. The output offset represents an error in that the output is non-zero when both the X and Y inputs are zero. An adjustment is usually available for forcing this to zero, although the offset is sensitive to termperature and power supply variations. The input offset, or feedthrough as it is sometimes called, is the output error which exists when one of the inputs is zero but the other is not zero. This offset varies with the magnitude of the input voltage and the frequency. Linearity error is usually measured by putting a fixed voltage on one input and plotting the output as a function of the second input, and noting the non-linearity. The dynamic characteristics of a multiplier concern the change in its parameters with frequency. A plot can be obtained which shows the effect of frequency on the magnitude and phase of the output. To obtain this, one of the inputs is usually held at a fixed d.c. voltage. Small signal response represents the frequency at which the output falls 3 decibels below the d.c. level. The slew rate and settling time parameters are similar to those of an operational amplifier. They are measured by applying a fixed voltage on one of the multiplier inputs and a step function on the other.

The only other application of the operational amplifier to be described in this section is the sample and hold circuit. It is used in systems where the output is required to follow the input signal during certain periods. called the sample time, and then to maintain the last stored stage during a second period called the hold time. This is shown in Fig. 7.17. Fig. 7.18 shows two possible circuit configurations. They both perform the same function. During the sample time transistor $TR1$ is on and the output follows the input. At the commencement of the hold period $TR1$ is turned off. However, there is a finite time, denoted by the aperture time, before this occurs so that capacitor C_1 still follows the input and gives rise to the aperture error. When $TR1$ turns off, the voltage on C_1 remains unchanged (leakage being ignored here) and the output is therefore fixed at this level. At the end of the hold time $TR1$ is again turned on. However, there is again a delay, called the acquisition time, before C_1 can adjust to this level.

Although both the sample and hold circuits shown in Fig. 7.18 are similar in operation they differ in detail. The inverting circuit reduces the loading on the input and is less demanding as regards the CMRR of the amplifier. For this circuit the acquisition time required to reach x per cent of the input voltage at the end of the hold period is given by:

$$t = C_1 R_2 \ln \left(\frac{1}{1 - x} \right) \tag{7.7}$$

The non-inverting sample and hold circuit has an acquisition time constant of C_1R where R is the series impedance of $TR1$ and the signal source. It can be made very short hence the circuit is generally faster in operation. However the capacitor current now flows via the source and $TR1$ so that an amplifier with good CMRR is required.

The drift in a sample and hold circuit should be the minimum possible to prevent voltage change during the hold period. This is caused by charge loss due to amplifier input current and leakage through $TR1$ when it is off. Low input current (FET) amplifiers, however, have low slew rates and a transistor with low leakage also has a large on resistance. Therefore operating speed is reduced in both instances. Increasing capacitor size to try to reduce voltage reductions during the hold time will also increase the acquisition time. Therefore the design of a sample and hold circuit needs careful trade-offs between conflicting parameters.

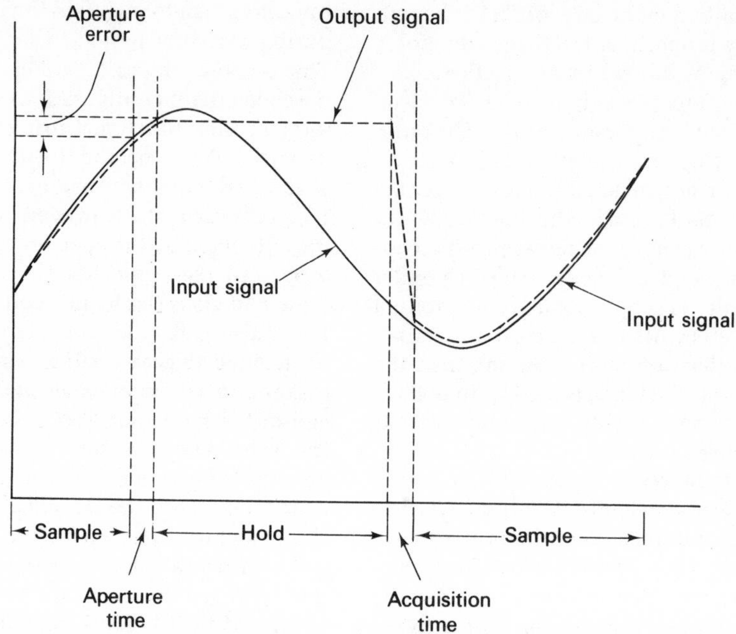

Fig. 7.17. Sample and hold system waveforms.

Several circuit modifications, such as feedback, are used in practice to improve its performance.

7.6 Interfacing analogue and digital systems

Very few industrial systems are totally digital or totally analogue. Therefore a requirement often arises for interfacing analogue and digital systems. Devices which do this are called digital to analogue (D to A) or analogue to digital (A to D) converters, and their construction and operation are described in this section.

7.6.1 *Digital to analogue converters*

Since the digital input, by its very nature, can only change in discrete steps, the analogue output will also change in this way. This is illustrated in Fig. 7.19. The analogue output voltage can be to any suitable scale and the magnitude of each of the individual steps is dependent on the number of bits used to represent the digital information. In many cases both positive and negative polarity inputs need to be catered for and converters which do this are called bipolar.

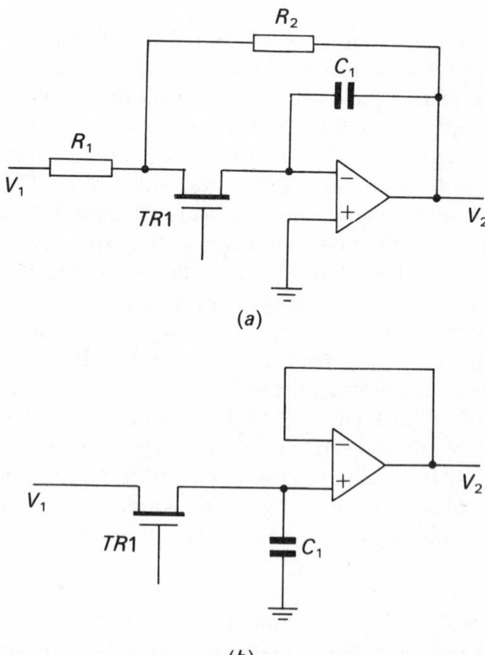

Fig. 7.18. Sample and hold circuits; (*a*) inverting, (*b*) non-inverting.

Fig. 7.20 (a) shows the construction of a common form of D to A converter. A three bit converter is illustrated. The digital input at any bit position will cause its associated switch to change over and connect the resistor to the reference supply V_{REF}. This results in a current into the summing junction of the amplifier, and an output voltage. The magnitude of this voltage is proportional to the current into the summing junction and therefore inversely proportional to the value of the resistor in the leg. If the resistors are weighted in binary code, as shown in Fig. 7.20 (a), then the output voltage is proportional to the binary value of the digital input.

Although the weighted resistor ladder network converter shown in Fig. 7.20 (a) is simple in concept it has several disadvantages. The accuracy of the output is dependent on the absolute accuracy of the reference voltage V_{REF} and the value of the resistors. Furthermore the resistors need to track each other closely with termperature in order to maintain this accuracy. Since the resistor values double between successive legs they can become very large. Not only is it difficult to make stable resistors in large values but it becomes nearly impossible to ensure that resistors of greatly differing values track each other closely over a wide temperature range. Resistors are also limited at the low value end by the fact that their impedance needs to be several orders larger than that of the switches in order to minimize their effect on conversion accuracy. Fig. 7.20 (b) shows the resistor weighting required for a converter which operates in BCD code. Two decades are shown and from this it is evident that the problems associated with large valued resistors, and with wide spreads in values through the network, are even greater. The

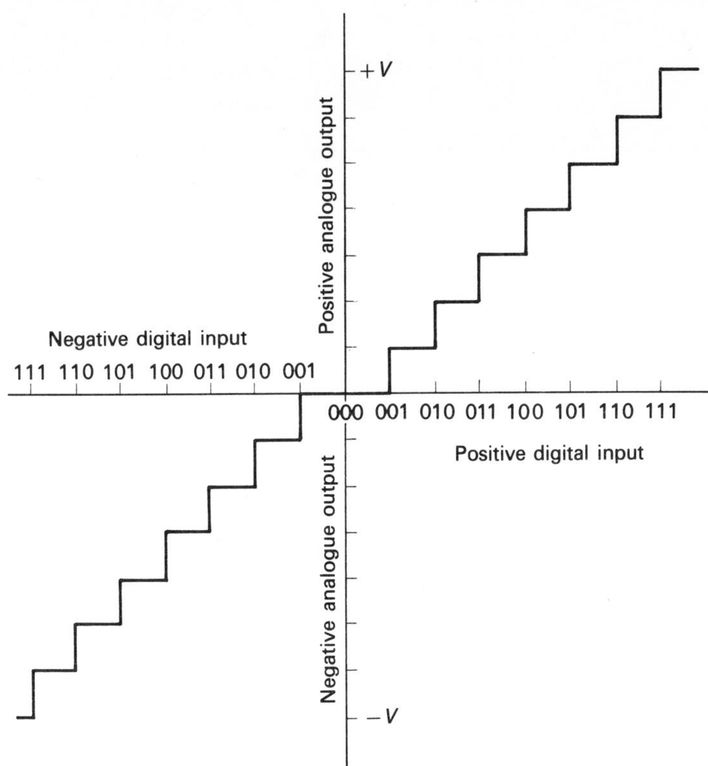

Fig. 7.19. Input–output waveform for D to A converter.

BCD converter is less efficient than the pure binary code converter due to the presence of redundant states. For example the two decade BCD converter of Fig. 7.20 (*b*) is capable of covering the full scale output with 100 steps. This means a resolution of 1 in 100 or 1 per cent. However eight bits are used so that in pure binary code an equivalent converter would have a resolution of 1 in 2^8 or 0.391 per cent.

Fig. 7.21 (*a*) shows an alternative D to A converter which uses a *R*–2*R* resistor network. It works on the principle that, since the input to the amplifier is at virtual earth, each node divides the current coming up the leg by two, so that the current at the summing junction is binary weighted according to the number of junctions passed through. BCD weighting can be incorporated as in Fig. 7.21 (*b*). Although the

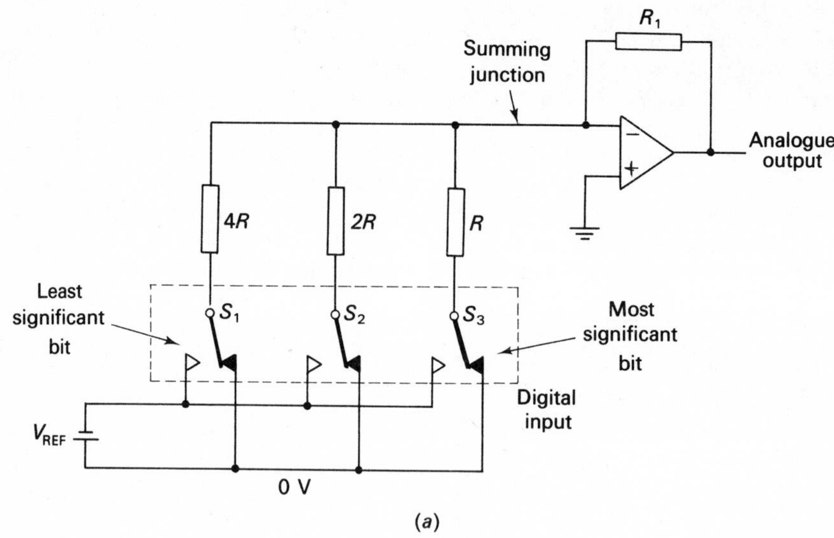

(*a*)

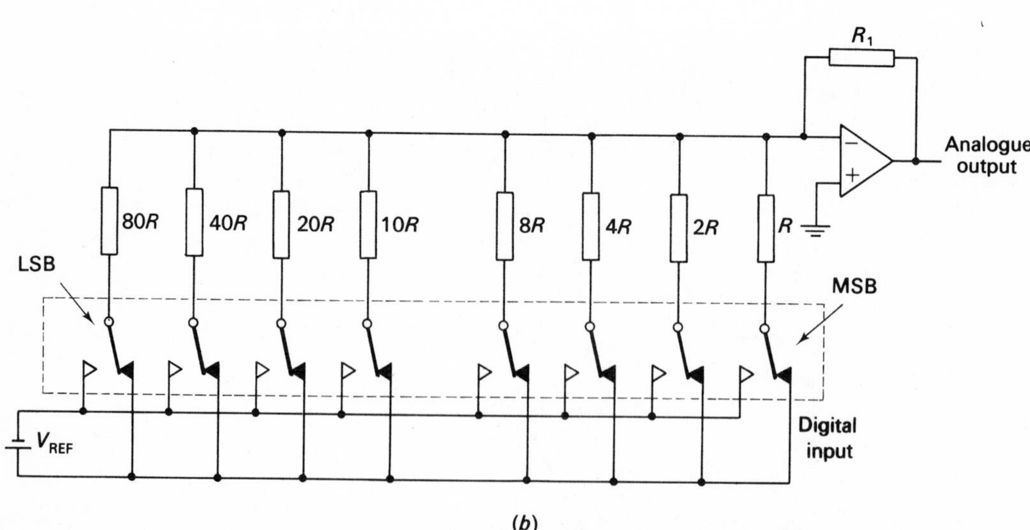

(*b*)

Fig. 7.20. Weighted ladder network D to A converter; (*a*) binary code, (*b*) BCD code.

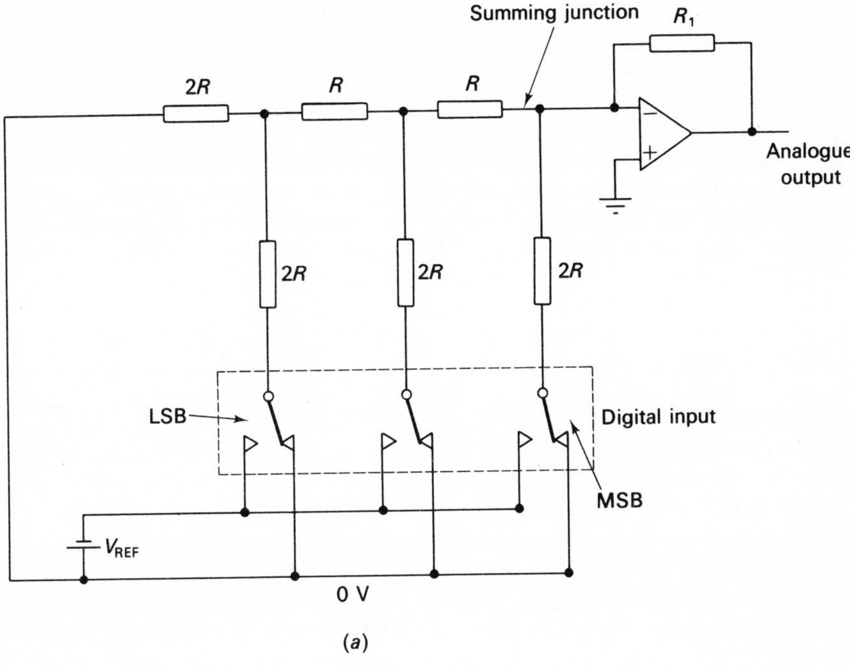

(a)

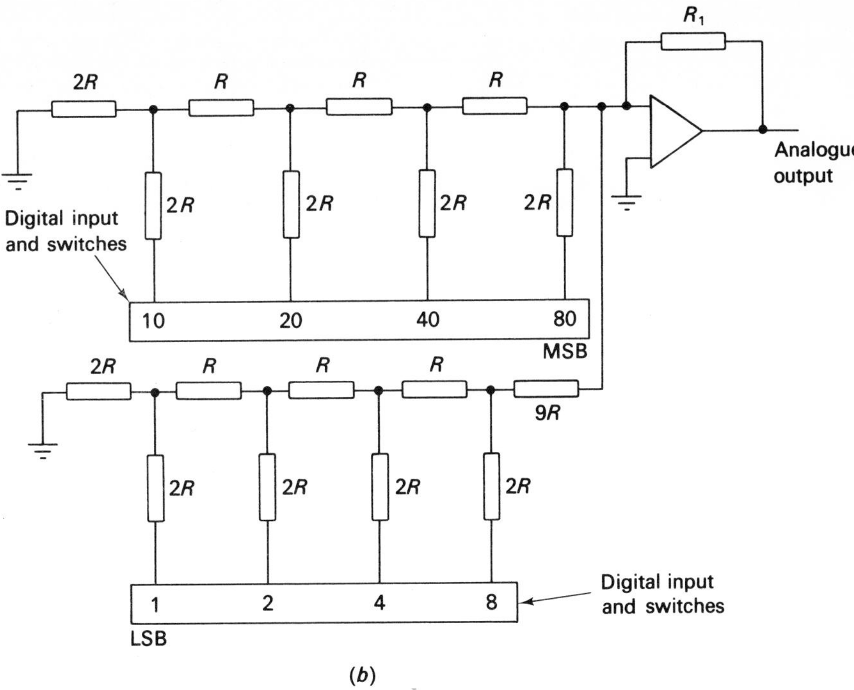

(b)

Fig. 7.21. R–2R ladder D to A converter; (a) binary code, (b) BCD code.

R–$2R$ ladder network requires twice as many resistors as a binary weighted network the accuracy of this converter depends on the relative values of the resistors and not on their absolute value. Furthermore only two different resistor values are used so that they can be chosen to give good resistor characteristics and can be made to track each other closely with temperature. A further advantage of the R–$2R$ network is that the impedance seen by the operational amplifier does not vary with the digital signal input, so that problems relating to the variations of the amplifier characteristics are avoided.

The D to A converters discussed so far are voltage output types. A current output D to A converter is shown in Fig. 7.22. The transistors are all biased on and since they operated in an unsaturated mode this converter is capable of high speeds. The diodes are connected via the digital switches to ground so that with no digital input the output voltage is negligible. A digital signal at any bit position will send the cathode of its associated diode to the logic 1 state making it reverse biased. Current can now flow via the resistor and its transistor to the output resistor R_1 to produce the output analogue voltage. The magnitude of this voltage is clearly dependent on the digital input and since the resistors are binary weighted the converter operates in binary code. Generally the circuit shown in Fig. 7.22 is capable of producing an output of a few volts only. For larger values an amplifier needs to be incorporated. This circuit also represents a simplified version of a current output converter. A commercially available circuit will be more complex since compensation is required for temperature drifts and for changes in transistor base–emitter volt drops due to current variations.

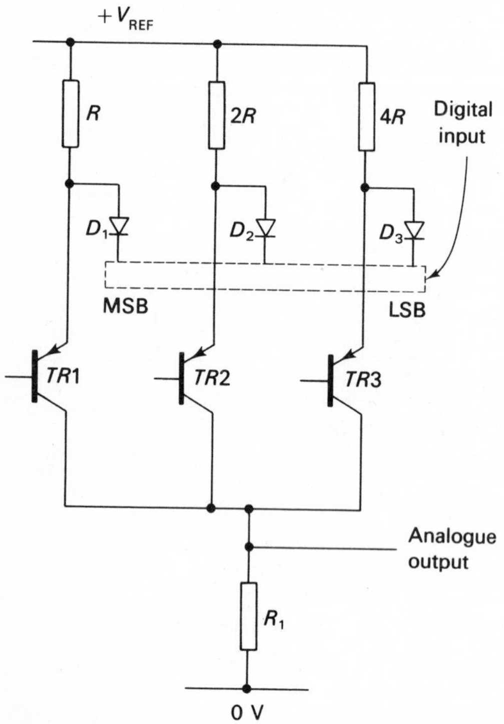

Fig. 7.22. Current output binary weighted D to A converter.

A variety of different switch configurations can be used for the digital input circuits of D to A converters. Fig. 7.23 (*a*) shows a simple transistor switch and Fig. 7.23 (*b*) gives its output characteristic. Although the ideal switch should be capable of operating between V_{REF} and zero volts the simple transistor switch goes between V_1 and V_2. The drop V_2 is accounted for by the saturation current through $TR1$ and ($V_{REF} - V_1$) is the voltage loss across R_1. This loss can be reduced by making R_1 small but then the current through $TR1$, when it is on, increases and so does V_2. The complementary bipolar switch shown in Fig. 7.23 (*c*) operates closer to the ideal characteristic since it presents a low impedance in both directions. The same considerations apply for the complementary unipolar switch of Fig. 7.23 (*d*). The transistors should present a low series impedance when conducting. There is now no offset voltage as in a bipolar device.

The conventional D to A converters can be modified to give two other modes of operation. The first is called a bipolar D to A and it was introduced at the start of this section. The output polarity of a D to A converter depends on the polarity of the reference voltage and on whether the input is supplied to the positive or negative terminal of the amplifier. To operate in a bipolar mode it is therefore necessary to switch the polarity of V_{REF} whenever a negative digit input is detected. The second D to A variation is a multiplying converter. The magnitude of the output voltage is proportional to the product of the reference voltage and the digital input. Normally this reference voltage is fixed in magnitude, but if it were made to vary proportional to a second input then the converter would be performing a multiplying function. If reference V_{REF} varied in proportion to an analogue signal then the converter is a hybrid multiplier since its output is proportional to the product of an analogue and digital signal. If V_{REF} is obtained as an output of a previous D to A converter which has a fixed reference input then the two converters behave as a digital multiplier since the two digital inputs to the converters are multiplied together.

A D to A converter which is very different in operation from the ladder network type is

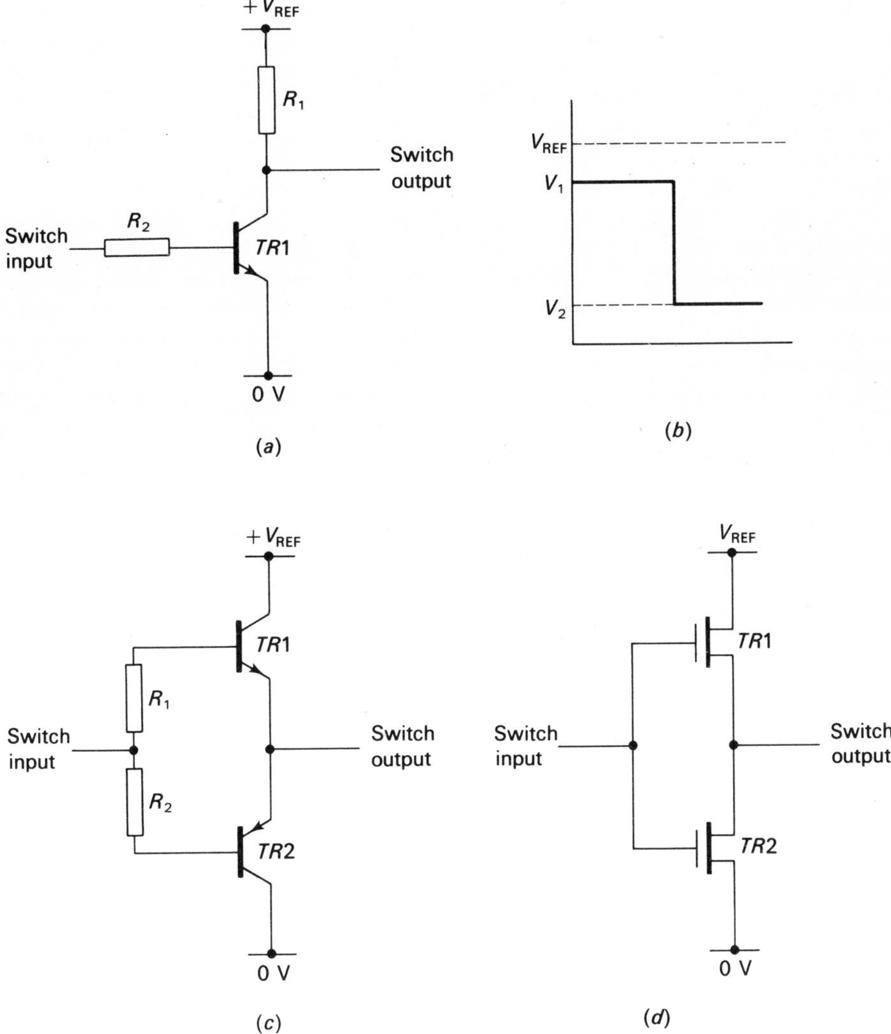

Fig. 7.23. D to A converter switches; (a) single transistor, (b) waveforms for switch illustrated in (a), (c) complementary bipolar, (d) unipolar.

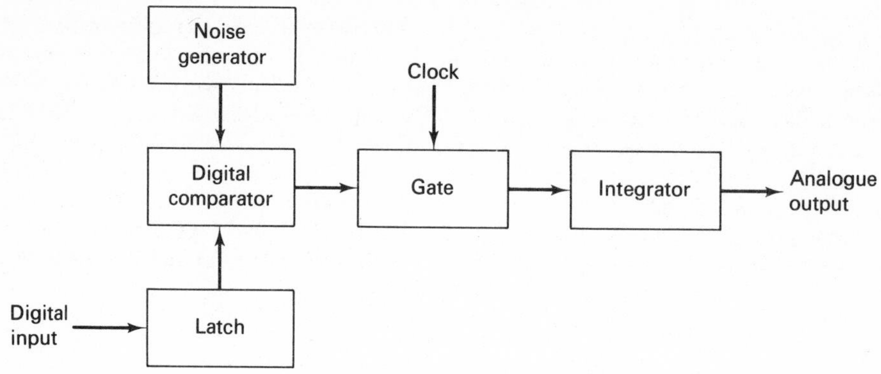

Fig. 7.24. Stochastic D to A converter.

shown in Fig. 7.24. It is based on the use of probability and is called a stochastic converter. It consists of a random noise generator and a latch which holds the digital input. The outputs from these are in the form of a binary number and they are compared in the binary comparator. When the latch output exceeds that from the noise generator a gate is enabled and clock pulses pass to the integrator. The output from the integrator represents the analogue output which is proportional to the digital input. Clearly the larger the digital input the greater the probability that it will exceed the random noise generator output giving a larger number of pulses to the integrator, and vice versa. The advantage of the stochastic converter is that there are no precision resistor ladder networks so that it is easier to manufacture in monolithic technology. It also has good accuracy. Its disadvantage is that it has a long conversion period.

7.6.2 *D to A converter characteristics*

Several parameters can be used to determine the characteristics of D to A converters. A few of these are discussed in this section. The 'resolution' of a converter has already been introduced earlier. It represents the number of steps which are used to cover the full scale output of the converter, and it is directly related to its number of bits. Therefore a 10 bit converter has 1024 steps and a resolution of less than 0.1 per cent. The 'linearity' of a D to A converter is measured in terms of the maximum deviation of the output from the best straight line drawn through it. This is shown in Fig. 7.25. For good linearity the value of L should be less than, or equal to, half the magnitude of the least significant bit. Linearity in a D to A converter depends on the accuracy of the resistors and the voltage drops in the switches. Since these change with temperature linearity is also temperature dependent.

The 'accuracy' of a converter is the shift of the analogue output voltage from the ideal value for any digital input. It is dependent on the accuracy of the reference voltage. Converter 'settling time' is defined in a similar manner to that of an operational amplifier. It is the time

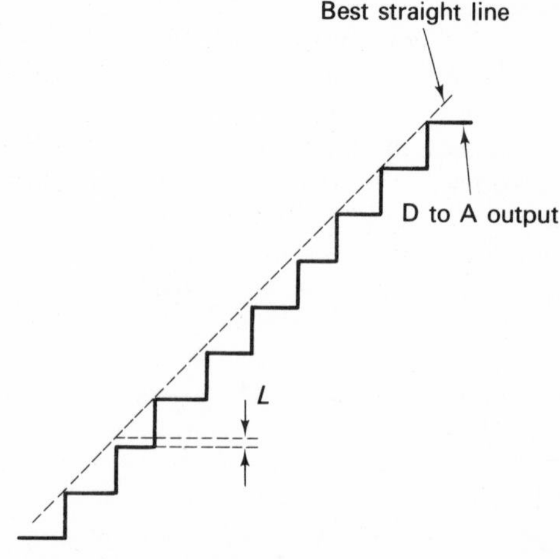

Fig. 7.25. Illustration of linearity in a D to A converter.

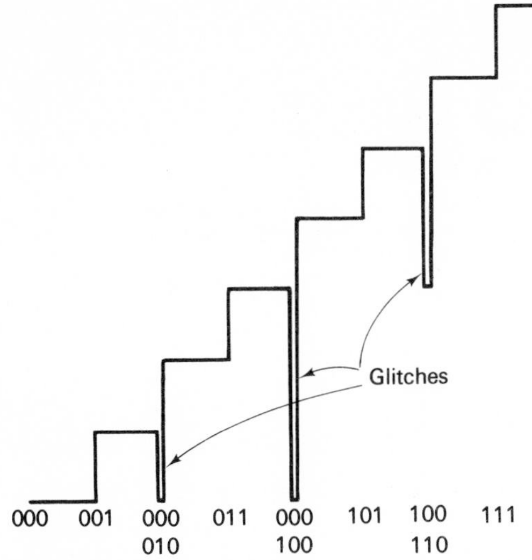

Fig. 7.26. Illustration of 'glitches' in a D to A converter.

needed to reach within a defined band of the final output value. This time is dependent on the type of switches used, the resistor characteristics and the output amplifier.

The output of an A to D converter is not usually in the comparatively smooth form shown in Fig. 7.19. The waveform is superimposed with overshoots and noise spikes. In addition it is also subjected to violent dips or 'glitches'. These are caused by the fact that the switches used have unequal turn on and turn off times. Therefore, for example, when switching from a digital signal of 011 to 100 it is possible for the output to pass momentarily through 000 if the switches go from 1 to 0 sooner than when going from 0 to 1. Fig. 7.26 illustrates the glitches which occur on a three bit converter when stepping up through its full scale. The size of the glitches can be reduced by using faster switches and their effect can be smoothed out by reducing the amplifier slew rate and by filtering. However this also reduces the overall response time. Sample and hold circuits can also be used to overcome glitch effects since the voltage can be held until the glitch has died down. Unfortunately the acquisition time of the sample and hold circuit now reduces the speed of response of the overall converter. The 'temperature stability' of the converter is dependent on the stability of the reference voltage, resistors, switches and amplifier. Converters are available which have internal reference sources and amplifiers although in some converters these must be provided externally. Where they are external components it is usual to specify the converter stability on the assumption that there is negligible change in the reference or amplifier offset with temperature.

7.6.3 *Analogue to digital converters*

There are many different techniques which have been used for converting from an analogue to a digital signal. They each have various advantages relating to cost, precision, stability, conversion speed and so on. In this section a few of these techniques are described.

Fig. 7.27 shows an A to D converter which uses a conventional D to A converter in a closed loop configuration for ramp generation. Assume that the analogue input is V_1 and that the counter is initially reset to zero. The D to A output is also zero and since this is less than V_1 the output from the comparator enables gate G_1 and allows the clock pulses to reach the counter. The counter therefore operates up to time t_1 when the D to A output exceeds V_1. The comparator now changes state and the gate is disabled stopping further clock pulses from reaching the counter. The output from the counter is held and it represents the digital value of the analogue input signal. On the next reset pulse the above operation recommences. It should be noted that this form of converter must have a finite error, no matter how small the D to A converter steps are made. This is due to the fact that the comparator switches when the D to A output exceeds that of the analogue input and not when it equals it.

The D to A converter in Fig. 7.27 can be replaced by a linear ramp generator. In this case the feedback from the counter to the generator is not included. In either event this form of converter is slow since it needs to count up from zero on each reset pulse. Its speed can be increased by using an up–down counter. In this instance the counter is not reset to zero at the start of the count period but it counts up or

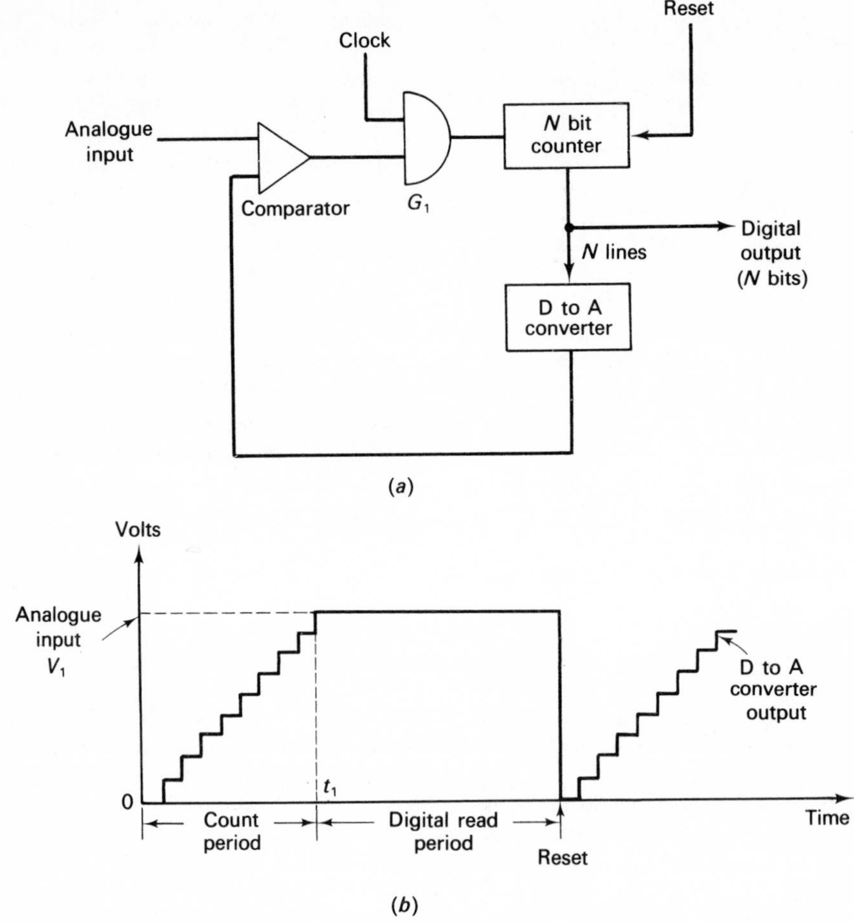

Fig. 7.27. A to D converter using a D to A converter for ramp generation; (a) circuit schematic, (b) waveforms.

down depending on the direction in which the analogue input has changed. The device is now called a tracking or servo A to D converter.

An alternative A to D conversion method is shown in Fig. 7.28. Its operation can be followed by means of the waveforms. Initially the counter and switch logic are reset. This closes S_1 so that the unknown analogue input V_1 is fed into the integrator. Since the integrator output is greater than zero the comparator output enables gate G_1 and allows clock pulses to reach the counter. This counts until a predetermined number of clock pulses are reached, as sensed by the switch logic, which gives a fixed time t_1.

The value of V_M is now given by

$$V_M = \frac{V_1 t_1}{\tau} \tag{7.8}$$

where τ is the time constant of the integrator.

The switch logic now closes S_2 and opens S_1. The reference input is of reverse polarity to the analogue input voltage so that the integrator output decreases from V_M until after further time t_2 it falls below zero. The comparator now switches over and disables G_1 so that no further clock pulses can reach the counter. The output voltage from the integrator, which is zero, now forms the equation

$$0 = V_M - \frac{V_{REF}\, t_2}{\tau}$$

or substituting for V_M this gives:

$$\frac{V_1\, t_1}{\tau} = \frac{V_{REF}\, t_2}{\tau}$$

$$\text{or } V_1 = \frac{V_{REF}\, t_2}{t_1} \quad (7.9)$$

Since V_{REF} and t_1 are constants, the counter reading, which is proportional to t_2, gives the value of the analogue input signal. The dual ramp converter gives excellent accuracy. It eliminates propagation errors in the circuits and compensates for changes in clock frequency and integrator time constant since these affect both the ramps equally. The converter also compensates for comparator offset currents and voltages since two zero crossings are involved and therefore these cancel out.

For fast analogue to digital conversion a successive approximation technique is often used. This principle is shown in Fig. 7.29 (*a*) for a three bit system. The most significant bit of the

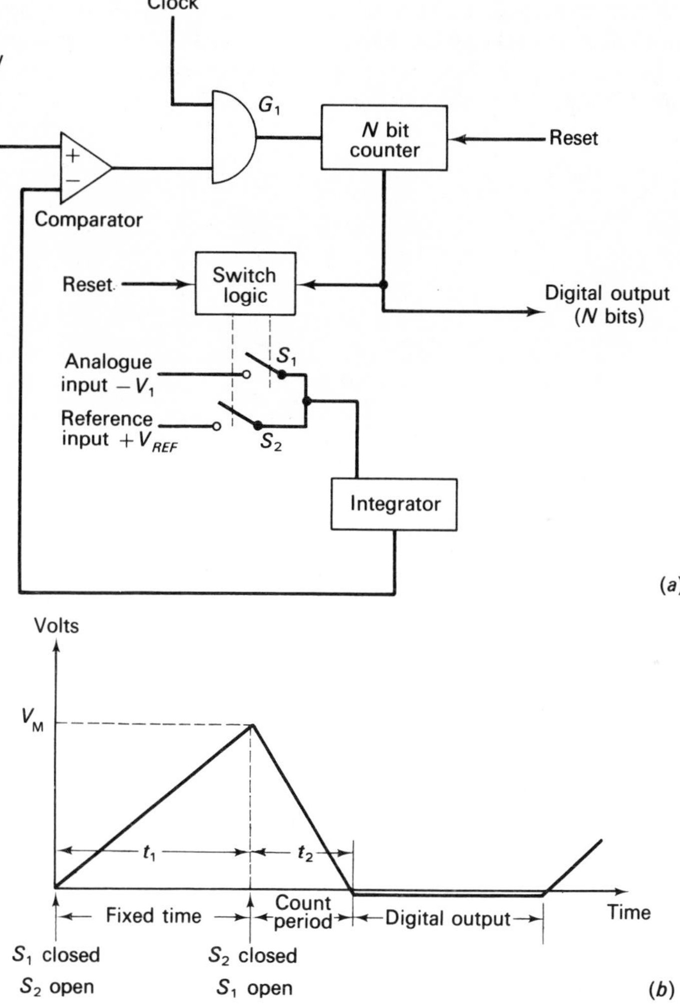

Fig. 7.28. A to D converter using an integrator ramp generator (dual ramp converter); (*a*) circuit schematic, (*b*) waveforms.

register (Fig. 7.29 (b)) is initially set to a logic 1 by the controller. This is fed to the D to A converter whose output is compared with the analogue input. If it is less than the input then path (1) in Fig. 7.29 (a) is followed and the second bit of the register is also set to a logic 1. If the D to A converter output is less than the analogue input then path (2) is followed and the most significant bit is reset while the second bit is set to a logic 1. Proceeding along path (1) the output from the D to A for an input of 110 is compared with the analogue signal, and if it is less the registers are changed to 111, otherwise they are set to 101. Therefore the successive approximation technique works by halving the range of possible values each time and comparing this with the input. The total number of comparisons required for any conversion is equal to the number of bits. The successive approximation technique is therefore very fast although it requires relatively complex logic circuitry.

There are three types of errors which occur in A to D converter systems. These are: (1) Quantization error, which is the smallest analogue input to which an output signal can be approximated. It has a maximum value equal to ± LSB and is therefore dependent on the number of bits. (2) Sampling error. This is caused by the fact that if the conversion time is relatively long the input signal could have changed in value. Therefore for a given conversion rate and accuracy the maximum a.c. signal frequency is fixed. (3) Electronic error.

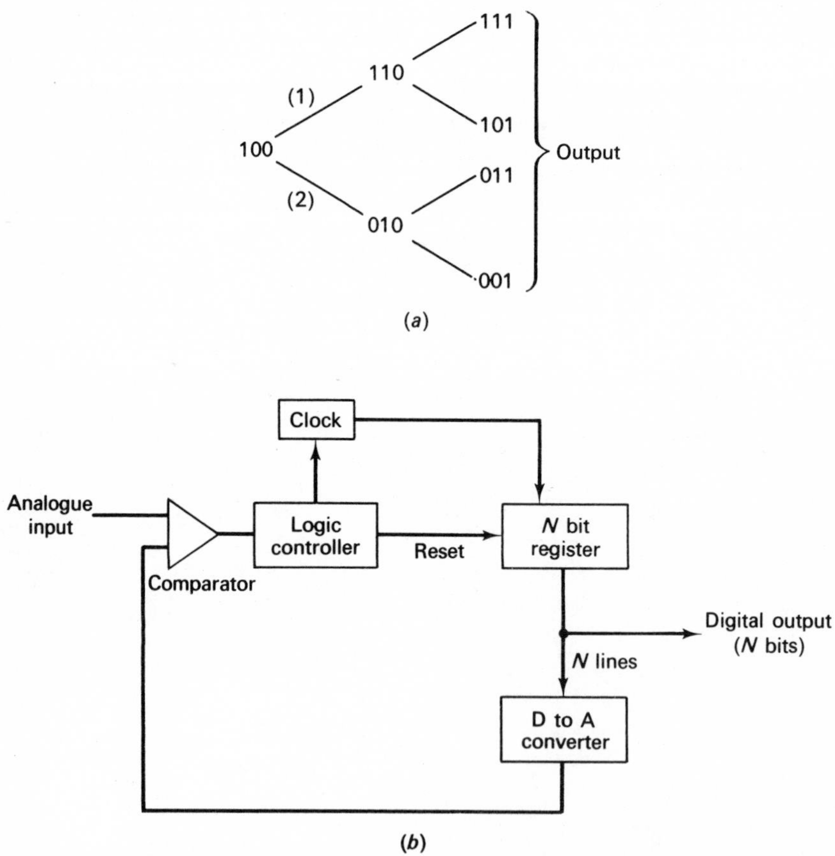

Fig. 7.29. Successive approximation A to D converter; (a) operating tree, (b) circuit diagram.

This is error which is contributed by the various circuitry through which the analogue signal needs to pass.

7.6.4. *Voltage to frequency converters*

Circuits which convert between voltage and frequency are also a form of A to D (or D to A) converter. Fig. 7.30 (*a*) shows a simple circuit by which this may be achieved. *TR*1 operates as an elementary constant current circuit so that the capacitor charging rate is proportional to the analogue input. When the capacitor voltage exceeds V_{REF} the comparator switches and triggers the monostable. This produces an output pulse of a fixed duration and at the same time momentarily turns on *TR*2 so that C_1 is discharged. The cycle then repeats so that the output consists of a string of pulses whose fre-

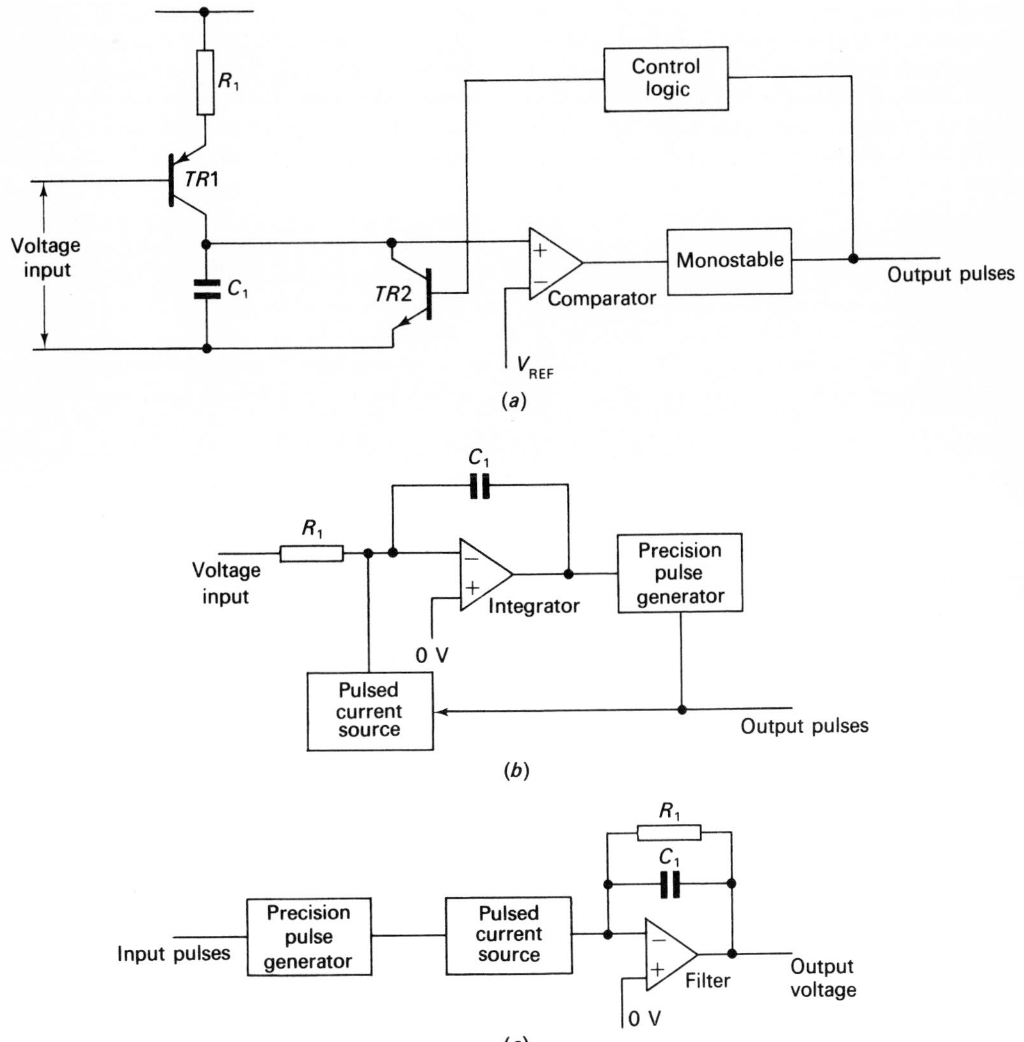

Fig. 7.30. Conversion between voltage and frequency; (*a*) simple V to *f* converter, (*b*) more accurate V of *f* converter, (*c*) *f* to V converter.

quency is proportional to the magnitude of the input voltage. Fig. 7.30 (*b*) shows an alternative voltage to frequency (*V* to *f*) converter which has better linearity. It is a closed loop system in which the current input from the voltage source is balanced by the current drawn out from the summing junction of the integrator by the pulsed current source operated by the precision pulse generator. This circuit can also be reversed as in Fig. 7.30 (*c*) to give an *f* to *V* converter. The operational amplifier and feedback components now act as a low pass filter to provide an analogue output which is an average of the pulse train input.

V to *f* systems represent a slow method for converting between analogue and digital signals. The signal is in serial form and a conversion time, say of 1 second, is needed to obtain a parallel reading from this. However it is capable of very good resolution and linearity. For example a *V* to *f* converter which runs at 10 kilohertz has a resolution of 1 in 10 000 for a conversion time of one second. This is equivalent to approximately 13 bits of a conventional converter. For 100 kilohertz the resolution equals 16 bits. However as the frequency increases so the linearity decreases. A better solution to increased resolution is to increase the conversion time assuming that this is permissible for the application concerned.

7.7. Voltage regulators

All electronic equipment require a power supply so that from this aspect an integrated circuit voltage regulator is a very important component. The function of this integrated circuit is to accept a relatively unregulated d.c. input voltage and to produce a d.c. output which has a low ripple content and good stability under conditions of variable load, changing input voltage and fluctuating temperature. It is assumed that for a.c. inputs the voltage regulator is preceded by a voltage conversion and rectification stage so as to produce a d.c. level with the required voltage and ripple tolerance. In addition to providing a regulated output voltage the integrated circuit often has such additional features as a current limit which serves to protect the external circuitry, and the regulator itself, under fault conditions.

7.7.1 *Construction of a voltage regulator*

The elements of a linear series regulator are shown in Fig. 7.31 (*a*). A reference voltage V_{REF} feeds into one terminal of the error amplifier and this is compared with a proportion of the output voltage. The output of the amplifier drives the series regulating element whose resistance is varied so as to produce an almost constant output voltage which is compensated for fluctuations in the output load current and the unregulated input voltage. The value of the output voltage can clearly be altered by selecting the ratio of R_1 to R_2 but it can never go above the reference voltage $V_{REF.}$ Fig. 7.31 (*b*) shows a simple circuit diagram for this basic regulator. The series regulating element consists of transistors *TR*6 and *TR*5 and resistor R_2 which together form a Darlington stage. Resistors R_3 and R_4 constitute the feedback components which can be selected to change the output voltage. The remaining devices are part of the error amplifier. This is basically a differential stage in which transistors *TR*1 and *TR*2 are used as collector loads to give higher gain.

It is common practice to build into the voltage regulator several protective features to prevent damage to the circuit being supplied, and to the regulator itself, in the event of a malfunction. A circuit diagram for such a system is shown in Fig. 7.32. Transistor *TR*3 is the series regulating element and resistors R_2, R_3 and the comparator form part of the feedback regulating loop. *TR*3 is fed via a current source and all the regulating and protection devices function by diverting this current from its base. These consist of transistor *TR*2 and the current sensing resistor R_1 for current limit, *TR*1 for thermal shutdown protection, and diode D_1 for safe area operation. A description of these devices is postponed until section 7.7.2.

A linear series regulator represents the simplest circuit configuration and is the most widely used system for supplying a constant voltage to a variable load. Where the load current is relatively constant a shunt regulator is also used. Fig. 7.33 shows a simple circuit arrangement. A proportion of the output voltage is fed back and compared with the reference voltage in the error amplifier, and this signal then controls the conduction of transistor *TR*1.

The output voltage is regulated by $TR1$ which varies the current, and therefore the voltage drop, in R_1. The advantages of the shunt regulator include an inherent voltage and current limiting feature although it is more complex and has a slower response than the series regulator.

External components are often required with monolithic voltage regulators. For instance for large power loads an external transistor may be added as a series regulator since most of the power loss occurs in this device. External potentiometers or resistors are required for adjustable voltage regulators in order to fix their

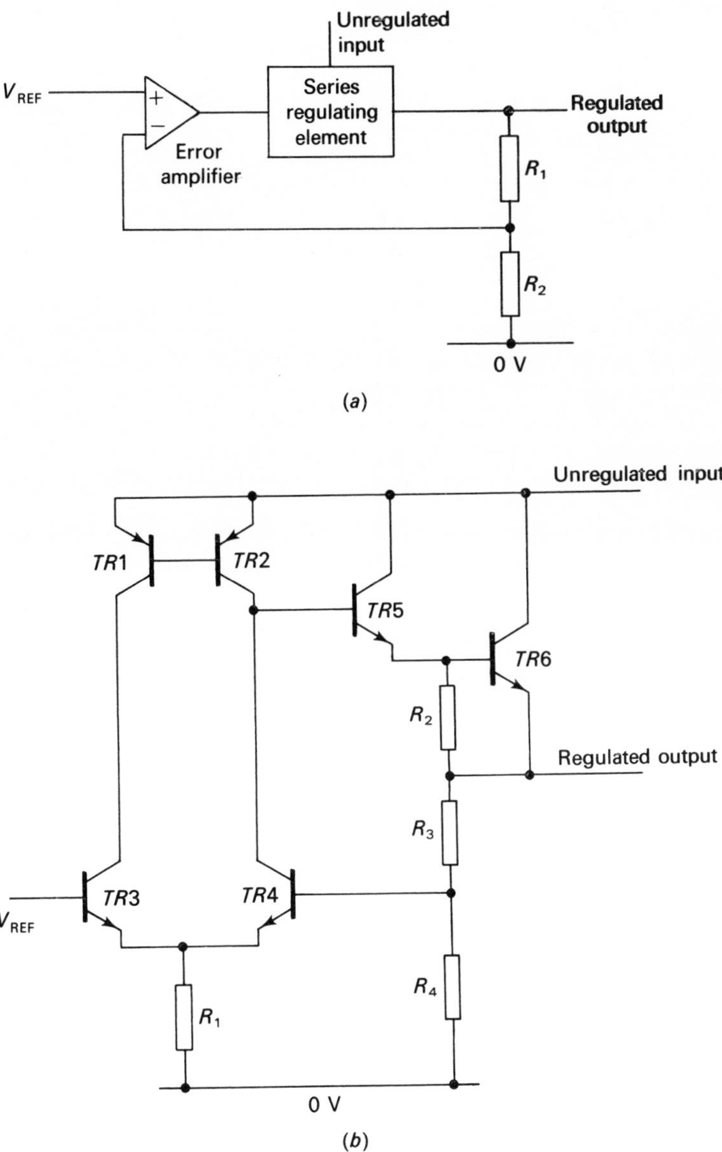

Fig. 7.31. Linear series regulator; (a) functional schematic, (b) circuit diagram.

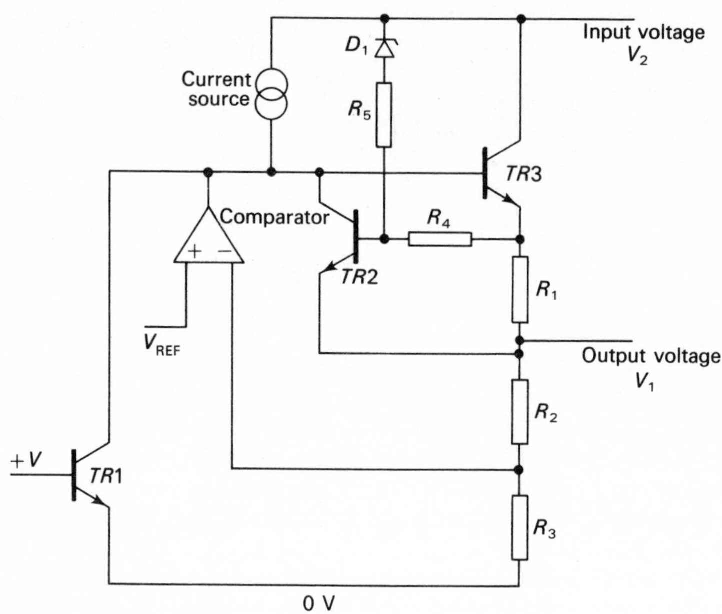

Fig. 7.32. Series linear regulator with current limit, thermal shutdown and safe area protection.

voltage output, and capacitors are added for stabilization, ripple reduction, noise reduction and improved transient response. In order to obtain improved voltage regulation an external differential amplifier may sometimes be used. Similarly an external reference voltage may be required to obtain higher or lower voltage outputs than that provided by the on-chip source, or for better stability.

7.7.2 *Voltage regulator characteristics*

Several characteristics are used to define the performance of a voltage regulator. These are:

(1) Load regulation, which is a measure of the maximum percentage change of the output voltage for a given percentage change of current over the operating range. This indicates the regulator's capability to cope with varying loads.

(2) Line regulation, which is the ratio of maximum percentage change of the output voltage for a specified percentage change of the input voltage. It indicates the ability of the regulator to tolerate input voltage fluctuations.

(3) Ripple rejection. This is the ratio of the peak to peak input voltage to peak to peak output voltage.

(4) Standby current, which is the current drawn by the regulator, excluding the load or any external components which may be used.

(5) Temperature drift. This is measured in terms of the temperature coefficient of the regulator. It is the percentage change in the output voltage for each degree centigrade change of the ambient temperature, over a specific range. Temperature drift is the single most important cause of output voltage change in a monolithic regulator. It is primarily caused by the change of the reference voltage and, as a secondary factor, by the drift in the error amplifier.

(6) Dropout voltage. This is the difference between the input and output voltages below which the circuit does not regulate against input voltage reductions.

Fig. 7.34 (*a*) shows the curves for load regulation. The characteristics are seen to improve with temperature due to the reduction in transistor junction volt drops. Fig. 7.34 also shows the current limit characteristics for conventional and foldback systems. These can be explained with reference to the circuit of Fig. 7.32. When $(V_2–V_1)$ is less than the zener voltage of diode D_1 there is no current in resistor R_5 and this circuit has no effect. Now as the current through resistor R_1 increases the base drive voltage to *TR*2 also increases until at a certain value it is sufficiently large to turn on *TR*2 and to prevent any increase of current into the base of *TR*3. The output current is therefore limited. As temperature increases the base emitter voltage of *TR*2 decreases so that it turns on at a lower load current. Now if the value of $(V_2 - V_1)$ exceeds the zener voltage of D_1 current flows in R_5 and R_4, the magnitude of this current being proportional to the magnitude of $(V_2 - V_1)$. This causes a drop in R_4 which adds to the voltage across R_1 and therefore causes TR_2 to turn on at lower absolute values of load current. This is called foldback current limiting and is shown in Fig. 7.34 (*c*). Alternatively it can be presented as the safe area curves of Fig. 7.34 (*d*). In either event it indicates that as more and more voltage is dropped across the regulator, either due to an increase in the input voltage or a reduction in the output voltage, then the maximum permissible load current is reduced in order to maintain the dissipation of the device at a constant value.

Since the temperature failure mode is the most prevalent in voltage regulators some devices use a direct temperature method to prevent overheating. This is done by transistor *TR*1 in Fig. 7.32. *TR*1 has a low voltage, of about 0.4 volts, applied to its base, and this is insufficient to turn it on at room temperature. *TR*1 is mounted close to *TR*3 on the silicon chip so that it heats up as the power dissipation increases. At a predetermined temperature the base–emitter voltage of *TR*1 will have fallen to a low value, and it turns on diverting base current from *TR*3, and switching off the regulator. Generally hysteresis is built into the circuit so that there is a difference of a few degrees between the turn off and re-connection temperatures, in order to prevent high frequency thermal oscillations.

7.8 Consumer integrated circuits

Integrated circuits are used in a wide range of configurations, for consumer equipment, in applications ranging from washing machines to automobiles. Many of these are digital in performance and primarily function as logic or storage elements. In this section only a few linear applications are described.

Integrated circuits for use in radio and television receivers are probably the best known. The basic function carried out in both these is that of signal amplification. General purpose operational amplifiers may be used for this and their gain and frequency response modified by external components. However it is usual to design special purpose amplifiers for this use since the demand, in terms of volume is large. There is very little standardization in this area so that each manufacturer tends to market his own group of integrated circuits. The amplifiers can be divided into several types.

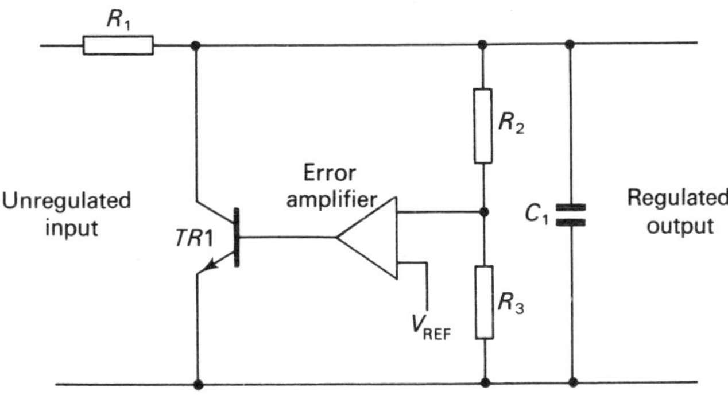

Fig. 7.33. Shunt regulator.

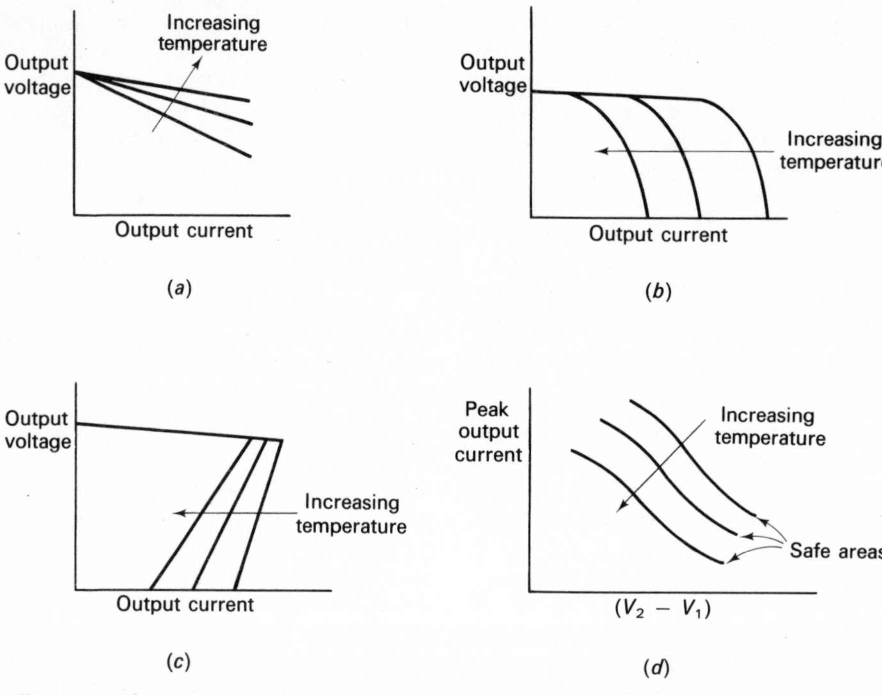

Fig. 7.34. Characteristics of a voltage regulator; (*a*) without current limit, (*b*) with conventional current limit, (*c*) with foldback current limit, (*d*) safe area curves.

(1) Audio amplifiers. These can be low power pre-amplifiers (up to 50 mW), medium power (up to 0.5 W) and high power (above 0.5 W). Class *A* techniques are generally used for low power outputs and Class *AB* push–pull for large power outputs. These circuits generally require relatively large capacitors which are difficult to incorporate in monolithic form and are added externally to the package. For very high powers hybrid circuits are preferred and these have the added advantage that external components can be incorporated as chips in the same package.

(2) Radio frequency and intermediate frequency amplifiers. These are required to operate at high frequencies with a large power gain per stage. In monolithic technology there is no need to economize on the number of transistors used as is done in discrete designs, so that multiple transistor arrangements are common. The main types of configurations used are cascode, long tail pair and feedback, as shown in Fig. 7.35. The cascode circuit has low noise, high stability and ease of interstage matching. The long tail pair arrangement has low noise, high stability, non-saturating action and fast recovery from an overdrive. The advantages of the feedback circuit are low noise, low power consumption and large signal handling capability.

(3) Wideband amplifiers. These are often also called broadband, baseband or video amplifiers. Their prime characteristic is an almost flat frequency response from d.c. to very high frequencies. For this they require low harmonic distortion, gain stability and low phase distortion. The advantages of monolithic technology as applied to these amplifiers is that the parasitic effects of long interconnection leads are avoided and that it is possible to diffuse thermal compensating elements into the chip so as to give good gain stability.

Radio and TV systems have been partitioned in many ways by different manufacturers, and accordingly there are various different types of integrated circuits in use. Fig. 7.36 shows a few typical examples in which each block

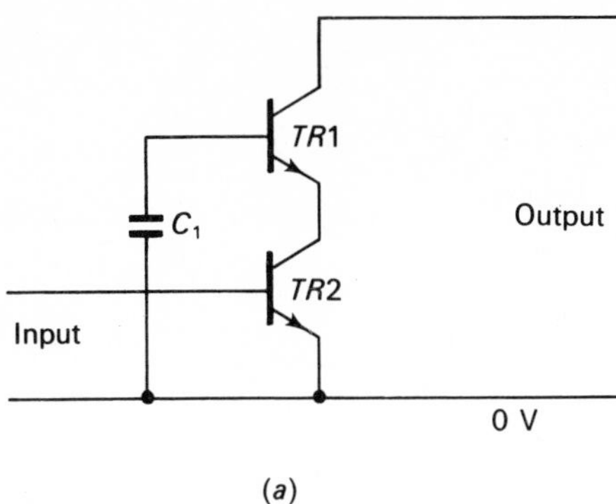

(a)

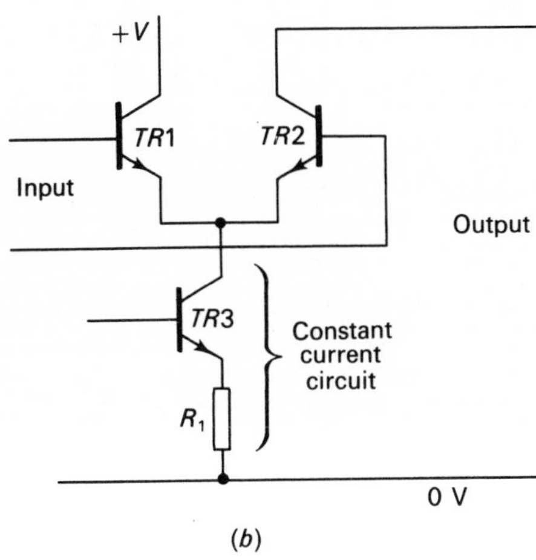

(b)

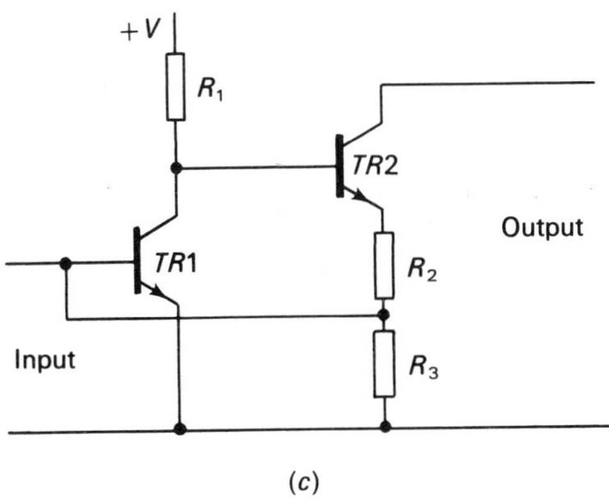

(c)

Fig. 7.35. R.F. and I.F. amplifier arrangements, (a) cascode, (b) long tail pair, (c) feedback.

represents an integrated circuit. Amplitude modulated (A.M.) receivers can be based on a conventional superhet system or they can use phase locked loop techniques, described in section 7.9. Generally a separate audio amplifier chip is needed for large output powers. Although several detection techniques may be used, for monolithic frequency modulated (F.M.) receivers the quadrature (or coincidence) system is the most popular since it requires only a single inductor. Fig. 7.36 (c) shows its operating principle. The input from the intermediate frequency (I.F.) amplifier is a square wave due to the limiting action in a previous stage. This is restored to a sine wave by the tuned circuit and both these signals are fed to the coincidence detector. When the carrier is modulated by an audio frequency (A.F.) signal a phase shift is introduced in the tuned circuit and this results in an output from the coincidence detector which is a series of pulses giving a mean value proportional to the modulating frequency. This is integrated to provide the A.F. output. For both monochromatic and colour TV receivers integrated circuits are used in all parts except for the power line scan outputs and the tuner, which needs to work at very high frequencies with low noise generation. Once again there is very little standardization of integrated circuits and this is accentuated by the fact that for colour the PAL, NTSC and SECAM systems are all in use in various countries.

A.M. and F.M. receivers are used for automobiles. In addition cars are becoming extremely sophisticated with the use of automatic seat belt interlocks, automatic anti-skid braking, pollution control and, perhaps in the future, radar control and collision avoidance. Also a host of digital circuits are used for such things as clocks, ignition, electronic speedometers and seat belt interlocks.

7.9 Phase locked loops

The principle involved in phase locked loops was known almost half a century ago, but it is only recently, since it first appeared as an integrated circuit, that the device has been extensively used.

Fig. 7.37 shows the basic elements of a phase locked loop. The phase detector is really a

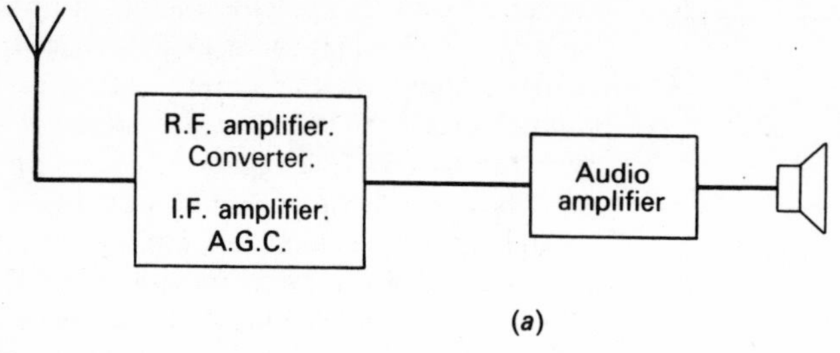

(a)

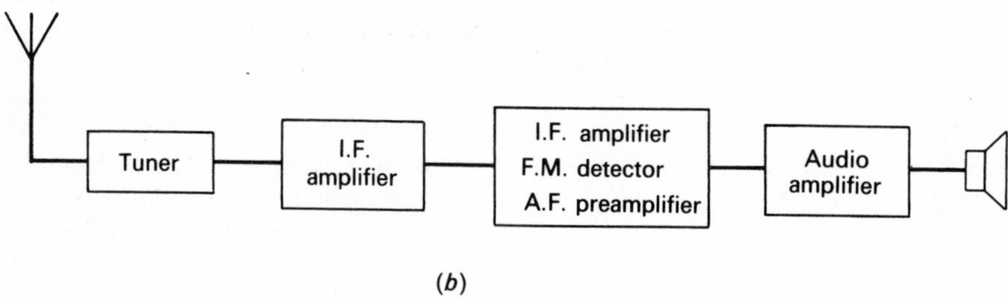

(b)

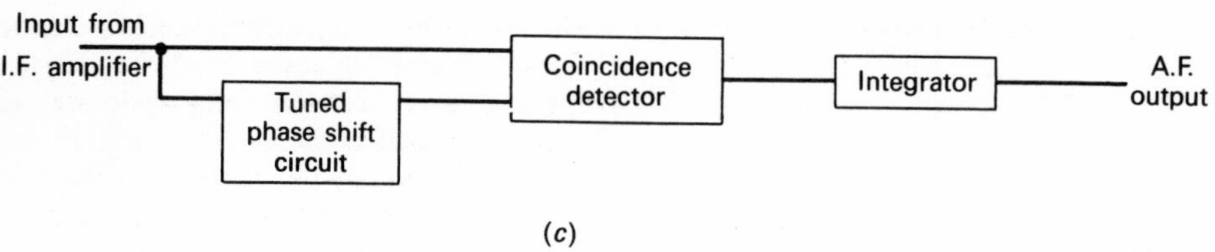

(c)

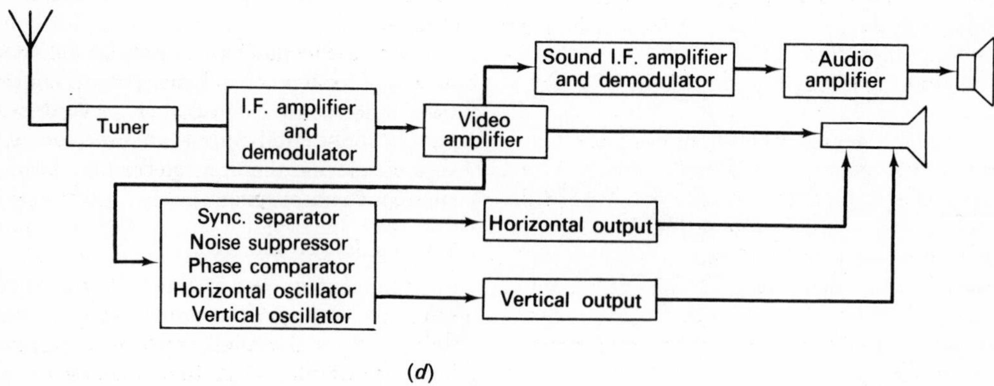

(d)

Fig. 7.36. Radio and TV system; (a) A.M. radio, (b) F.M. radio, (c) quadrature F.M. detector, (d) monochrome TV.

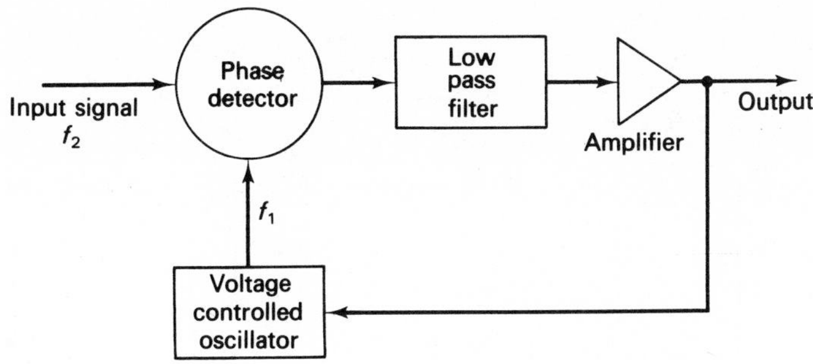

Fig. 7.37. Block schematic of a phase locked loop.

mixer whose output is determined by the frequency and phase of the two inputs. With no input signal the voltage controlled oscillator runs at some fixed frequency called its free running frequency. This can usually be selected by means of timing components external to the integrated circuit package. For an input signal of frequency f_2 the phase detector produces an output equal to $f_2 \pm f_1$. The low pass filter only allows the difference frequency $f_2 - f_1$ to go through. This signal is amplified and appears as the output of the phase locked loop. In addition the signal is fed back to the voltage controlled oscillator such that it drives the frequency f_1 towards f_2. This feedback action continues until f_1 is equal to f_2 and the loop is now locked to the input signal. However, there is still a phase difference between f_2 and f_1 and this produces a d.c. output signal which is just sufficient to drive the oscillator away from its free running frequency and to keep it locked to f_2.

Once the phase locked loop has locked onto the input signal it will follow it closely with a delay determined by the time constant of the loop. If however the input signal drifts very far from the free running frequency of the oscillator then the phase locked loop will fall out of lock since the design of the oscillator limits the range over which it can vary. This is defined as the lock range.

The lock range of the phase locked loop must be differentiated from its capture range. Initially, before the loop is in lock, the output from the phase detector will give a signal $f_2 - f_1$, which is relatively large and falls outside the band edge of the low phase filter. There is therefore no output and the voltage controlled oscillator stays in its free running mode. As f_2 approaches f_1 the value of the phase detector output decreases until eventually it falls inside the band edge of the filter. This is the capture frequency of the phase locked loop and $2(f_2 - f_1)$ is the capture range. Once this state is achieved the output from the filter drives f_1 towards f_2 in a regenerative loop so that the system is rapidly locked to the input signal. The capture range and the speed of capture are both reduced by the time constants involved in the filter. However, this same time constant also dampens the response of the loop to spurious signals so that its noise rejection capability is increased.

The phase locked loop has found acceptance in a wide range of industrial and consumer applications. For instance it offers a very simple means for demodulating a frequency modulated signal. If the input to the loop is a F.M. signal then the output will be such as to vary the oscillator frequency in tune with the input. This output will therefore directly represent the demodulated output of the F.M. signal. The linearity of the system is now directly related to the linearity of the voltage controlled oscillator. Fig. 7.38 illustrates another frequency synthesis application in which the output signal is a multiple N of the input where N is programmable. Fig. 7.38 (*a*) shows the most basic

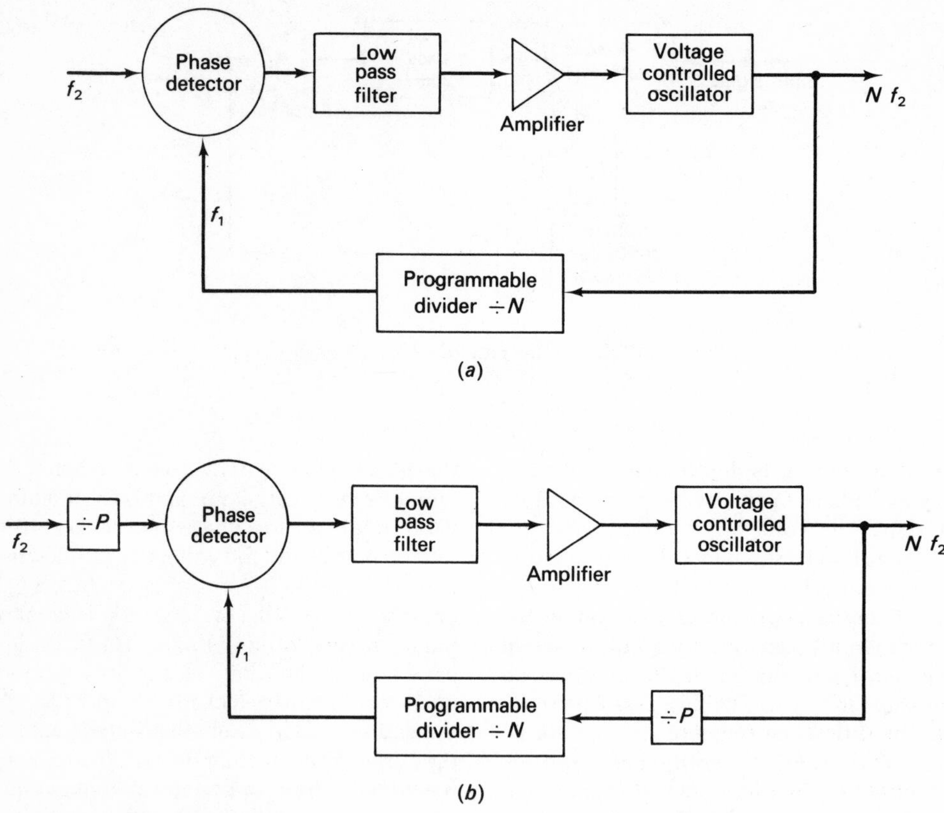

Fig. 7.38. Frequency synthesis using a phase locked loop; (*a*) basic system, (*b*) prescaler system.

arrangement with a programmable divider in the feedback loop. Using present day technology the oscillator can be designed to run at very high frequencies. However, the programmable counter must also be made to run at this frequency and since it is relatively complex this can be expensive. Fig. 7.38 (*b*) shows an alternative solution in which prescalers are used to divide the input and feedback frequencies by a fixed amount P. This also represents the fraction at which the programmable divider runs compared to the output frequency.

Bibliography

Morton L. Topfer, *Thick-Film Microelectronics,* Van Nostrand Reinhold Company, 1971, 210 pages.

The book gives a very good general introduction to thick-film hybrid circuits. Topics covered include the design of the circuit layout, technology of materials, details of assembly techniques for printing, firing, bonding and packaging, and a discussion on typical applications.

J.R.A. Beale, E.T. Emms, R.A. Hilbourne, *Microelectronics*, Taylor & Francis Ltd, 1971, 310 pages.

The book essentially covers the technology of monolithic integrated circuits. Following a detailed discussion on the physics of semiconductor devices the processes involved in the manufacture of monolithic integrated circuits are described and analysed. Chapters cover assembly and final testing, and circuit design techniques and mask making are described in detail. Although two chapters of this book are devoted to the applications of digital and linear circuits, the treatment is not very good compared to the other books given in this bibliography.

T.D. Towers, *Elements of Linear Microelectronics*, Butterworth & Co., Ltd, 1973, 108 pages.

The book gives a good overall introduction to the more commonly used linear integrated circuits. The treatment is essentially in the form of a user's guide to commercially available devices and the emphasis is on consumer components. Topics covered include operational amplifiers, audio amplifiers, radio and intermediate frequency amplifiers, wideband amplifiers, voltage regulators, amplitude modulated radio receivers, F.M. radio receivers, and television receivers.

Eugene R. Hnatek, *A User's Handbook of Integrated Circuits,* John Wiley & Son, Inc., 1973, 449 pages.

The book presents an excellent introduction to the whole field of integrated circuits although it does not mention many of the newer components such as microprocessors and charge transfer devices. The processes involved in the manufacture of bipolar and unipolar monolithic circuits are described in detail and the application of digital circuits is also adequately covered. The emphasis is on the more established components. Linear integrated circuits receive a comparatively narrow treatment with only operational amplifiers and voltage regulators being described in detail and the book concludes with a chapter on packaging and interconnection techniques.

Frederick H. Edwards, *Principles of Switching Circuits*, The Massachusetts Institute of Technology, 1973.

This is an analytical book relating to digital circuits. The treatment is up to date and concise. Topics covered include switching algebra, minimization techniques and sequential circuit analysis.

John Allison, *Electronic Integrated Circuits*, McGraw-Hill Book Company Ltd, 1975, 139 pages.

The book gives a good introduction to the technology of monolithic integrated circuits and the material covered is relatively up to date. Applications of integrated circuits are only briefly introduced in one short chapter. The topics covered in the technology chapters include the preparation of semiconductor slices, oxide layer growth and photolithography, monolithic circuit components, isolation, interconnection and packaging.

Glossary of acronyms

ALU	Arithmetic logic unit
BBD	Bucket brigade device
BCD	Binary coded decimal
BRM	Binary rate multiplier
CAM	Content addressable memory
CCD	Charge coupled device
CDI	Collector diffusion isolation
CHL	Current hogging logic
CML	Common mode logic
CMOS	Complementary metal–oxide-semiconductor
CMRR	Common mode rejection ratio
CPU	Central processor unit
DCTL	Direct coupled transistor logic
DIL	Dual-in-line
DMA	Direct memory access
DMOS	Double diffused metal–oxide-semiconductor
DTL	Diode transistor logic
ECL	Emitter coupled logic
EROM	Electrically programmable read only memory
FAMOS	Floating gate avalanche injection metal–oxide-semiconductor
FET	Field effect transistor
FIFO	First-in first-out
FPM	Fixed program memory
FROM	Field programmable read only memory
IIL (or I^2L)	Integrated injection logic
J–FET	Junction field effect transistor
LIFO	Last-in first-out
LPTTL	Low power transistor–transistor logic
LPTTL–S	Low power transistor–transistor logic – Schottky
LSB	Least significant bit
LSI	Large scale integration
MAOS	Metal–alumina–oxide-semiconductor
MNOS	Metal–nitride–oxide-semiconductor
MOS	Metal–oxide semiconductor
MOS–FET	Metal–oxide semiconductor field effect transistor
MPU	Microprocessor unit
MSB	Most significant bit
MSI	Medium scale integration
NMOS	n-channel metal–oxide-semiconductor
PLA	Programmable logic array
PMOS	p-channel metal–oxide-semiconductor
PPL	Phase locked loop
RAM	Random access memory
RMM	Read mostly memory
ROM	Read only memory
RTL	Resistor transistor logic
SFL	Substrate fed logic
SOI	Silicon on insulator
SOS	Silicon on sapphire
SSI	Small scale integration
TTL	Transistor–transistor logic
TTL–S	Transistor–transistor logic – Schottky
UART	Universal asynchronous receiver transmitter
ULA	Uncommitted logic array
VATE	Vertical anisotrophic etch
VIP	V isolation with polysilicon backfill

Glossary of terms

Address: The designation of the location in *memory* where information is stored.

Air abrasive trimming: A technique for reducing the cross sectional area of thick film resistors by blasting them with a jet of air containing fine abrasive particles.

Architecture: The structure of a system. The *hardware* architecture specifies its construction in terms of components whereas the *software* architecture specifies its user characteristics.

Bias: It is the voltage or current which is applied to a component in order to put it into a specified operating mode, e.g. conducting or non-conducting.

Binary: A mathematical system which has a base of two.

Bipolar: A semiconductor having both holes and electrons as charge carriers.

Black box: A device whose performance is specified in terms of input and output signals, and whose internal structure is undefined.

Bonding: Techniques for making an electrical connection between two conductors.

Bulk memory: A memory system having the capability of storing very large amounts of information and working at relatively slow speeds. This memory usually acts as a *peripheral* to the CPU and smaller, faster memory. It is also called mass memory and backing memory.

Characteristic: The parameters which define the performance of a circuit or component under specified conditions.

Chip: An unencapsulated component. These are available as chip semiconductors, also called *dice*, chip capacitors or chip resistors.

Chip density: Used in connection with an integrated circuit chip it refers to the number of circuit elements which can be made into the chip.

Clock: An electrical signal which occurs at a specified periodic interval and which determines the timing of the circuit to which it is applied.

Combinational logic: A logic circuit whose output is determined by the combination of the input signals and not by the sequence in which they occur.

Control memory: Memory which stores the user's *instructions*. These determine the way in which the circuit operates. This is also called a *program memory*.

Custom integrated circuit: An integrated circuit which has been designed for a specific customer application and is sold exclusively for use by that customer.

Decimal: A mathematical system which has a base of ten.

Depletion mode: A mode of operation in which the semiconductor device is on, i.e. can conduct current in the absence of a signal at its control terminal.

Dice: Plural of *die*.

Die: See chip.

Diffusion: A technique for introducing *impurities* into the semiconductor. Under the influence of high temperatures these impurity atoms gradually work their way into the crystal lattice of the semiconductor.

Dopant: An additive to a semiconductor designed to give it certain electrical properties. This is also called an *impurity*.

Driver: An electronic component designed to supply a high voltage or current output. It is also called a load-driver.

Dynamic: A mode of operation for semiconductor circuits in which the electrical signal is stored on capacitors. Since this signal will leak away with time a dynamic circuit needs to be periodically *refreshed*.

Electrically programmed read only memory: A *field programmed read only memory* which is programmed by means of electrical signals.

Enhancement mode: A mode of operation in which the semiconductor device is off, i.e. cannot conduct current in the absence of a signal at its control terminal.

Epitaxy: A relatively thin layer of semiconductor, with a closely defined chemical structure, which is formed on the surface of the semiconductor *substrate*.

Field programmed read only memory: A *read only memory* which can be programmed after it leaves the manufacturer's plant.

Functional adjustment: Adjustment of components in a system while it is performing the function for which it was designed.

Hardware: The parts of a system which are built from components. See also *software.*

Hermetic: A package or seal designed to protect its contents from the effects of adverse environment such as moisture and chemicals.

Hybrid integrated circuit: A *substrate*, containing miniature active and passive devices, usually in chip form or as *printed components.*

Impurity: See *dopant.*

Ink: See *paste.*

Instructions: A sequence of electrical signals, usually stored in *control memory*, which determine the way in which the system operates.

Inverter: A device for changing the polarity of the signal, for example positive voltage to negative voltage and logic 0 to logic 1.

Ion implantation: A technique for introducing *impurities* into the semiconductor surface by accelerating them to a high velocity and then bombarding them onto the semiconductor.

Isolation: The technique used to electrically separate different parts of a system on a semiconductor die.

Large scale integration: A semiconductor die which has a large number of circuit elements formed in it. Usually LSI devices have more than 100 gates.

Loading: The power taken from the output terminal of an integrated circuit. Generally the output voltage is fixed and the loading is stated as a current.

Logic families: Groups of logic circuits which resemble each other in electrical characteristics, such as *loading*, and operating speed, but which perform different functions.

Medium scale integration: A semiconductor die which has a medium number of circuit elements formed in it. Usually MSI devices have between 10 and 100 gates.

Memory: A store for electrical signals, which may be analogue or digital.

Memory array: A group of *memory cells* usually arranged in a matrix so that any cell can be *addressed* to obtain its information.

Memory cell: A simple memory element capable of storing one unit of electrical signal, e.g. one logic bit or one analogue voltage level.

Microelectronics: Covers the whole field of miniature electronic components and equipment. Often the term is used synonamously with integrated circuits, but it has a much wider meaning.

Monolithic integrated circuit: A system in which all circuit elements are formed into a single semiconductor die.

Negative logic: A digital convention in which the logic 0 state is represented by a higher or a more positive voltage than the logic 1 state.

Noise: Unwanted electrical signals.

Off-the-shelf: Devices which are not *custom*. This is also called *standard*.

Oxide: Term often used for the silicon dioxide layer on the semiconductor die.

Parasitic capacitance: Unwanted capacitance which is unavoidably formed when the system is built.

Paste: Material used for making *printed components.*

Positive logic: A digital convention in which the logic 0 state is represented by a lower or more negative voltage than the logic 1 state.

Printed components: Devices formed on a *substrate* by making the desired pattern in *paste* and then *processing* further such as firing at a high temperature. Printed components can be conductors, resistors or dielectrics.

Process: The manufacturing steps used in the production of an integrated circuit.

Program memory: See *control memory.*

Random access memory: An array of *read–write memory* cells arranged such that the access time to the information stored in any *memory cell* is the same.

Rating: The maximum electrical, mechanical or environmental stress which the device can stand without being destroyed.

Read mostly memory: A memory which is designed to be read more often than it is written into. Generally the time required to read information is much less than the time needed to write information into a *memory cell*. This is also called a *reprogrammable read only memory.*

Read only memory: A memory which has information written into it only once, usually by the manufacturer, and thereafter this information is read out only.

Read–write memory: A memory which is designed to be read and written into with equal frequency. The time required to read and write are very similar. Generally a read write memory is also *random access*.

Refresh: The operation of recharging the capacitors of a *dynamic* system to compensate for charge leakage.

Sequential logic: A logic circuit whose output is determined by the sequence in which the input signals occur. A *clock* is generally used as the timing element in a sequential logic circuit.

Signal: The electrical representation of the required information. This can either be digital, for example a positive or negative voltage level, or analogue, for example a continuous variation of voltage over several levels.

Slice: A semiconductor material containing many dice prior to their separation into individual die. A slice is also called a *wafer*.

Small scale integration: A semiconductor die which has a small number of circuit elements formed in it. Usually SSI devices have less than 10 gates.

Software: The parts of a system which are not built from components. Software generally resides in the system as electrical signals.

Standard integrated circuit: See *off-the-shelf*.

Static: A system which can store information indefinitely so long as the power supply is maintained. Static circuits do not therefore require to be *refreshed*.

Substrate: The base material on which the integrated circuit is formed. For a monolithic circuit this is a semiconductor, whereas for a hybrid circuit it is usually alumina or glass.

Unipolar: A system having either holes or electrons as charge carriers within the semiconductor, but not both at the same time.

Volatile memory: A memory whose information is lost when the power supply is disconnected.

Wafer: See *slice*.

Index